Circadian Rhythm Sleep-Wake Disorders

R. Robert Auger
Editor

Circadian Rhythm Sleep-Wake Disorders

An Evidence-Based Guide for Clinicians and Investigators

Springer

Editor
R. Robert Auger
Mayo Center for Sleep Medicine
Department of Psychiatry and Psychology
Mayo Clinic College of Medicine
Rochester, MN
USA

ISBN 978-3-030-43802-9 ISBN 978-3-030-43803-6 (eBook)
https://doi.org/10.1007/978-3-030-43803-6

This Springer imprint is published by the registered company Springer Nature Switzerland AG
The registered company address is: Gewerbestrasse 11, 6330 Cham, Switzerland

Preface

Identification and treatment of circadian rhythm sleep-wake disorders (CRSWDs) can be a fascinating yet vexing clinical encounter in the Sleep Medicine realm. While circadian-based basic science research, chronobiological research of healthy humans (including simulated CRSWD studies), and chronotherapeutic research are quite sophisticated and abundant, there is a relative dearth of clinical research. In my point of view, one major reason for the outpacing of CRSWD clinical research pertains to the disparate sources of information and related lack of interdisciplinary input that is necessary for holistic care. This publication makes an attempt to bridge this gap by including authorities from numerous relevant disciplines to inform clinical practice.

Accordingly, the text begins with a scientific background on mammalian circadian rhythms and progresses to a discussion of chronobiological protocols that will allow the reader to critically review relevant literature. Following an introduction to CRSWDs, there are two chapters that describe assessment tools for clinicians' consideration and separate chapters dedicated to individual CRSWDs. Recognizing that delayed sleep-wake phase disorder has a particularly relevant impact among adolescents, a chapter is included that highlights the school start time debate, which has significant relevance given the recent legislation signed by the Governor of California. Subsequent chapters from lighting research experts may help clinicians to standardize implementation of treatments and should also serve to optimize comparative clinical trials. A final chapter discusses the impact that specified lighting environments may have in the workplace and other settings.

It is ultimately my hope that this book will serve as an immediate reference for Sleep Medicine clinicians but that it will also serve as an impetus to address clinical research deficiencies by encouraging cross-talk between various scientific, medical, and educational disciplines.

Rochester, MN, USA R. Robert Auger

Acknowledgments

I would like to acknowledge the esteemed authors who assisted with this project. I feel fortunate that their extraordinary efforts and insights are contained within this singular text (a true treasure trove of circadian knowledge!). I would also like to extend my gratitude to my colleagues in the Mayo Center for Sleep Medicine, with whom I feel privileged to work. I work in an environment that is collegial, encouraging, and filled with intelligent mentors and peers who enrich the clinical experience on a daily basis.

I would also be remiss if I did not acknowledge my wonderful family—my lovely wife Tracy, my son Josh, and my daughter Emily. From them I derive happiness and meaning, and they are the perfect antidote to life's deadlines and stressors. I also love and admire my parents Ray and Judy Ann. They remain exemplars of integrity and intellect and taught me most of what I know about careful writing and editing.

Finally, I would like to acknowledge Ms. Lori Solmonson, who always provides expert, cheerful, and thoughtful technical assistance, which makes projects such as this easier to complete.

Contents

Contributors

R. Robert Auger, MD, FAASM, FAPA, DABPN Mayo Center for Sleep Medicine and Department of Psychiatry and Psychology, Mayo Clinic College of Medicine, Rochester, MN, USA

Matthew R. Brown, BS Department of Physiology and Biomedical Engineering, Mayo Clinic School of Medicine, Rochester, MN, USA

Gregory S. Carter, MD, PhD Department of Neurology and Neurotherapeutics, The University of Texas Southwestern Medical Center at Dallas, Dallas, TX, USA

Jonathan Emens, MD, FAASM, DFAPA Department of Psychiatry, Oregon Health and Science University, Division of Mental Health and Clinical Neurosciences, VA Portland Health Care System, Portland, OR, USA

Mariana G. Figueiro, PhD Lighting Research Center, Rensselaer Polytechnic Institute, Troy, NY, USA

Danielle Goldfarb, MD University of Arizona College of Medicine, Phoenix, AZ, USA

Cathy Goldstein, MS, MD Michael S. Aldrich Sleep Disorders Center, Department of Neurology, University of Michigan Health System, Ann Arbor, MI, USA

Bhanu Prakash Kolla, MD, MRCPsych Center for Sleep Medicine and Department of Psychiatry and Psychology, Mayo Clinic, Rochester, MN, USA

Vincent A. LaBarbera, MD Department of Neurology, Rhode Island Hospital/ Warren Alpert Medical School of Brown University, Providence, RI, USA

Elliott Kyung Lee, MD, FRPC(C), DABSM Department of Psychiatry, Royal Ottawa Mental Health Center and Institute for Mental Health Research, University of Ottawa, Ottawa, ON, Canada

Meghna P. Mansukhani, MD Center for Sleep Medicine, Mayo Clinic, Rochester, MN, USA

Aleksey V. Matveyenko, PhD Department of Physiology and Biomedical Engineering, Mayo Clinic School of Medicine, Rochester, MN, USA

Department of Medicine, Division of Endocrinology, Metabolism, Diabetes, and Nutrition, Mayo Clinic School of Medicine, Rochester, MN, USA

Eric J. Olson, MD Department of Medicine, Division of Pulmonary and Critical Care Medicine, and Center for Sleep Medicine, Mayo Clinic, Rochester, MN, USA

Cinthya Pena-Orbea, MD Center for Sleep Medicine, Mayo Clinic, Rochester, MN, USA

Mark S. Rea, PhD Lighting Research Center, Rensselaer Polytechnic Institute, Troy, NY, USA

Alok Sachdeva, MD Michael S. Aldrich Sleep Disorders Center, Department of Neurology, University of Michigan Health System, Ann Arbor, MI, USA

Katherine M. Sharkey, MD, PhD Department of Medicine and Psychiatry and Human Behavior, Alpert Medical School of Brown University, Providence, RI, USA

Kyla L. Wahlstrom, PhD Department of Organizational Leadership, Policy and Development, College of Education and Human Development, University of Minnesota, Minneapolis, MN, USA

Chapter 1
Biological Timekeeping: Scientific Background

Matthew R. Brown and Aleksey V. Matveyenko

Introduction

The timekeeper, often underappreciated, is an individual or device in competition, experimentation, and daily life that records and regulates the initiation and termination of key processes. In physiology, the timekeeper or pacemaker is crucial for maintaining delicate temporal organization of molecular and physiological events that dictate the survival of an organism. Throughout the human body, there are a myriad of biological rhythms that oscillate at rates as short as 1–2 seconds (i.e., sinoatrial node of the heart) and as long as weeks to months (i.e., estrous cycle) that are essential to life. A circadian rhythm is the term for a process that is under the control of our internal pacemakers with a period of approximately 24 hours [1]. Rhythms shorter than a day are referred to as ultradian rhythms, while rhythms longer than a day are called infradian rhythms [2]. The word circadian is derived from the Latin words *circa* ("approximately") and *dies* ("day") [1]. As such, circadian rhythms adapt to, follow, and oscillate with a period nearly matching the Earth's 24-hour rotation with respect to the sun.

It is becoming clear that nearly all fundamental physiological processes are under the control of the circadian system [3]. From the autonomic control of blood pressure to the coordination of hormonal secretion, investigators are starting to

M. R. Brown
Department of Physiology and Biomedical Engineering, Mayo Clinic School of Medicine, Rochester, MN, USA

A. V. Matveyenko (✉)
Department of Physiology and Biomedical Engineering, Mayo Clinic School of Medicine, Rochester, MN, USA

Department of Medicine, Division of Endocrinology, Metabolism, Diabetes, and Nutrition, Mayo Clinic School of Medicine, Rochester, MN, USA
e-mail: Matveyenko.Aleksey@mayo.edu

R. R. Auger (ed.), *Circadian Rhythm Sleep-Wake Disorders*,
https://doi.org/10.1007/978-3-030-43803-6_1

unravel the importance of the circadian system for anticipating and optimizing the timing of various physiological functions [4]. In concert with these findings, recent clinical and preclinical research have begun to shed light on the detrimental consequences associated with chronic disruption of circadian rhythms [5–9]. The information technology revolution of the twenty-first century has inadvertently created a "24/7" environment that does not abide the natural rhythms of the Earth. In 2018, Nielsen estimated American adults spent an average of 11 hours per day exposed to artificial light emitted from electronic screens, an increase of 25% over the past 5 years [10]. This time does not even include the time that is spent using these devices at work or school which are becoming ever more digital. Additionally, shift work is becoming increasingly common in our connected world. It is estimated by the International Agency for Research on Cancer that 15–30% of the Western population is exposed to rotating shift work [11]. Taken together, modern humans are exposed to stressors that negatively regulate the circadian system. Due to its fundamental role in essential physiological processes, disruption of one's circadian rhythm has been linked to an increasing risk of developing cancer [12], type 2 diabetes [4, 13], neurological disorders [14, 15], and cardiovascular disease [16, 17], among other adverse health consequences. Therefore, it is critical to further our understanding of the mechanisms underlying the circadian system to develop new therapies and strategies for maintenance of the circadian clock in our twenty-first-century environment.

History of Chronobiology

The concept of timekeeping has permeated civilization for millennia; however, endogenous biological clocks were not truly appreciated until the eighteenth century [18]. Although ancient physicians such as *Galen* and *Hippocrates* noted that conditions like fevers exhibited periodic 24-hour rhythms, they unlikely appreciated that these rhythms were controlled, in part, by an endogenous timekeeper which persists independent of environmental cues [19]. In 1729, French astronomer Jean Jacques d'Ortous de Mairan was the first to realize that daily rhythms were intrinsically controlled and were not simply a response to the rhythmic light cycle of the Earth [18]. He observed that *Mimosa* leaves moved with a 24-hour cycle, even when placed in constant darkness, and therefore concluded that the movements could not have been affected by the light-dark cycle. Following de Mairan's work, many confirmed his findings and demonstrated that the circadian rhythms in plants were independent of other rhythmic external stimuli. However, it was still unclear exactly what internal factors drove the maintenance of the circadian rhythm. With the explosion of Mendelian genetics in the early 1900s, Dr. Erwin Bunning, a German botanist, crossed bean plants with various period lengths and demonstrated that the period length of the offspring was an intermediate of the previous generation [20]. As a result, Bunning inferred that endogenous circadian clocks were "heritable" from the genetic code. Bunning's pioneering work unraveling the interaction between intrinsic clocks and light sensitivity in plants set the stage for an explosion in the understanding of circadian rhythms throughout the rest of the twentieth century.

The modern field of chronobiology is an ever-growing discipline which traverses a broad range of basic and clinical research [21]. Nevertheless, the diversity in the field today mirrors its composition upon its founding in the early 1960s [22]. At the time, the field was led by Dr. Colin Pittendrigh and Dr. Jurgen Aschoff, often considered the fathers of chronobiology. Both pioneered early studies defining the properties and characteristics of circadian rhythms [23]. Most notably, Pittendrigh developed a parametric model describing how circadian rhythms can be reset or disrupted by a singular pulse of light [24]. Aschoff, meanwhile, showed that humans have intrinsic clocks and that these clocks could be synchronized by external cues. He coined the term *zeitgeber*, a German word meaning time-giver, to describe the environmental stimuli that can synchronize internal clocks [25]. While Aschoff, Pittendrigh, and others were unraveling the underpinnings of the biological clock, there was a push to apply these principles to improve human health. Dr. Franz Halberg, a physician at the University of Minnesota, led this charge at the time by pioneering novel preclinical and clinical studies that actively considered circadian time as a biological variable [1]. In one of his early studies, Halberg measured the circadian variability in the temperature of oral tumors, as a metric of their circadian metabolism, in order to optimize radiation therapy treatment [26]. He observed that patients who received treatment at peak tumor temperature were twice as likely to be cancer-free 2 years following radiation. Despite their diverse approaches and methods, these individuals clearly understood the importance in maintaining robust circadian rhythms and provided the guiding principles being applied today to improve human health.

Measurement and Assessment of Circadian Rhythms

Circadian rhythms, like any other biological or physical rhythm, consist of predictable patterns of oscillations over a finite element of time [22]. In order to describe and compare how circadian patterns of motor activity, temperature, food intake, and other rhythmic behaviors vary between individuals or organisms, one must appreciate the distinct features of a rhythm. To identify whether a process is truly a circadian rhythm, the period or frequency must be assessed. The period is the time to complete one cycle, while the frequency, also known as the inverse period, is the number of cycles completed over a unit of time. In many scientific disciplines, Fourier analysis, pioneered by Joseph Fourier in the late eighteenth century, is used to quantify the various frequency components of a signal [27]. Simply put, Fourier analysis transforms a discrete or continuous rhythmic signal into a sum of sinusoidal functions with varying frequencies. The relative contribution of each frequency to the signal can then be visualized by calculating their spectral densities, otherwise known as a periodogram. In circadian biology, modified periodogram analysis such as the chi-squared periodogram, which utilizes least squares regression, is used to measure the dominant period of physiological rhythms [28]. Periodogram analysis also reveals the strength or robustness of the rhythm. Essentially, periods that are "neater" and more consistent will have a larger power because the period has a greater contribution to the overall sinusoidal signal. The power is different than the amplitude of the signal as the latter only quantifies the relative deviation from the mean to the peak value, while the former is not affected by signal deviation and relies solely on the contribution of the various frequency components of the rhythm. Although the rhythm's period may describe whether processes occur under the control of the circadian system, it does not explain why they occur and why they are optimized to occur at different times of the day. The secretion of cortisol and melatonin, for example, both occur with a robust 24-hour period; however, their peak secretion occurs at different times of the circadian cycle [29]. Typically, plasma cortisol levels rise throughout the solar day and begin to fall just as melatonin levels begin to rise in anticipation of the biological night. The relationship of the opposing rhythms can be quantified by the phase, which describes the initial time (or angle) at the signal's origin. The lag in the phase of the melatonin rhythm relative to the cortisol rhythm is referred to as the phase difference.

The use of melatonin and cortisol secretion as markers of circadian timekeeping exemplifies how the field of chronobiology still mainly relies on indirect, noninvasive methods such as hormonal secretion, internal temperature, and voluntary locomotor activity to "keep time" *in vivo* [30, 31]. With the development of mobile, noninvasive devices for monitoring of voluntary locomotor activity, assessment of circadian activity rhythms is most commonly used to indirectly measure circadian pacemaking in humans [32]. Activity rhythms are typically visualized by means of an actogram, which vertically organizes each day of behavior allowing for the efficient visual analysis of the period and phase of this circadian output [33]. Nevertheless, novel methods to directly measure circadian pacemaking are currently being explored. Recent preclinical animal studies visualizing fluorescently tagged circadian genes have started to allow for the longitudinal and direct measurement of circadian rhythms in pacemakers of mice; however, it is unclear whether such techniques will become applicable to humans [34]. Advances in imaging technology such as BOLD fMRI show promise, as they may be able to directly quantify biological timekeeping activity via blood flow to the pacemaking region [35]. Moreover, novel computational methods using integrated transcriptomic data are also being explored in humans to estimate the internal circadian time from a single blood or tissue sample [31, 36, 37]. Taken together, these metrics and tools will be essential to allow for the direct assessment of circadian rhythms in humans under various physiological and pathophysiological conditions.

Properties of Circadian Rhythms

Over the past 300 years, endogenous circadian rhythms have been assessed in a diverse group of plants, insects, birds, fish, amphibians, and mammals. Although the underlying molecular mechanisms that encode circadian processes may not be completely conserved between species, the essential properties of their circadian rhythms remain. The three key properties that define a circadian rhythm are (1) persistence under constant conditions with a free-running period of approximately 24 hours, (2) ability to entrain to a 24-hour cycle via rhythmic environmental stimuli, and (3) free-running periods which are temperature compensated such that an organism maintains an approximately 24-hour circadian period regardless of temperature fluctuations [22].

Circadian Rhythms Persist Under Constant Conditions

Jean Jacques d'Ortous de Mairan's initial observation that the *Mimosa* plant maintained a 24-hour period in constant darkness suggested that circadian rhythms persist with a free-running period (FRP) of ~24 hours. The FRP is the period of activity in constant conditions and is also referred to as tau (t) or the inverse/frequency of the FRP [38]. An intrinsic FRP in organisms is critical for biological timekeeping because it ensures that the "clock keeps ticking" regardless of external stimuli. Nevertheless, it was unclear for many years whether this was conserved in higher mammals such as humans. In order to test this hypothesis, subjects were placed in an underground bunker devoid of any timing cues or potential zeitgebers [39] and were asked to maintain a "regular" life for 3–4 weeks. The light intensity and internal temperature were carefully controlled from the outside. Upon recording activity, temperature, and urine rhythms, Aschoff's team calculated an average free-running period of ~25 hours in the subjects. Future studies led to uncertainty in the study's findings [40, 41], and researchers further attempted to refine measurements by desynchronizing subjects' sleep-wake cycles from their circadian pacemakers. Nathaniel Kleitman pioneered a technique whereby subjects are exposed to a 28-hour "day", such that the zeitgebers related to the sleep-wake schedule are equally distributed throughout the circadian cycle over the duration of the experiment [30]. Indirect measurement of the periods of activity, core temperature, plasma melatonin, and plasma cortisol all revealed an intrinsic period of 24.18 hours with 90% of measurements ranging between 24.00 and 24.35 hours. This study demonstrated that the intrinsic human pacemaker was precise and truly circadian.

Rhythmic Environmental Stimuli Can Entrain Circadian Rhythms

Although the internal timekeeper maintains an approximately 24-hour rhythm, environmental cues are required to continually tune the internal timekeeper to maintain a precise period and phase. Consider an intrinsic period of 24.18 hours described above, which is only 12 minutes longer than the Earth's 24-hour solar cycle. After only 1 week, individuals would be subjected to an ~1-hour phase shift which would progressively push one to a nocturnal state after a few months. Without tuning, our internal pacemakers would create an asynchronous and inefficient world that would overthrow our modern societal norms. Most typically, the Earth's light-dark cycle is considered the foremost entrainment agent; however, other agents such as food availability and external temperature have been also shown to act as zeitgebers for internal clocks [25, 42].

The paradigm describing how light tunes the period and phase of the biological clock was championed by Pittendrigh, leading to the development of a nonparametric, discrete model of synchronization [43]. The model suggests that the circadian pacemaker maintains equilibrium by shifting its phase in response to a stimulus of light outside of its current circadian phase. In order to visualize how pulses of light affect the phase of the pacemaker, phase response curves (PRCs) can be constructed describing whether the light pulse will advance, delay, or have little effect [44]. Intuitively, the effect of a light pulse during the active phase of a diurnal mammal will be minimal, while exposure to light during the night will significantly phase shift the FRP. This concept has been translated and repeatedly demonstrated in humans by exposing subjects to a 3–6-hour light stimulus at various times throughout the circadian cycle and subsequently measuring physiological rhythms in the absence of entrainment cues [45–48]. Researchers demonstrated that there is variability in response to the pulses of light; however, humans generally experience a phase delay when exposed to light early in sleep and a phase advance as the sleep phase approaches dawn [49]. Practically, PRCs have been a valuable tool in assessing the efficacy of chronotherapies for individuals exposed to circadian disruptions [50, 51].

Free-Running Periods Are Temperature Compensated

The light-dark cycle plays a dominant role in resetting and entraining the circadian clock; however, the biological pacemaker encodes these environmental cues through a series of biochemical reactions that are directly affected by the temperature of the reaction environment. Dr. Hans Kalmus first demonstrated that the rate of the circadian cycle in drosophila increased threefold with every 10 °C increase in temperature [52, 53]. Intuitively, a temperature-dependent clock would present significant challenges as the clock would run faster on a hot, summer day and slower on a cold, winter day, resulting in unreliable period measurements from 1 day or one season to the next. As such, further investigations on bees, mice, drosophila, and cultured human cells have refuted this initial finding [54–56]. In the presence of increasing ambient temperature, the FRP has been shown to remain relatively constant and, therefore, temperature compensated. Despite this finding, *in vitro* studies of chick pineal cells demonstrated that a heat pulse, similar to light pulses described above, could cause a phase shift of the circadian clock as measured by melatonin secretion [57]. Recently, *Buhr* et al. demonstrated that *ex vivo* tissue, absent of photic cues, can also be entrained to a new phase by a heat pulse while still maintaining an approximately 24-hour period [58]. Taken together, these findings revealed that an organism's innate circadian period is insensitive to temperature; however, it also demonstrates that other rhythmic environmental stimuli, such as the daily circadian rhythm in body temperature, play a role in setting the phase of the circadian pacemaker.

Organization and Entrainment of the Circadian System

Provided the knowledge that the circadian pacemaker can be tuned by external cues, it is clear that it is actively engaged with the surrounding environment rather than passively counting time [59]. In turn, the circadian system can be directly modeled by simple oscillator networks. An oscillator is a rhythmic signal that provides inputs into a system to initiate change (i.e., response to light pulse), records and measures feedback from the system, and institutes a delay in response to feedback from a system not at equilibrium. Early models of the circadian oscillator system considered various configurations including a single, master pacemaker that drives systemic circadian outputs, coupled master oscillators that drive dependent or independent peripheral outputs, and a hierarchical oscillator system that couples a master timekeeper to peripheral oscillators that control local circadian outputs [60, 61].

The discovery of the suprachiasmatic nuclei (SCN) of the hypothalamus as the master pacemaker in mammals by two independent teams in 1972 began to eliminate the concept of a coupled oscillator model [62, 63]. Lesioning of the SCN resulted in loss of circadian regulation of locomotor activity, drinking behavior, and cortisol rhythms. Electrical recordings of SCN neurons *in vivo* and *in vitro* demonstrated that the SCN functions as a self-sustained oscillator as the neurons have the ability to intrinsically maintain a rhythm in electrical activity regardless of external input [64, 65]. Additionally, transplantation studies that implanted the SCN into an arrhythmic animal were successful in restoring the circadian rhythmicity of the donor animals [66, 67]. Although the SCN was demonstrated as the master circadian pacemaker, peripheral cells and tissues were also found to have intrinsic circadian oscillations [68, 69]. In the absence of SCN input, *ex vivo* examination of the mammalian liver, lung, and skeletal muscle revealed that these tissues displayed a robust circadian oscillation in gene expression. Recent studies using a luciferase (light) reporter as a real-time marker of circadian rhythms in mice confirmed the existence of self-sustainable peripheral oscillators in nearly all peripheral tissues [70, 71]. The growing experimental evidence suggests that the circadian system exhibits a hierarchical oscillator structure whereby a master oscillator provides inputs and tunes a network of peripheral oscillators that control downstream physiological outputs.

Light Is Encoded at the Master Timekeeper by a Specialized Detection Mechanism

Light, as the primary zeitgeber, requires an efficient communication system to relay changes in the light-dark cycle to the SCN. Intrinsically photosensitive retinal ganglion cells (ipRGCs) contain specialized melanopsin receptors which are present in only ~1% of RGCs [72, 73]. Melanopsin is a photopigment sensitive to short wavelengths of light with a peak absorption at ~480 nm corresponding to blue/green light [74]. Activation of melanopsin leads to an intracellular signaling cascade causing ipRGCs to depolarize, initiating an action potential that travels down the retinohypothalamic tract (RHT) and directly innervates the SCN [75]. Lesioning the RHT causes animals to free-run, despite a rhythmic light-dark cycle, exemplifying its role in entrainment [76]. Simply put, the fundamental role of the RHT is to indicate the start and end of the day in order to entrain the master oscillator to a consistent phase response.

The symbiotic relationship between the SCN and RHT is the key to maintaining precise timing and is best illustrated by the monosynaptic connection between them [77]. Increase in the activity of the RHT at dawn, for instance, is directly received by SCN neurons. Inappropriate exposure to light stimuli due to jetlag or shift work can induce spiking activity of the RHT which results in a phase shift of SCN neuronal activity, as described by the PRC previously discussed. Synaptic communication between the RHT and SCN occurs via presynaptic release of the excitatory neurotransmitter glutamate [78]. The released glutamate is received postsynaptically by N-methyl-D-aspartate (NMDA) receptors, causing an influx of intracellular Ca^{2+} into SCN neurons and a cyclic adenosine monophosphate (cAMP)-dependent signaling cascade that resets and entrains the pacemaker to the light input [79]. Thus, a light stimulus can be quickly detected and communicated by ipRGCs to SCN neurons via an elegant and efficient signal transduction system.

Organization of the Suprachiasmatic Nucleus (SCN): The Master Timekeeper

The SCN is located in the ventral periventricular zone of the hypothalamus, dorsal to the optic chiasm, and is composed of approximately 20,000 neurons that have a circadian pattern of electrical activity [80]. Division of SCN neurons by amplitude of electrical activity, neuropeptide expression, and afferent inputs reveals two distinct subpopulations [81]. The core SCN neurons receive the majority of photic input, fire with low-amplitude rhythms that can be easily reset by environmental stimuli, and express a variety of neuropeptides including vasoactive intestinal peptide (VIP) and gastrin-releasing peptide (GRP) [82, 83]. Moreover, light-induced rhythmic oscillations occur exclusively in core neurons, suggesting that the core is primarily responsible for sensory processing of environmental cues and entrainment of the circadian system to the light-dark cycle [84, 85]. In contrast, shell SCN neurons primarily express arginine vasopressin (AVP), experience high-amplitude, self-sustaining oscillations, and receive little photic input [83]. These features highlight the shell's role in relaying SCN outputs to modulate the clock in peripheral tissue oscillators [86]. Despite these differences, communication between the core and shell SCN neurons is critical for maintaining the central pacemaker's 24-hour circadian rhythm. In an *ex vivo* SCN preparation, separation of the shell and core neurons caused the shell SCN to desynchronize [87]. Elimination of VIP in the SCN, a key marker of the core, abolished behavioral rhythms and highlighted the downstream role the substance plays in entraining SCN outputs to light stimuli [88]. In combination with studies tracing neuronal circuits [83], it is clear that signaling from the core to the shell is necessary for the entrainment of the master circadian pacemaker.

SCN-Dependent Outputs Entrain Peripheral Oscillators

The hierarchical oscillator model of the circadian system suggests that outputs from the master pacemaker can modulate peripheral circadian oscillators. The SCN directly and indirectly entrains peripheral clocks through a combination of synaptic and neuroendocrine outputs [89]. Projections from SCN neurons mainly innervate the hypothalamic paraventricular nucleus (PVN). The PVN serves as the hypothalamic hub for hormonal and autonomic control and thus effectively relays SCN outputs for the circadian control of physiological outputs. Pre-autonomic neurons in the PVN are directly innervated by the SCN and subsequently project to sympathetic and parasympathetic motor nuclei controlling several organs including the heart, pancreas, and liver [90–92]. Lesioning the connection between the SCN and PVN was effective in eliminating the circadian rhythmicity in heart rate suggesting that the SCN plays a direct role in autonomic regulation of cardiac, as well as other peripheral rhythms [91].

Additionally, the PVN relays SCN outputs to the pineal gland for the circadian regulation of melatonin secretion. The pineal gland is indirectly controlled through a multisynaptic connection. SCN outputs travel from the PVN to the intermediolateral cell column and finally to the superior cervical ganglion (SCG) which innervates the pineal gland [93]. During the day, the SCN provides an inhibitory signal (via γ-aminobutyric acid [GABA]) to this pathway in order to suppress melatonin secretion [94]. These inhibitory signals suppress the expression and activity of the enzymes responsible for synthesizing melatonin from serotonin, arylalkylamine N-acetyltransferase (AANAT) and methionine adenosyltransferase 2a (MAT2A) [95]. Conversely, glutamatergic output from the SCN to the PVN has been demonstrated to be responsible for enhanced melatonin synthesis and secretion during the night in mammals [96]. Importantly, the nightly increase in melatonin secretion has been shown to functionally impact a wide range of tissues and processes and is considered a potential zeitgeber for peripheral tissue oscillators [97–100].

The hypothalamic-pituitary-adrenal (HPA) axis also receives SCN output. In a series of experiments by Kalsbeek and Buijis, they demonstrated that the SCN mediates its effect on the HPA by secretion of the neuropeptide vasopressin during an organism's inactive cycle. In turn, this inhibits the secretion of corticotrophin-releasing hormone (CRH) and vasopressin from PVN neurons [101–104]. During an organism's active cycle when the SCN ceases to secrete vasopressin, CRH and vasopressin subsequently induce the secretion of adrenocorticotropic hormone (ACTH) from the anterior pituitary, which can then act directly on the adrenal cortex, modulating the release of glucocorticoids, catecholamines, and mineralocorticoids [105–107]. Similar to the role of melatonin, the daytime increase in stress hormone (i.e., corticosterone) secretion from the adrenal cortex has been proposed as an SCN-dependent mechanism of peripheral oscillator entrainment [108]. Through local control of its synaptic output, the SCN exerts its role as master timekeeper by controlling downstream autonomic and endocrine outputs to entrain peripheral clocks.

Non-photic Entrainment of Peripheral Oscillators

Although the SCN is fundamentally required to entrain peripheral tissues to the light-dark cycle via direct and indirect pathways, peripheral circadian rhythms are thought to also be modulated and entrained by SCN-independent mechanisms. The timing of a variety of rhythmic behaviors such as food intake and exercise has been demonstrated to phase shift peripheral oscillators while leaving SCN rhythms unaffected [42, 109, 110]. One of the strongest non-photic entrainment agents is food, specifically the timing of feeding patterns, and is likely due to feeding's role in mammalian survival [111]. Precise timing of feeding patterns or time-restricted feeding (tRF) can act as a zeitgeber for peripheral food entrainable oscillators (FEO). The concept that FEOs are independent of the SCN was confirmed in a seminal study where tRF during the inactive cycle of a mouse shifted the phase of the liver, pancreas, heart, and kidney while leaving the phase of the SCN unchanged [42]. Current research is building on this finding to identify the signals that mediate feeding-dependent phase responses. Centrally, FEOs seem to depend in part on orexin, a neuropeptide regulating appetite, because ablation of orexin-expressing neurons effectively prohibited food anticipatory activity in response to tRF [112]. Dopamine and ghrelin have also been implicated as responsible for food anticipation in response to tRF [113, 114].

In peripheral tissues, nutritional signals have been thoroughly investigated as the potential zeitgeber for FEOs. By feeding nutritionally homogeneous food to mice during their inactive phase, it was discovered that a combination of glucose and protein (casein) resulted in a significant phase advance in the liver that was not observed with mono-nutrient diets, suggesting that a balanced diet is required for entrainment of peripheral oscillators [115]. Parallel studies also demonstrated that glucose controls peripheral clock oscillations via an AMP-activated protein kinase (AMPK) signaling cascade [116, 117]. Hormones responsible for the systemic response to feeding and fasting such as insulin, incretins, and glucagon exhibit a robust circadian rhythm and thus have been suggested as potential entrainment agents for FEOs [118–120].

In addition to feeding, timing of exercise has also been implicated in entraining peripheral tissue oscillators. In humans, exposure to an acute bout of high-intensity exercise during the evening caused a significant phase shift in melatonin onset relative to the baseline melatonin onset [121]. Additional studies exposing subjects to various exercise routines have confirmed the potential role of exercise in non-photic entrainment [122]. In animal studies, wheel-running during the inactive phase caused a phase advance in the liver and kidney clocks without affecting SCN rhythms [123]. The exercise-induced entrainment was likely coupled to an SCN-independent (exercise-dependent) secretion of corticosterone and/or catecholamines. Nevertheless, other exercise-induced signals such as local hypoxia and acute inflammation have been demonstrated to modulate the peripheral circadian clock [124, 125]. Overall, non-photic cues clearly play an important role in entraining peripheral tissues to the rhythmic environment for the optimization of physiological outputs.

Molecular Basis for Circadian Rhythms

Thus far, we have considered central and peripheral circadian rhythms as cellular and physiological outputs in response to various environmental cues. This is an overly simplified view of the circadian system which ignores the intracellular mechanisms that transduce an external stimulus into a physiological response. The circadian system is centrally and peripherally encoded by a transcriptional-translational negative feedback loop (TTFL) that was first proposed by Hall, Robash, and Young [126]. Their work deciphering the molecular clock in drosophila was recognized with the Nobel Prize in Physiology or Medicine in 2017. The TTFL is an elegant and efficient negative feedback loop that encodes and maintains an organism's circadian rhythm [127]. In short, the TTFL comprises an integrated circuit whereby the positive limb initiates the transcription of a negative limb that, upon translation, translocates back into the nucleus to negatively regulate its own transcription. The discovery of this approximately 24-hour oscillator network effectively described Bunning's initial observation that the inherited FRP is derived from the period of the molecular clock and encodes the circadian rhythm.

Organization of the Mammalian Molecular Clock

The components and structure of the mammalian molecular clock are highly conserved, allowing for the translatable study of this framework in various model organisms [128]. The positive limb of the mammalian TTFL is comprised of the core circadian transcription factors circadian locomotor output cycles kaput (CLOCK) and its heterodimer partner brain and muscle aryl hydrocarbon receptor nuclear translocator-like 1 (BMAL1), encoded by the aryl hydrocarbon receptor nuclear translocator-like (*ARNTL)* gene [127]. Genetic deletion of BMAL1 leads to loss of behavioral and physiological circadian rhythms because BMAL1 is the only mammalian circadian clock gene that does not have a compensatory paralogue [129]. In contrast, loss of CLOCK expression leads to mixed phenotypes because neuronal period-aryl hydrocarbon receptor nuclear translocator protein-single-minded protein (PAS) domain protein 2 (NPAS2) can compensate since it is a paralogue to CLOCK [130]. The CLOCK-BMAL1 heterodimer promotes transcription of genes *period 1–3* (*PER1, PER2, PER3*) and *cryptochrome 1–2* (*CRY1, CRY2*), comprising the negative limb of the TTFL [131, 132]. PER and CRY proteins, along with the stabilizing casein kinases 1δ and 1ε (CK1δ, CK1ε), join together upon translation and translocate into the nucleus where they interact with CLOCK-BMAL1 to inhibit their own transcription [133]. The TTFL is further stabilized by the nuclear receptor subfamily 1, group D, member 1 and 2 (REV-ERBα/β) and nuclear receptor subfamily 1, group F, member 1 (RORα/β) that, respectively, repress and activate the transcription of *ARNTL* as part of secondary regulatory loops [134]. Overall, the molecular oscillator ensures the generation of robust circadian rhythms by encoding a 24-hour period into the TTFL.

Outside of its role as the master molecular timekeeper, the CLOCK-BMAL1 heterodimer activates transcription of target genes by binding to E-box regions in deoxyribonucleic acid (DNA), which are conserved promoters coded by a palindromic CACGTG sequence [135]. The binding of the heterodimer also includes recruitment and interaction with a variety of epigenetic factors such as histone acetyltransferases (i.e., p300), the cAMP response element binding protein (CREB), and cell-specific enhancers (i.e., pancreatic and duodenal homeobox 1) to increase expression of target genes [135–138]. A seminal study by *Koike* et al. demonstrated that CLOCK and BMAL1 bind to more than 10,000 combined sites regulating ~3000 unique genes in the mammalian liver [139]. Highlighting the critical role of the circadian architecture, these genes were significantly enriched for pathways regulating liver metabolism, cancer development, and insulin signaling. The negative repressors, PER1/PER2 and CRY1/CRY2, were demonstrated to bind to more than 12,000 and 25,000 liver sites, respectively. Consistent with this study, CRY1–CRY2 are thought to regulate nuclear receptors such as the glucocorticoid receptor, which are critical for non-photic entrainment [140]. PER1/PER2, meanwhile, have been shown to directly control cell proliferation and lipid metabolism; however, additional work is needed to fully investigate their downstream role in cellular physiology [141, 142]. Overall, the circadian control of molecular processes effectively transduces environmental cues into precisely timed physiological outputs.

Entrainment of the Molecular Clock

The entrainment of central and peripheral pacemakers is mediated by both photic and non-photic cues. These zeitgebers cause a distinct phase response by modulating the molecular clock through intricate intracellular signaling cascades. The phase shifting effects of light on SCN neurons described previously can be directly attributed to the plasticity of the system's molecular framework. Glutamatergic signaling induced by a light stimulus elicits an influx of Ca^{2+} and activation of cAMP in SCN core neurons [143]. The transcription of *PER1* and *PER2* is directly stimulated at the core because the *PER1/PER2* promoter region contains a cAMP/ Ca^{2+} response element (CRE), in addition to the E-box element [144]. Communication from the core to the shell is mainly mediated by VIP, a potent adenylate cyclase activator, which leads to an activation of cAMP and, in turn, *PER1/PER2* expression within a few hours [145]. This allows the circadian system to quickly adapt to inappropriate light exposure (i.e., jet lag) because the expression of *PER1/PER2* will reset the TTFL to a new phase as it interacts with and inhibits the BMAL1-CLOCK heterodimer. Relative to the PRC, light exposure toward dawn while endogenous *PER1/PER2* expression is low will accelerate and advance the cycle, while light during the day has essentially no impact on *PER1/PER2*. Non-photic entrainment, meanwhile, has been shown to utilize both CRE-dependent and CRE-independent mechanisms to modulate the phase of peripheral oscillators [116]. Time-restricted feeding, for instance, has been demonstrated to effectively modulate the phase of CREB, resulting in a resetting of the hepatic clock [146]. Nevertheless, parallel studies have demonstrated that inhibition of mitogen-activated protein kinases and phosphoinositide 3-kinases prevents the entrainment of the liver oscillator by insulin, suggesting that these kinases are likely involved in the signaling cascade responsible for resetting peripheral clocks [119]. Moreover, the nutrient regulated AMP-activated protein kinase has also been demonstrated to directly regulate the stability of CRY1 *in vitro*, providing additional evidence that CRE-independent pathways may be required for entrainment of peripheral clocks [116]. Finally, the heat shock signaling pathway mediated by heat shock factor 1 has been shown to be necessary for the peripheral entrainment by both heat and feeding [58, 147]. Taken together, the phase response of circadian oscillators to entrainment cues is transduced via intracellular signaling cascades that allow the circadian system to quickly adapt to its environment, the details of which are only beginning to be fully appreciated.

Conclusion

In summary, the circadian system is an endogenous feature of organisms that precisely maintains an approximately 24-hour period, driven by a molecular negative feedback loop. The tight and stable control of intrinsic circadian rhythms allows for the anticipation of external stimuli such that relevant physiological outputs can be optimized for efficient and effective responses. This chapter highlights the extensive investigations over the last 300 years that delineate the physiological and molecular mechanisms which form the framework of the circadian system. Recent clinical and preclinical studies have confirmed that disruption of these mechanisms through genetic or environmental stressors produces a state of circadian misalignment which is associated with the development of a variety of diseases. As such, further investigation defining the cellular and molecular mechanisms controlling the circadian machinery is needed to identify potential therapeutic targets and to restore the body's temporal balance in our 24/7 world.

References

1. Halberg F. Circadian (about twenty-four-hour) rhythms in experimental medicine [abridged]. Proc R Soc Med. SAGE Publications. 1963;56:253.
2. Aschoff J. A survey on biological rhythms. In: Biological Rhythms. Boston: Springer; 1981. p. 3–10.
3. Matveyenko AV. Consideration for circadian physiology in rodent research. Bethesda: American Physiological Society; 2018.
4. Javeed N, Matveyenko AV. Circadian etiology of type 2 diabetes mellitus. Physiology. 2018;33(2):138–50.
5. Qian J, Dalla Man C, Morris CJ, Cobelli C, Scheer FA. Differential effects of the circadian system and circadian misalignment on insulin sensitivity and insulin secretion in humans. Diabetes Obes Metab. 2018;20(10):2481–5.
6. Sharma A, Laurenti MC, Dalla Man C, Varghese RT, Cobelli C, Rizza RA, et al. Glucose metabolism during rotational shift-work in healthcare workers. Diabetologia. 2017;60(8): 1483–90.
7. Davis S, Mirick DK, Stevens RG. Night shift work, light at night, and risk of breast cancer. J Natl Cancer Inst. 2001;93(20):1557–62.
8. Haus EL, Smolensky MH. Shift work and cancer risk: potential mechanistic roles of circadian disruption, light at night, and sleep deprivation. Sleep Med Rev. 2013;17(4):273–84.
9. Knutsson A, Jonsson B, Akerstedt T, Orth-Gomer K. Increased risk of ischaemic heart disease in shift workers. Lancet. 1986;328(8498):89–92.
10. Nielsen. Time flies: U.S. adults now spend nearly half a day interacting with media. Nielsen Insights. 2018. https://www.nielsen.com/us/en/insights/news/2018/time-flies-us-adults-now-spend-nearly-half-a-day-interacting-with-media.html. Accessed 6 Aug 2019.
11. Stevens RG, Hansen J, Costa G, Haus E, Kauppinen T, Aronson KJ, et al. Considerations of circadian impact for defining 'shift work'in cancer studies: IARC Working Group Report. Occup Environ Med. 2011;68(2):154–62.
12. Straif K, Baan R, Grosse Y, Secretan B, El Ghissassi F, Bouvard V, et al. Carcinogenicity of shift-work, painting, and fire-fighting. The Lancet Oncology. 2007;8(12):1065–6.
13. Pan A, Schernhammer ES, Sun Q, Hu FB. Rotating night shift work and risk of type 2 diabetes: two prospective cohort studies in women. PLoS Med. 2011;8(12):e1001141.
14. Musiek ES, Xiong DD, Holtzman DM. Sleep, circadian rhythms, and the pathogenesis of Alzheimer disease. Exp Mol Med. 2015;47(3):e148.
15. Wulff K, Gatti S, Wettstein JG, Foster RG. Sleep and circadian rhythm disruption in psychiatric and neurodegenerative disease. Nat Rev Neurosci. 2010;11(8):589.
16. Scheer FA, Hilton MF, Mantzoros CS, Shea SA. Adverse metabolic and cardiovascular consequences of circadian misalignment. Proc Natl Acad Sci U S A. 2009;106(11):4453–8.
17. Rüger M, Scheer FA. Effects of circadian disruption on the cardiometabolic system. Rev Endocr Metab Disord. 2009;10(4):245–60.
18. Roenneberg T, Merrow M. Circadian clocks—the fall and rise of physiology. Nat Rev Mol Cell Biol. 2005;6(12):965.
19. Pittendrigh CS. Temporal organization: reflections of a Darwinian clock-watcher. Annu Rev Physiol. 1993;55(1):17–54.
20. Chandrashekaran M. Erwin Bünning (1906–1990): a centennial homage. J Biosci. 2006;31(1):5–12.
21. Takahashi JS, Hong H-K, Ko CH, McDearmon EL. The genetics of mammalian circadian order and disorder: implications for physiology and disease. Nat Rev Genet. 2008;9(10):764.
22. Pittendrigh CS, editor. Circadian rhythms and the circadian organization of living systems. In: Cold Spring Harbor symposia on quantitative biology. Cold Spring Harbor, NY: Cold Spring Harbor Laboratory Press; 1960.
23. Daan S. Colin Pittendrigh, Jürgen Aschoff, and the natural entrainment of circadian systems. J Biol Rhythms. 2000;15(3):195–207.

24. Pittendrigh CS. Circadian systems: entrainment. In: Biological rhythms. Boston: Springer; 1981. p. 95–124.
25. Aschoff J, editor. Exogenous and endogenous components in circadian rhythms. In: Cold Spring Harbor symposia on quantitative biology. Cold Spring Harbor, NY: Cold Spring Harbor Laboratory Press; 1960.
26. Halberg F, Prem K, Halberg F, Norman C, Cornélissen G. Origins of timed cancer treatment: early marker rhythm-guided individualized chronochemotherapy. J Exp Ther Oncol. 2006;6(1):55–61.
27. Coppel W. JB Fourier—on the occasion of his two hundredth birthday. Am Math Mon. 1969;76(5):468–83.
28. Enright JT. Data analysis. In: Biological rhythms. Boston: Springer; 1981. p. 21–39.
29. Sharma M, Palacios-Bois J, Schwartz G, Iskandar H, Thakur M, Quirion R, et al. Circadian rhythms of melatonin and cortisol in aging. Biol Psychiatry. 1989;25(3):305–19.
30. Czeisler CA, Duffy JF, Shanahan TL, Brown EN, Mitchell JF, Rimmer DW, et al. Stability, precision, and near-24-hour period of the human circadian pacemaker. Science. 1999;284(5423):2177–81.
31. Decoursey PJ, Pius S, Sandlin C, Wethey D, Schull J. Relationship of circadian temperature and activity rhythms in two rodent species. Physiol Behav. 1998;65(3):457–63.
32. Bellone GJ, Plano SA, Cardinali DP, Chada DP, Vigo DE, Golombek DA. Comparative analysis of actigraphy performance in healthy young subjects. Sleep Sci. 2016;9(4):272–9.
33. Lieberman HR, Wurtman JJ, Teicher MH. Circadian rhythms of activity in healthy young and elderly humans. Neurobiol Aging. 1989;10(3):259–65.
34. Mei L, Fan Y, Lv X, Welsh DK, Zhan C, Zhang EE. Long-term in vivo recording of circadian rhythms in brains of freely moving mice. Proc Natl Acad Sci U S A. 2018;115(16):4276–81.
35. McGlashan EM, Poudel GR, Vidafar P, Drummond SP, Cain SW. Imaging individual differences in the response of the human suprachiasmatic area to light. Front Neurol. 2018;9:1022.
36. Wittenbrink N, Ananthasubramaniam B, Münch M, Koller B, Maier B, Weschke C, et al. High-accuracy determination of internal circadian time from a single blood sample. J Clin Invest. 2018;128(9):3826–39.
37. Ruben MD, Wu G, Smith DF, Schmidt RE, Francey LJ, Lee YY, et al. A database of tissue-specific rhythmically expressed human genes has potential applications in circadian medicine. Sci Transl Med. 2018;10(458):eaat8806.
38. Pittendrigh CS, Daan S. A functional analysis of circadian pacemakers in nocturnal rodents. J Comp Physiol. 1976;106(3):223–52.
39. Aschoff J. Circadian rhythms in man. Science. 1965;148(3676):1427–32.
40. Proll J, Wever RA. The circadian system of man, results of experiments under temporal isolation. XII und 276 Seiten, 181 Abb. Springer Verlag, New York, Heidelberg, Berlin (West) 1979. Preis: 98,—DM. Food/Nahrung. 1981;25(7):708–709.
41. Campbell SS, Dawson D, Zulley J. When the human circadian system is caught napping: evidence for endogenous rhythms close to 24 hours. Sleep. 1993;16(7):638–40.
42. Damiola F, Le Minh N, Preitner N, Kornmann B, Fleury-Olela F, Schibler U. Restricted feeding uncouples circadian oscillators in peripheral tissues from the central pacemaker in the suprachiasmatic nucleus. Genes Dev. 2000;14(23):2950–61.
43. Pittendrigh CS, Minis DH. The entrainment of circadian oscillations by light and their role as photoperiodic clocks. Am Nat. 1964;98(902):261–94.
44. Johnson CH. Forty years of PRCs-what have we learned? Chronobiol Int. 1999;16(6):711–43.
45. Khalsa SBS, Jewett ME, Cajochen C, Czeisler CA. A phase response curve to single bright light pulses in human subjects. J Physiol. 2003;549(3):945–52.
46. Minors DS, Waterhouse JM, Wirz-Justice A. A human phase-response curve to light. Neurosci Lett. 1991;133(1):36–40.
47. Honma K, Honma S, Wada T. Phase-dependent shift of free-running human circadian rhythms in response to a single bright pulse. Experientia. 1987;43(11–12):1205–7.
48. Czeisler CA, Kronauer RE, Allan JS, Duffy JF, Jewett ME, Brown EN, et al. Bright light induction of strong (type 0) resetting of the human circadian pacemaker. Science. 1989;244(4910):1328–33.

49. Boivin DB, Duffy JF, Kronauer RE, Czeisler CA. Sensitivity of the human circadian pacemaker to moderately bright light. J Biol Rhythms. 1994;9(3–4):315–31.
50. Ancoli-Israel S, Martin JL, Kripke DF, Marler M, Klauber MR. Effect of light treatment on sleep and circadian rhythms in demented nursing home patients. J Am Geriatr Soc. 2002;50(2):282–9.
51. Eastman CI, Gazda CJ, Burgess HJ, Crowley SJ, Fogg LF. Advancing circadian rhythms before eastward flight: a strategy to prevent or reduce jet lag. Sleep. 2005;28(1):33–44.
52. Pittendrigh CS. On temperature independence in the clock system controlling emergence time in Drosophila. Proc Nat Acad Sci U S A. 1954;40(10):1018–29.
53. Kalmus H. Periodizität und autochronie (ideochronie) als zeitregelnde eigenschaffen der organismen. Biologia generalis. 1935;11:93–114.
54. Sweeney BM, Hastings JW, editors. Effects of temperature upon diurnal rhythms. In: Cold Spring Harbor symposia on quantitative biology. Cold Spring Harbor, NY: Cold Spring Harbor Laboratory Press; 1960.
55. Wahl O. Neue untersuchungen über das zeitgedächtnis der bienen. Z Vgl Physiol. 1932;16(3):529–89.
56. Tsuchiya Y, Akashi M, Nishida E. Temperature compensation and temperature resetting of circadian rhythms in mammalian cultured fibroblasts. Genes Cells. 2003;8(8):713–20.
57. Barrett RK, Takahashi JS. Temperature compensation and temperature entrainment of the chick pineal cell circadian clock. J Neurosci. 1995;15(8):5681–92.
58. Buhr ED, Yoo S-H, Takahashi JS. Temperature as a universal resetting cue for mammalian circadian oscillators. Science. 2010;330(6002):379–85.
59. Bünning E, editor. Opening address: biological clocks. In: Cold Spring Harbor symposia on quantitative biology. Cold Spring Harbor, NY: Cold Spring Harbor Laboratory Press; 1960.
60. Winfree AT. Biological rhythms and the behavior of populations of coupled oscillators. J Theor Biol. 1967;16(1):15–42.
61. Pavlidis T. Populations of interacting oscillators and circadian rhythms. J Theor Biol. 1969;22(3):418–36.
62. Moore RY, Eichler VB. Loss of a circadian adrenal corticosterone rhythm following suprachiasmatic lesions in the rat. Brain Res. 1972;42:201–6.
63. Stephan FK, Zucker I. Circadian rhythms in drinking behavior and locomotor activity of rats are eliminated by hypothalamic lesions. Proc Nat Acad Sci U S A. 1972;69(6):1583–6.
64. Inouye S-I, Kawamura H. Persistence of circadian rhythmicity in a mammalian hypothalamic "island" containing the suprachiasmatic nucleus. Proc Nat Acad Sci U S A. 1979;76(11):5962–6.
65. Schaap J, Pennartz CM, Meijer JH. Electrophysiology of the circadian pacemaker in mammals. Chronobiol Int. 2003;20(2):171–88.
66. Sawaki Y, Nihonmatsu I, Kawamura H. Transplantation of the neonatal suprachiasmatic nuclei into rats with complete bilateral suprachiasmatic lesions. Neurosci Res. 1984;1(1):67–72.
67. Drucker-Colín R, Aguilar-Roblero R, García-Hernández F, Fernández-Cancino F, Rattoni FB. Fetal suprachiasmatic nucleus transplants: diurnal rhythm recovery of lesioned rats. Brain Res. 1984;311(2):353–7.
68. Balsalobre A, Damiola F, Schibler U. A serum shock induces circadian gene expression in mammalian tissue culture cells. Cell. 1998;93(6):929–37.
69. Yamazaki S, Numano R, Abe M, Hida A, Takahashi R-I, Ueda M, et al. Resetting central and peripheral circadian oscillators in transgenic rats. Science. 2000;288(5466):682–5.
70. Konturek P, Brzozowski T, Konturek S. Gut clock: implication of circadian rhythms in the gastrointestinal tract. J Physiol Pharmacol. 2011;62(2):139–50.
71. Yoo S-H, Yamazaki S, Lowrey PL, Shimomura K, Ko CH, Buhr ED, et al. PERIOD2::LUCIFERASE real-time reporting of circadian dynamics reveals persistent circadian oscillations in mouse peripheral tissues. Proc Nat Acad Sci U S A. 2004;101(15):5339–46.
72. Freedman MS, Lucas RJ, Soni B, von Schantz M, Muñoz M, David-Gray Z, et al. Regulation of mammalian circadian behavior by non-rod, non-cone, ocular photoreceptors. Science. 1999;284(5413):502–4.

73. Provencio I, Jiang G, Willem J, Hayes WP, Rollag MD. Melanopsin: an opsin in melanophores, brain, and eye. Proc Nat Acad Sci U S A. 1998;95(1):340–5.
74. Takahashi JS, DeCoursey PJ, Bauman L, Menaker M. Spectral sensitivity of a novel photoreceptive system mediating entrainment of mammalian circadian rhythms. Nature. 1984;308(5955):186.
75. Berson DM, Dunn FA, Takao M. Phototransduction by retinal ganglion cells that set the circadian clock. Science. 2002;295(5557):1070–3.
76. Johnson RF, Moore RY, Morin LP. Loss of entrainment and anatomical plasticity after lesions of the hamster retinohypothalamic tract. Brain Res. 1988;460(2):297–313.
77. Cahill GM, Menaker M. Responses of the suprachiasmatic nucleus to retinohypothalamic tract volleys in a slice preparation of the mouse hypothalamus. Brain Res. 1989;479(1):65–75.
78. Ding JM, Chen D, Weber ET, Faiman LE, Rea MA, Gillette MU. Resetting the biological clock: mediation of nocturnal circadian shifts by glutamate and NO. Science. 1994;266(5191):1713–7.
79. Wang L, Schroeder A, Loh D, Smith D, Lin K, Han J, et al. Role for the NR2B subunit of the N-methyl-D-aspartate receptor in mediating light input to the circadian system. Eur J Neurosci. 2008;27(7):1771–9.
80. Schwartz WJ, Gross RA, Morton MT. The suprachiasmatic nuclei contain a tetrodotoxin-resistant circadian pacemaker. Proc Nat Acad Sci U S A. 1987;84(6):1694–8.
81. Leak RK, Moore RY. Topographic organization of suprachiasmatic nucleus projection neurons. J Comp Neurol. 2001;433(3):312–34.
82. Pulivarthy SR, Tanaka N, Welsh DK, De Haro L, Verma IM, Panda S. Reciprocity between phase shifts and amplitude changes in the mammalian circadian clock. Proc Nat Acad Sci U S A. 2007;104(51):20356–61.
83. Yan L, Karatsoreos I, LeSauter J, Welsh D, Kay S, Foley D, et al., editors. Exploring spatiotemporal organization of SCN circuits. In: Cold Spring Harbor symposia on quantitative biology. Cold Spring Harbor Laboratory Press; 2007.
84. Schwartz W, Carpino A Jr, De la Iglesia H, Baler R, Klein D, Nakabeppu Y, et al. Differential regulation of fos family genes in the ventrolateral and dorsomedial subdivisions of the rat suprachiasmatic nucleus. Neuroscience. 2000;98(3):535–47.
85. Hamada T, Antle MC, Silver R. Temporal and spatial expression patterns of canonical clock genes and clock-controlled genes in the suprachiasmatic nucleus. Eur J Neurosci. 2004;19(7):1741–8.
86. Evans JA, Suen T-C, Callif BL, Mitchell AS, Castanon-Cervantes O, Baker KM, et al. Shell neurons of the master circadian clock coordinate the phase of tissue clocks throughout the brain and body. BMC Biol. 2015;13(1):43.
87. Yamaguchi S, Isejima H, Matsuo T, Okura R, Yagita K, Kobayashi M, et al. Synchronization of cellular clocks in the suprachiasmatic nucleus. Science. 2003;302(5649):1408–12.
88. Aton SJ, Colwell CS, Harmar AJ, Waschek J, Herzog ED. Vasoactive intestinal polypeptide mediates circadian rhythmicity and synchrony in mammalian clock neurons. Nat Neurosci. 2005;8(4):476.
89. Dibner C, Schibler U, Albrecht U. The mammalian circadian timing system: organization and coordination of central and peripheral clocks. Annu Rev Physiol. 2010;72:517–49.
90. Buijs RM, Chun SJ, Niijima A, Romijn HJ, Nagai K. Parasympathetic and sympathetic control of the pancreas: a role for the suprachiasmatic nucleus and other hypothalamic centers that are involved in the regulation of food intake. J Comp Neurol. 2001;431(4):405–23.
91. Scheer F, Ter Horst G, van Der Vliet J, Buijs R. Physiological and anatomic evidence for regulation of the heart by suprachiasmatic nucleus in rats. Am J Physiol Heart Circ Physiol. 2001;280(3):H1391–9.
92. la Fleur SE, Kalsbeek A, Wortel J, Buijs RM. Polysynaptic neural pathways between the hypothalamus, including the suprachiasmatic nucleus, and the liver. Brain Res. 2000;871(1):50–6.
93. Klein D, Smoot R, Weller J, Higa S, Markey S, Creed G, et al. Lesions of the paraventricular nucleus area of the hypothalamus disrupt the suprachiasmatic→ spinal cord circuit in the melatonin rhythm generating system. Brain Res Bull. 1983;10(5):647–52.

94. Kalsbeek A, Garidou ML, Palm IF, Van Der Vliet J, Simonneaux V, Pévet P, et al. Melatonin sees the light: blocking GABA-ergic transmission in the paraventricular nucleus induces daytime secretion of melatonin. Eur J Neurosci. 2000;12(9):3146–54.
95. Kim J-S, Coon SL, Blackshaw S, Cepko CL, Møller M, Mukda S, et al. Methionine adenosyltransferase: adrenergic-cAMP mechanism regulates a daily rhythm in pineal expression. J Biol Chem. 2005;280(1):677–84.
96. Perreau-Lenz S, Kalsbeek A, Pévet P, Buijs RM. Glutamatergic clock output stimulates melatonin synthesis at night. Eur J Neurosci. 2004;19(2):318–24.
97. Costes S, Boss M, Thomas AP, Matveyenko AV. Activation of melatonin signaling promotes β-cell survival and function. Mol Endocrinol. 2015;29(5):682–92.
98. Thomas AP, Hoang J, Vongbunyong K, Nguyen A, Rakshit K, Matveyenko AV. Administration of melatonin and metformin prevents deleterious effects of circadian disruption and obesity in male rats. Endocrinology. 2016;157(12):4720–31.
99. Ahluwalia A, Brzozowska IM, Hoa N, Jones MK, Tarnawski AS. Melatonin signaling in mitochondria extends beyond neurons and neuroprotection: implications for angiogenesis and cardio/gastroprotection. Proc Nat Acad Sci U S A. 2018;115(9):E1942–3.
100. Pevet P, Challet E. Melatonin: both master clock output and internal time-giver in the circadian clocks network. J Physiol Paris. 2011;105(4–6):170–82.
101. Kalsbeek A, Buijs RM. Output pathways of the mammalian suprachiasmatic nucleus: coding circadian time by transmitter selection and specific targeting. Cell Tissue Res. 2002;309(1):109–18.
102. Buijs RM, Kalsbeek A. Hypothalamic integration of central and peripheral clocks. Nat Rev Neurosci. 2001;2(7):521.
103. Waite EJ, McKenna M, Kershaw Y, Walker JJ, Cho K, Piggins HD, et al. Ultradian corticosterone secretion is maintained in the absence of circadian cues. Eur J Neurosci. 2012;36(8):3142–50.
104. Buijs RM, Kalsbeek A, van der Woude TP, van Heerikhuize JJ, Shinn S. Suprachiasmatic nucleus lesion increases corticosterone secretion. Am J Physiol. 1993;264(6):R1186–92.
105. Ulrich-Lai YM, Herman JP. Neural regulation of endocrine and autonomic stress responses. Nat Rev Neurosci. 2009;10(6):397.
106. Valenta LJ, Elias AN, Eisenberg H. ACTH stimulation of adrenal epinephrine and norepinephrine release. Horm Res. 1986;23(1):16–20.
107. Kem DC, Weinberger MH, Gomez-Sanchez C, Kramer NJ, Lerman R, Furuyama S, et al. Circadian rhythm of plasma aldosterone concentration in patients with primary aldosteronism. J Clin Invest. 1973;52(9):2272–7.
108. Balsalobre A, Brown SA, Marcacci L, Tronche F, Kellendonk C, Reichardt HM, et al. Resetting of circadian time in peripheral tissues by glucocorticoid signaling. Science. 2000;289(5488):2344–7.
109. Stokkan K-A, Yamazaki S, Tei H, Sakaki Y, Menaker M. Entrainment of the circadian clock in the liver by feeding. Science. 2001;291(5503):490–3.
110. Edgar DM, Dement WC. Regularly scheduled voluntary exercise synchronizes the mouse circadian clock. Am J Physiol. 1991;261(4):R928–R33.
111. Piggins HD, Bechtold DA. Circadian rhythms: feeding time. Elife. 2015;4:e08166.
112. Akiyama M, Yuasa T, Hayasaka N, Horikawa K, Sakurai T, Shibata S. Reduced food anticipatory activity in genetically orexin (hypocretin) neuron-ablated mice. Eur J Neurosci. 2004;20(11):3054–62.
113. LeSauter J, Hoque N, Weintraub M, Pfaff DW, Silver R. Stomach ghrelin-secreting cells as food-entrainable circadian clocks. Proc Nat Acad Sci U S A. 2009;106:13582–7. https://doi.org/10.1073/pnas.0906426106.
114. Gallardo CM, Darvas M, Oviatt M, Chang CH, Michalik M, Huddy TF, et al. Dopamine receptor 1 neurons in the dorsal striatum regulate food anticipatory circadian activity rhythms in mice. Elife. 2014;3:e03781.
115. Hirao A, Tahara Y, Kimura I, Shibata S. A balanced diet is necessary for proper entrainment signals of the mouse liver clock. PLoS One. 2009;4(9):e6909.

116. Lamia KA, Sachdeva UM, DiTacchio L, Williams EC, Alvarez JG, Egan DF, et al. AMPK regulates the circadian clock by cryptochrome phosphorylation and degradation. Science. 2009;326(5951):437–40.
117. Qian J, Block GD, Colwell CS, Matveyenko AV. Consequences of exposure to light at night on the pancreatic islet circadian clock and function in rats. Diabetes. 2013;62:3469–78. https://doi.org/10.2337/db12-1543.
118. Gil-Lozano M, Mingomataj EL, Wu WK, Ridout SA, Brubaker PL. Circadian secretion of the intestinal hormone, glucagon-like peptide-1, by the rodent L-cell. Diabetes. 2014;63:3674–85. https://doi.org/10.2337/db13-1501.
119. Yamajuku D, Inagaki T, Haruma T, Okubo S, Kataoka Y, Kobayashi S, et al. Real-time monitoring in three-dimensional hepatocytes reveals that insulin acts as a synchronizer for liver clock. Sci Rep. 2012;2:439.
120. Sun X, Dang F, Zhang D, Yuan Y, Zhang C, Wu Y, et al. Glucagon-CREB/CRTC2 signaling cascade regulates hepatic BMAL1 protein. J Biol Chem. 2015;290(4):2189–97.
121. Buxton OM, Lee CW, L'Hermite-Balériaux M, Turek FW, Van Cauter E. Exercise elicits phase shifts and acute alterations of melatonin that vary with circadian phase. Am J Physiol. 2003;284(3):R714–R24.
122. Yamanaka Y, Hashimoto S, Masubuchi S, Natsubori A, Nishide S-Y, Honma S, et al. Differential regulation of circadian melatonin rhythm and sleep-wake cycle by bright lights and nonphotic time cues in humans. Am J Physiol. 2014;307(5):R546–R57.
123. Tahara Y, Shiraishi T, Kikuchi Y, Haraguchi A, Kuriki D, Sasaki H, et al. Entrainment of the mouse circadian clock by sub-acute physical and psychological stress. Sci Rep. 2015;5: 11417.
124. Javeed N, Rakshit K, Matveyenko A. Assessment of proinflammatory mediators of beta-cell circadian clock dysfunction in diabetes. Diabetes. 2018;67((Suppl 1):193-OR.
125. Wu Y, Tang D, Liu N, Xiong W, Huang H, Li Y, et al. Reciprocal regulation between the circadian clock and hypoxia signaling at the genome level in mammals. Cell Metab. 2017;25(1):73–85.
126. Burki T. Nobel Prize awarded for discoveries in circadian rhythm. Lancet. 2017;390(10104):e25.
127. Dunlap JC. Molecular bases for circadian clocks. Cell. 1999;96(2):271–90.
128. Takahashi JS. Transcriptional architecture of the mammalian circadian clock. Nat Rev Genet. 2017;18(3):164.
129. Bunger MK, Wilsbacher LD, Moran SM, Clendenin C, Radcliffe LA, Hogenesch JB, et al. Mop3 is an essential component of the master circadian pacemaker in mammals. Cell. 2000;103(7):1009–17.
130. DeBruyne JP, Weaver DR, Reppert SM. CLOCK and NPAS2 have overlapping roles in the suprachiasmatic circadian clock. Nat Neurosci. 2007;10(5):543.
131. Zheng B, Albrecht U, Kaasik K, Sage M, Lu W, Vaishnav S, et al. Nonredundant roles of the mPer1 and mPer2 genes in the mammalian circadian clock. Cell. 2001;105(5):683–94.
132. Vitaterna MH, Selby CP, Todo T, Niwa H, Thompson C, Fruechte EM, et al. Differential regulation of mammalian period genes and circadian rhythmicity by cryptochromes 1 and 2. Proc Nat Acad Sci U S A. 1999;96(21):12114–9.
133. Lee C, Etchegaray J-P, Cagampang FR, Loudon AS, Reppert SM. Posttranslational mechanisms regulate the mammalian circadian clock. Cell. 2001;107(7):855–67.
134. Forman BM, Chen J, Blumberg B, Kliewer SA, Henshaw R, Ong ES, et al. Cross-talk among ROR alpha 1 and the Rev-erb family of orphan nuclear receptors. Mol Endocrinol. 1994;8(9):1253–61.
135. Ripperger JA, Schibler U. Rhythmic CLOCK-BMAL1 binding to multiple E-box motifs drives circadian Dbp transcription and chromatin transitions. Nat Genet. 2006;38(3):369.
136. Hosoda H, Asano H, Ito M, Kato H, Iwamoto T, Suzuki A, et al. CBP/p300 is a cell type-specific modulator of CLOCK/BMAL1-mediated transcription. Mol Brain. 2009;2(1):34.
137. Etchegaray J-P, Lee C, Wade PA, Reppert SM. Rhythmic histone acetylation underlies transcription in the mammalian circadian clock. Nature. 2003;421(6919):177.

138. Glick E, Leshkowitz D, Walker MD. Transcription factor BETA2 acts cooperatively with E2A and PDX1 to activate the insulin gene promoter. J Biol Chem. 2000;275(3):2199–204.
139. Koike N, Yoo S-H, Huang H-C, Kumar V, Lee C, Kim T-K, et al. Transcriptional architecture and chromatin landscape of the core circadian clock in mammals. Science. 2012;338(6105):349–54.
140. Lamia KA, Papp SJ, Ruth TY, Barish GD, Uhlenhaut NH, Jonker JW, et al. Cryptochromes mediate rhythmic repression of the glucocorticoid receptor. Nature. 2011;480(7378):552.
141. Yang X, Wood PA, Ansell CM, Quiton DFT, Oh E-Y, Du-Quiton J, et al. The circadian clock gene Per1 suppresses cancer cell proliferation and tumor growth at specific times of day. Chronobiol Int. 2009;26(7):1323–39.
142. Grimaldi B, Bellet MM, Katada S, Astarita G, Hirayama J, Amin RH, et al. PER2 controls lipid metabolism by direct regulation of PPARγ. Cell Metab. 2010;12(5):509–20.
143. Kuhlman SJ, Silver R, Le Sauter J, Bult-Ito A, McMahon DG. Phase resetting light pulses induce Per1 and persistent spike activity in a subpopulation of biological clock neurons. J Neurosci. 2003;23(4):1441–50.
144. Obrietan K, Impey S, Smith D, Athos J, Storm DR. Circadian regulation of cAMP response element-mediated gene expression in the suprachiasmatic nuclei. J Biol Chem. 1999;274(25):17748–56.
145. Yan L, Silver R. Differential induction and localization of mPer1 and mPer2 during advancing and delaying phase shifts. Eur J Neurosci. 2002;16(8):1531–40.
146. Vollmers C, Gill S, DiTacchio L, Pulivarthy SR, Le HD, Panda S. Time of feeding and the intrinsic circadian clock drive rhythms in hepatic gene expression. Proc Nat Acad Sci U S A. 2009;106(50):21453–8.
147. Reinke H, Saini C, Fleury-Olela F, Dibner C, Benjamin IJ, Schibler U. Differential display of DNA-binding proteins reveals heat-shock factor 1 as a circadian transcription factor. Genes Dev. 2008;22(3):331–45.

Chapter 2
Review of Protocols and Terminology to Enhance Understanding of Circadian-Based Literature

Vincent A. LaBarbera and Katherine M. Sharkey

The rhythms of life have intrigued humankind for millennia—since the first observations of the periodicity of parasitic fevers during the time of Hippocrates—through today, including the awarding of the 2017 Nobel Prize in Physiology or Medicine for the isolation and characterization of the clock gene "period" and its molecular regulation of rhythms in the fruit fly *Drosophila* [1, 2].

A biological rhythm can be defined as the recurrence of an event "within a biological system at more-or-less regular intervals" [3, 4]. The intervals may be on the order of one cycle per millisecond or per years and may occur at the level of the cell within an organism or even at the population level. Biological rhythms may be considered *exogenous*, meaning arising as a response to a periodic input coming from outside of the biological unit, or *endogenous*, that is those which arise from within.

The word *circadian* is derived from the Latin words "circa" and "diem," translating to "about a day." The term dates to 1959 and is attributed to Professor Franz Halberg, a leading circadian researcher from the University of Minnesota. Halberg also notably coined the term *chronobiology* or the study of time as it relates to biological processes. In addition to *circadian*, the description of other "circa" rhythms are attributed to Halberg: *circatidal* (in relation to the natural rhythm of the oceans' tides), *circalunar* (in relation to the approximately monthly rhythm of the moon's orbit around the Earth), and *circannual* (in relation to the yearly revolution of the Earth around our sun). *Circhoral* is used to describe an approximately hourly rhythm, the most well-studied being episodic hormone secretion [4].

V. A. LaBarbera
Department of Neurology, Rhode Island Hospital/Warren Alpert Medical School of Brown University, Providence, RI, USA

K. M. Sharkey (✉)
Department of Medicine and Psychiatry and Human Behavior, Alpert Medical School of Brown University, Providence, RI, USA
e-mail: katherine_sharkey@brown.edu

R. R. Auger (ed.), *Circadian Rhythm Sleep-Wake Disorders*,
https://doi.org/10.1007/978-3-030-43803-6_2

Circadian rhythms run independently of exogenous factors and are driven by an internal biological clock. Several examples of circadian rhythms have been documented throughout the plant and animal kingdoms. Notable examples in humans include the sleep-wake cycle, the rhythmicity of core body temperature, and hormonal cycling. For cycles with durations shorter and longer than 24 hours, the terms *ultradian* and *infradian*, respectively, are used. For instance, rapid-eye-movement and non-rapid-eye-movement (REM-NREM) sleep cycles that occur at ~90–120-minute intervals during sleep are examples of ultradian rhythms, whereas the human menstrual cycle, which lasts approximately 28 days, is an infradian rhythm.

The *suprachiasmatic nuclei* (SCN) of the anterior hypothalamus comprise the dominant clock in the brain and serve as the prime driver of circadian rhythms in mammals. Environmental lighting conditions are relayed to the SCN through the *retinohypothalamic tract* via unique receptors, the *intrinsically photosensitive retinal ganglion cells* (ipRGCs). Separate from the rods and cones that relay information to the visual cortex for image formation, the ipRGCs, along with input from the intergeniculate leaflet of the thalamus and the midbrain raphe nuclei, convey both photic and nonphotic information to the SCN circadian clock. The SCN, in turn, regulate peripheral clocks in cells throughout the brain and the rest of the body to produce downstream circadian rhythms in virtually every aspect of physiology and behavior, from circulation of immune cells to hormonal cycling to digestion and metabolism to cognition and mood regulation. Interestingly, there can be a hierarchical entrainment of multiple oscillations within an organism that follow an established sequence laid forth by the primary pacemaker. These secondary oscillations maintain an entrained cycle, but only due to the initial entrainment of the system pacemaker [3].

Circadian rhythms play an important role in our sleep-wake cycles. Sleep and wakefulness coordinate and counteract one another via two processes. One is the circadian process, which drives diurnal species to be behaviorally active during the solar day and to sleep in the dark at night. The other is the homeostatic process, which is dependent on the length of time that an individual has been awake, and aims to equilibrate the physiologic need for sleep with sleep initiation and maintenance. In other words, the longer an individual is awake, the greater the homeostatic drive for sleep; the converse is also true, in that if an individual has had recent sleep, the homeostatic drive for sleep is lessened [5].

The properties of circadian rhythms are analogous to the terminology describing harmonic oscillations in the field of physics. Circadian rhythms are represented schematically by the sine wave function (Fig. 2.1). In the case of the circadian rhythm of sleep propensity, the range from neutral sleepiness to the highest or lowest level of sleepiness, similar to the maximum displacement from equilibrium, is called the amplitude. Amplitude of sleep propensity can be mutable, depending on variables such as age, gender, or the individual's sleep state (i.e., NREM vs REM sleep). The duration of the rhythm from nadir to nadir, or peak to peak, is termed the period or tau (τ). This also may be thought of as the intrinsic duration of the internal clock in "free-running conditions," such as in no-light environments without time cues. In humans, period length is nearly 24 hours.

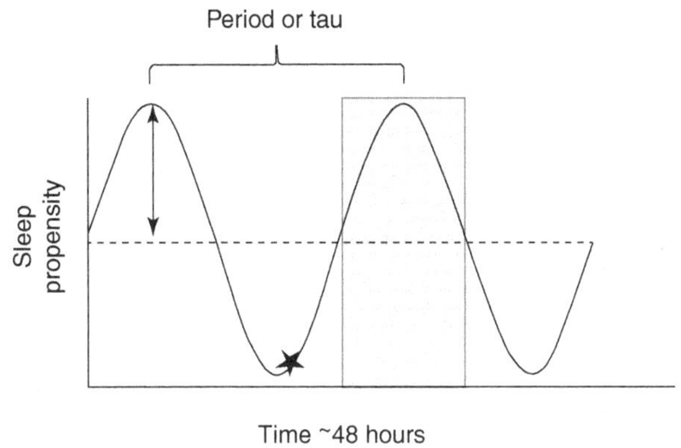

Fig. 2.1 This figure depicts an example of the circadian rhythm of sleep propensity across two cycles represented by a sine wave. The gray-shaded box indicates usual sleep time, the double-sided arrow depicts amplitude, and the black star notes a specific phase position, in this case, just after the nadir of sleepiness

To describe the current state of a particular circadian measurement, such as sleep, temperature, and melatonin secretion, among others, one uses the term "*phase*," which identifies the state at a specific instance in time. Commonly measured phase positions depend on the circadian marker of interest. For example, a conventional marker of temperature phase is the minimum, whereas for hormonal secretion the onset or peak is frequently used. The phase angle (ψ) describes the duration of time between two circadian rhythms, for example, an individual's sleep onset, midpoint, or offset in relation to the time of the temperature minimum, or the onset or peak of melatonin secretion.

Individuals can be catagorized by *chronotype*, i.e., a phenotype describing the person's tendency to adhere to a particular periodicity. "Larks" describe individuals with a tendency and/or preference to awaken early and retire early, whereas "owls" have and/or prefer later sleep patterns. Most individuals have a neutral chronotype landing somewhere in between owls and larks [6]. When an individual's chronotype is out of sync with their desired or required daily work, school, or social schedule, there are implications regarding the perpetuation of insomnia or the experience of daytime sleepiness.

Because most individuals' internal body clocks have a period length (tau) that is not precisely 24 hours in length, circadian rhythms must be synchronized or "entrained," daily. These small adjustments occur mostly in response to the natural light-dark cycle, which is the strongest "zeitgeber" or "time giver" to the biological clock. Entrainment allows the organism to align the internal clock with external time cues, including light-dark patterns. If the light-dark cycle changes, the circadian rhythms shift gradually to re-entrain with the new cycle. However, in the absence of time cues, these rhythms "free run," thereby only cycling based on endogenous periodicity. Since the endogenous tau is usually close, but not equal, to 24 hours, if entrainment is not achieved, the internal circadian rhythms may become uncoupled from external time cues. The most common cause of free-running circadian rhythms is a lack of photic stimulation of sufficient strength to entrain the SCN, such as may occur among people with blindness.

Circadian rhythms are also vulnerable to misalignment, wherein the phase angle between the endogenous propensity for sleep or wake and external time cues and schedules is not synchronized. Common causes of circadian misalignment are jet lag, daylight savings time, and night shift work. In these cases, the internal clock needs to readjust or "phase shift" to achieve realignment. An individual can "phase advance," which refers to circadian rhythms resetting to an earlier time, as is required for most eastward travel or "springing forward," or "phase delay," where the rhythms must entrain to later time cues. Since most individuals have a period length (tau) slightly longer than 24 hours, it is generally easier to phase delay than to phase advance. Exposure to the external light-dark cycle is the strongest zeitgeber to promote re-entrainment and correct misalignment. Night shift workers represent a special case of circadian misalignment because they continue to be exposed to a light-dark cycle that conflicts with re-entrainment.

Phase shifting to an earlier time, that is, phase advancing, is best accomplished with exposure to morning bright light, whereas exposure to evening bright light will facilitate a phase delay. The phase shifts produced by a stimulus (zeitgebers such as light exposure, exercise, or exogenous melatonin administration) at a specific circadian phase can be described using a phase response curve (PRC, See Fig. 2.2). A PRC is created by plotting the circadian phase shifts produced across multiple trials of zeitgeber exposure at different circadian phases. The resultant PRC can help researchers predict the magnitude and direction of a phase shift in response to zeitgeber exposure across the 24-hour circadian cycle. There is a robust literature on phase response to various stimuli, with several phase response curves documented to exogenous melatonin, light of variable intensity, duration, and wavelength, as well as physical activity [7–18]. One of the most potent zeitgebers is photic stimulation that does not necessarily come from natural light. For example, bright artificial lights and light of short wavelength (i.e., blue-green) have significant effects on the biological clock.

Protocols

Various protocols can be used to measure circadian rhythms. Oftentimes, research schema can measure circadian rhythms with more precision than can be done clinically. Two such protocols are the constant routine (CR) and the forced desynchrony (FD) [5].

Constant routine protocols aim to mitigate factors that impede the researcher's or clinician's attempt to measure endogenous circadian rhythms. Constant routines (CRs) minimize variables that influence the outcome of interest, literally by keeping the data collection conditions as constant as possible. There are many published CR protocols that impose strict limits on factors that can mask the output of the clock, including eating/feeding, movement, exercise, postural changes, light exposure, cognitive load, and knowledge of clock time e.g., [19, 20]. For example, the pattern of melatonin secretion is affected by light exposure and can be measured with more

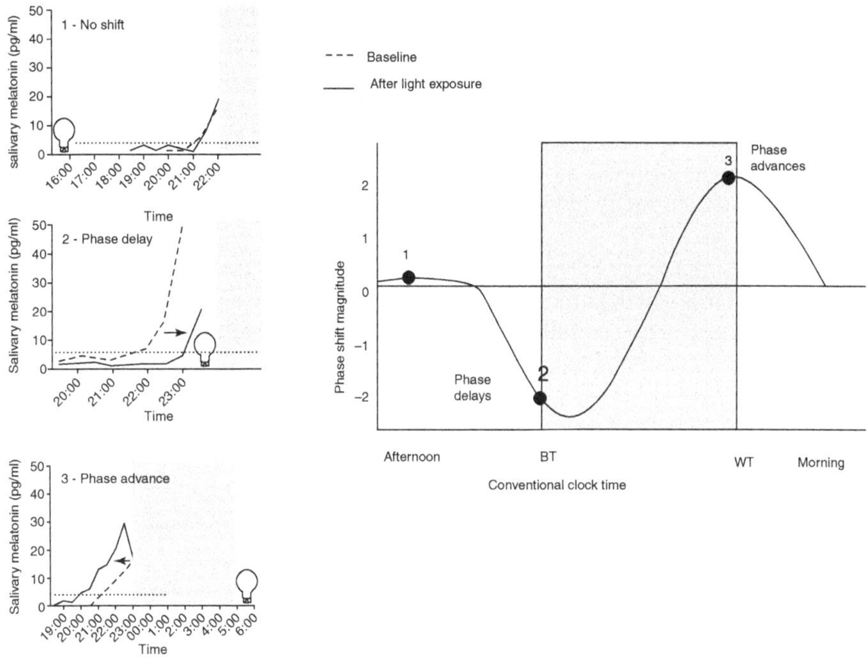

Fig. 2.2 An example phase response curve to light: This schematic illustrates how a phase response curve (PRC) is derived and indicates the expected circadian response to bright light exposure at various times of day. A PRC plots the results of multiple experimental trials; in each trial, the baseline phase position is measured and then the stimulus is presented at a specific time of day. A second phase position measure is obtained, and the difference between the two phase measures is calculated and plotted on the PRC with the stimulus time on the x-axis and the magnitude of the difference in the two phase measures on the y-axis. By convention, phase advances are plotted as positive and phase delays are plotted as negative, that is, below the 0 (no shift) line. The three small panels on the left show the expected phase shift in dim light melatonin onset (DLMO) at the three time-points plotted on the large PRC. No shift is expected when light is presented in the afternoon (point 1), a delay in DLMO is anticipated with bright light exposure in the late evening and early part of the subjective night (point 2), and a phase advance in DLMO is expected when light exposure occurs in the late part of the subjective night and early morning (point 3). There are many specific PRCs to light that detail the intensity and duration of light as well as other study parameters, including references [3, 8, 11–14]. (Graphic by Alexander Callahan)

fidelity by providing closely monitored uniform conditions in which samples for melatonin measurement are collected only in dim and/or long wavelength light. Many CR protocols require that participants maintain a wakeful state in a reclined position to attenuate the effects of sleep or activity on measurements of the circadian rhythm of body temperature. In other CR protocols, regimentation of food and liquid intake to regularly spaced meals, or even intravenous nutrition, have also been used. These protocols may require participants to stay in a laboratory setting for one or more circadian cycles to determine the endogenous circadian phase position, amplitude, etc.

Other less commonly used modifications of the constant routine protocol are the constant bed rest protocol and multiple nap protocol [5]. Constant bed rest allows patients to choose when and for how long they sleep. This protocol is oftentimes less burdensome on the patient, and circadian rhythms such as sleep propensity or REM sleep can be measured. The multiple nap protocol schedules longer naps across the day to suppress the homeostatic drive for sleep, thereby demonstrating rhythms that are normally hidden, such as subjective sleepiness in association with circadian cycles.

Forced desynchrony (FD) is another circadian rhythm measurement technique used mostly for research purposes. FD protocols aim to uncouple the two processes that control sleep and wakefulness: the circadian process and the homeostatic process. In a healthy, entrained individual, the endogenous circadian rhythm and homeostatic drive are highly coordinated, resulting in consolidated wakefulness during the day and sleep at night. This makes it impossible to distinguish whether an outcome of interest is related to the homeostatic drive, circadian phase, or both.

In an FD protocol, the sleep-wake schedule, and thus by definition the light-dark cycle, is altered such that the research participant's circadian cycle cannot entrain to the manipulation of their sleep timing. For example, a "20-hour day" FD protocol is comprised of 13.3 hours of wake and 6.7 hours of sleep opportunities, such that FD Day 2 starts after just 20 hours and when there are 4 hours remaining in the first calendar day. A 20-hour day is outside the human "range of entrainment." This means that the internal clock cannot adjust to a cycle that is so different from its endogenous cycle length. In these circumstances, the internal clock will continue its circadian rhythms on its endogenous tau and become desynchronized from the sleep-wake cycle imposed by the FD protocol. After several cycles, the circadian pacemaker and the homeostatic drive become uncoupled, so that they are running in parallel, each maintaining its own properties. Over time in an FD protocol, scheduled sleep and wakefulness will be allocated across all phases of the endogenous circadian rhythm, and variables of interest can be measured with respect to each parameter. In this manner, the homeostatic drive for sleep, which increases across wakefulness and diminishes with sleep, and the circadian drive for sleep, which is low during the typical "day" and increases when sleep occurs at night, become dissociated and are less confounding on the dependent variable being measured. Common dependent variables, such as sleepiness, metabolism, performance tasks and reaction time, mood, etc., can be measured upon awakening, thereby deducing the effect of circadian rhythm while controlling for the homeostatic drive, as ideally the drive to sleep would have dissipated during the preceding sleep opportunity.

Astute observers have utilized the concepts described in this chapter in their quest to characterize circadian behaviors and phenomena witnessed in the natural world for centuries. In recent decades, scientists and clinicians have applied these concepts to study the organization of fundamental physiological processes (e.g., eating/feeding, metabolism, hormonal regulation, growth, sleep, and wake) and have gained a deeper understanding of complex pathophysiology that is impacted by circadian rhythms.

References

1. Refinetti R. Early research on circadian rhythms. In: Refinetti R, editor. Circadian physiology. 3rd ed. Boca Raton: CRC Press, Taylor & Francis Group; 2016. p. 3–32.
2. Callaway E, Ledford H. Medicine Nobel awarded for work on circadian clocks. Nature. 2017;550(7674):18.
3. Moore-Ede MC, Sulzman FM, Fuller CA. The clocks that time us: physiology of the circadian timing system. Cambridge, UK: Harvard University Press; 1982.
4. Aschoff J. Handbook of behavioral neurobiology, v.4 biological rhythms. New York: Plenum Press; 1981.
5. Wirz-Justice A. How to measure circadian rhythms in humans. Medicographia. 2007;29(1):84–90.
6. Horne JA, Ostberg O. A self-assessment questionnaire to determine morningness-eveningness in human circadian rhythms. Int J Chronobiol. 1976;4(2):97–110.
7. Wever RA. Light effects on human circadian rhythms. A review of recent experiments. J Biol Rhythms. 1989;4:161–85.
8. Boivin DB, Duffy JF, Kronauer RE, et al. Dose–response relationships for resetting of human circadian clock by light. Nature. 1996;379:540–2.
9. Brainard GC, Hanifin JP, Greeson JM, et al. Action spectrum for melatonin regulation in humans: evidence for a novel circadian photoreceptor. J Neurosci. 2001;21(16):6405–12.
10. Wright HR, Lack LC. Effect of light wavelength on suppression and phase delay of the melatonin rhythm. Chronobiol Int. 2001;18:801–8.
11. Czeisler CA, Kronauer RE, Allan JS, et al. Bright light induction of strong (type 0) resetting of the human circadian pacemaker. Science. 1989;244:1328–33.
12. Honma K, Honma S. A human phase response curve for bright light pulses. Jpn J Psychiatry Neurol. 1988;42:167–8.
13. Minors DS, Waterhouse JM, Wirz-Justice A. A human phase–response curve to light. Neurosci Lett. 1991;133:36–40.
14. Gooley JJ, Rajaratnam SM, Brainard GC, et al. Spectral responses of the human circadian system depend on irradiance and duration of exposure to light. Sci Transl Med. 2010;2:31ra33.
15. Lewy AJ, Ahmed S, Jackson JML, et al. Melatonin shifts human circadian rhythms according to a phase–response curve. Chronobiol Int. 1992;9:380–92.
16. Baehr EK, Fogg LF, Eastman CI. Intermittent bright light and exercise to entrain human circadian rhythms to night work. Am J Physiol. 1999;277:R1598–604.
17. Buxton OM, Lee CW, L'Hermite-Baleriaux M, et al. Exercise elicits phase shifts and acute alterations of melatonin that vary with circadian phase. Am J Physiol Regul Integr Comp Physiol. 2003;284:R714–24.
18. Eastman CI, Hoese EK, Youngstedt SD, et al. Phase-shifting human circadian rhythms with exercise during the night shift. Physiol Behav. 1995;58:1287–91.
19. Minors DS, Waterhouse JM. The use of constant routines in unmasking the endogenous component of human circadian rhythms. Chronobiol Int. 1984;1(3):205–16.
20. Duffy JF, Dijk DJ. Getting through to circadian oscillators: why use constant routines? J Biol Rhythms. 2002;17(1):4–13.

Chapter 3
Introduction to Circadian Rhythm Disorders

Elliott Kyung Lee

Definition

The term "circadian rhythm sleep/wake disorders" is used to encompass a wide variety of maladies in which there is misalignment of the endogenous circadian rhythm and the light/dark cycle that give rise to various sleep-wake complaints [1]. The word "circadian" is derived from two Latin terms—"circa" meaning about and "diem" which means "day," hence "about a day" [2]. When this term is applied to physiologic conditions, the implication is that these rhythms would continue in this near-24-hour cycle endogenously, i.e., in the absence of exogenous factors [3]. When synchronized to environmental cues such as the light/dark cycle, or social/activity cycles, this process is termed entrainment [4]. Each factor that can adjust or entrain this circadian rhythm is identified as a *zeitgeber*, which is German for "time givers" [2]. The history and evolution of these terms and concepts are described in Chap. 1.

In previous iterations of the International Classification of Sleep Disorders (ICSD), dyssynchrony between the circadian rhythm and the accompanying environment was identified simply as a circadian *sleep* disorder; the term "wake" was added to the most recent iteration to draw attention to the additional impairments in daytime function [5]. They are thought to occur because either (a) the external environment is not properly synchronized with the internal circadian system (e.g., jet lag, shift work) or (b) the circadian system itself is misaligned with the external environment (e.g., delayed sleep phase, advanced sleep phase, etc.) [4]. The clinical presentation of these disorders, however, is also influenced by numerous environmental, physiologic, psychologic, and social factors.

E. K. Lee (✉)
Department of Psychiatry, Royal Ottawa Mental Health Center and Institute for Mental Health Research, University of Ottawa, Ottawa, ON, Canada
e-mail: elliott.lee@theroyal.ca

R. R. Auger (ed.), *Circadian Rhythm Sleep-Wake Disorders*,
https://doi.org/10.1007/978-3-030-43803-6_3

Two major protocols have been developed to evaluate circadian rhythms [3, 6]. The first is a constant routine, during which subjects are kept awake for 24–48 hours in constant dim light conditions. The second is a forced desynchrony protocol, during which subjects alter the timing of their sleep and wake periods to prevent entrainment, while still preserving a 2:1 ratio of wake/sleep. The duration of the allotted sleep periods can vary widely under such protocols depending upon the investigative goals, ranging from as short as 7 minutes for sleep alternating with 14 minutes for wakefulness up to even a 42-hour day (i.e., a 14-hour sleep episode coupled with a 28-hour wake episode). Such protocols "desynchronize" physiologic circadian rhythms from the sleep/wake cycle to permit isolated analyses and are described in greater detail in Chap. 2. Their implementation consumes considerable resources and thus are not practical for routine clinical use.

Extensive research has determined that the human circadian cycle is controlled by the suprachiasmatic nuclei (SCN), which serves as the central biological clock ("master clock") or pacemaker, and is located in the anterior basal hypothalamus [4]. Under "free-running" conditions (i.e., in the absence of any cues or entrainment factors), these nuclei have an endogenous rhythm that averages 24.18 hours (range 23.47–24.64) [7, 8], i.e., usually slightly longer than the 24-hour day [2]. This genetically programmed rhythm is sometimes referred to as "tau" [9]. Further studies have shown that this intrinsic circadian period is slightly shorter in women (24.09 hours ± 0.2 hours) compared to men (24.19 hours ± 0.2 hours), which may have implications for sleep duration, insomnia symptoms and other circadian rhythm sleep/wake disorders [10] (see Chap. 9). Consequently, to function adequately in a 24-hour day, these nuclei require constant resetting or entrainment. This is accomplished by exposure to various environmental cues which serve as zeitgebers. These cues can reposition the circadian cycle forward or backward, depending on the type, timing, and intensity of exposure [1].

The generation of this circadian rhythmicity within SCN cells, as well as other cells in the body, is largely determined by numerous genes that are carefully regulated. These processes are described in more detail in Chap. 1. Central to this process are the positive transcription factors CLOCK and BMAL1 [11]. These proteins exit the nucleus and heterodimerize to act as an "on" switch for the beginning of the day by reentering the nucleus and binding to enhancer box (E-box) promoter elements of the PERIOD (PER1, PER2, PER3) and CRYPTOCHROME (CRY1, CRY2) gene families to induce their expressions [12]. PER and CRY proteins accumulate over the afternoon and peak in the evening in the extracellular space, before being phosphorylated. PER proteins phosphorylated by casein kinase enzymes are marked for proteosomal degradation. However, degradation is inhibited if CRY binds to PER1/PER 2, and this heterodimer is stabilized by PER3. In this instance, this PER-CRY-CK phosphorylated multicomplex is translocated to the cell nucleus in order to exert negative feedback on the expression of the CLOCK/BMAL1 complex [6]. Because of their integral role in maintaining circadian rhythmicity, variations, mutations, and/or polymorphisms in these genes may be linked to subsequent increases or decreases in circadian period length, which may underlie the development of some circadian rhythm sleep/wake disorders (Fig. 3.1) [12]. Circadian rhythm derangements have been implicated in a wide variety of disorders including cancers, mood disorders, neurodegenerative disease, cardiovascular diseases, endocrine difficulties, and gastrointestinal tract issues [3, 13].

Borbély initially proposed a two-process model to explain how the sleep/wake cycle is regulated, and this work laid the foundation for decades of additional research [14, 15]. The model proposes two theoretical primary influences, a longitudinal homeostatic drive, termed "Process S," and a circadian drive, termed "Process C." Normally the two work in concert to promote maximal wakefulness in

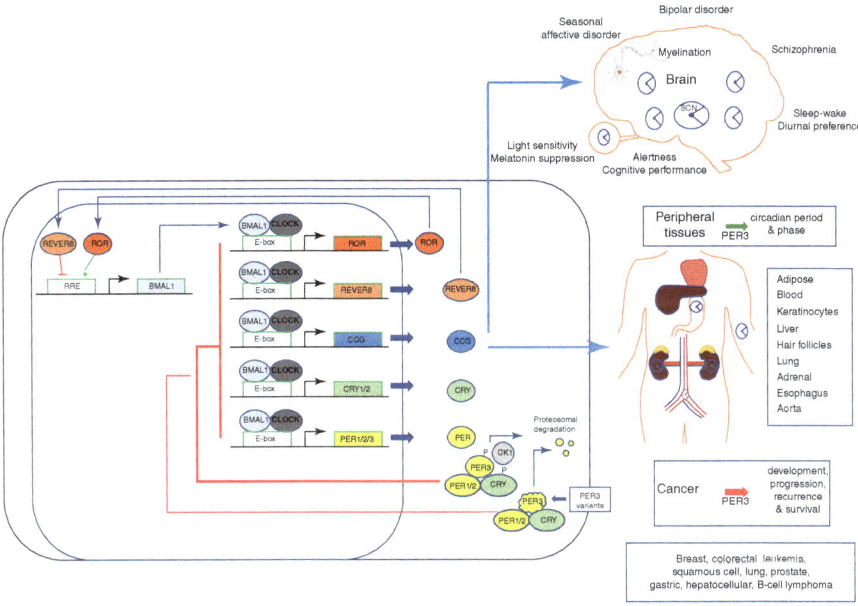

Fig. 3.1 The circadian clock consists of positive and negative integrated transcription and translation feedback loops. Transcription factors BMAL1 (also known as Arntl1 and Mop3) and CLOCK form a heterodimer that then binds to E-box motifs to promote the transcription of clock-controlled genes (CCGs), including CRY1/CRY2, PER1/PER2/PER3, REVERB and ROR, as well as others. CCGs go on to communicate the circadian timing period to a variety of other cellular processes throughout numerous peripheral tissues to maintain rhythmic cellular processes. The effects of CCGs may also underlie several circadian disturbances seen in numerous psychiatric disorders including mood and psychotic disorders, seasonal affective disorder, and diurnal preferences (circadian typology). ROR and REVERB proteins feedback and bind to ROR/REVERB response elements (RRE) to either enhance (ROR) or suppress (REVERB) expression of BMAL1. CRY and PER proteins dimerize in several combinations and subsequently translocate back into the nucleus to inhibit CLOCK/BMAL1 activity, resulting in autoregulation of PER and CRY, and other CCG activity. Alternatively, CRY/PER heterodimer proteins can also become phosphorylated by CK1 variants (e.g., CK1 delta, CK1epsilon) and thereby tagged for proteasomal degradation. Consequently, variants in PER and CRY genes can influence the stability of the circadian phase. For instance, PER3 variants (shown in the figure) can influence the expression of many CCGs and may underlie a wide range of phenotypes and disease conditions including numerous cancers. (Reprinted from Archer et al. [13], with permission from Elsevier)

the morning and increased propensity for sleep in the late evening. Process S posits a drive for sleep that increases with increased time spent in wakefulness [16]. An increase in adenosine over the course of a 24-hour period has been associated with an increased drive for sleep and likely is an important endogenous homeostatic sleep factor [17]. Other established markers include slow wave activity (SWA) in NREM sleep, as well as theta activity in wakefulness [11]. Only sleep can reduce this accumulated homeostatic drive [4]. While quality of sleep is driven by Process S, sleep quantity is more strongly influenced by circadian factors or Process C [16, 18]. This system has three components: the circadian oscillator in the SCN (approximating a rhythm of 24.18 hours), input pathways for other external/environmental stimuli (primarily light) to synchronize the SCN, and finally the output pathways from the SCN [4]. When properly aligned with Process S, this endogenous rhythm facilitates wakefulness during the day and continuous sleep at night. Subsequent to prolonged wakefulness, Process C will prevent sleep recovery at certain circadian phases, primarily during the second half of the waking day, and especially 2 hours before the habitual sleep onset time (sometimes referred to as "forbidden zones" or "wake maintenance zones"), even if homeostatic sleep drive is high (Fig. 3.2). These processes and their implications are described in greater detail in Chap. 11.

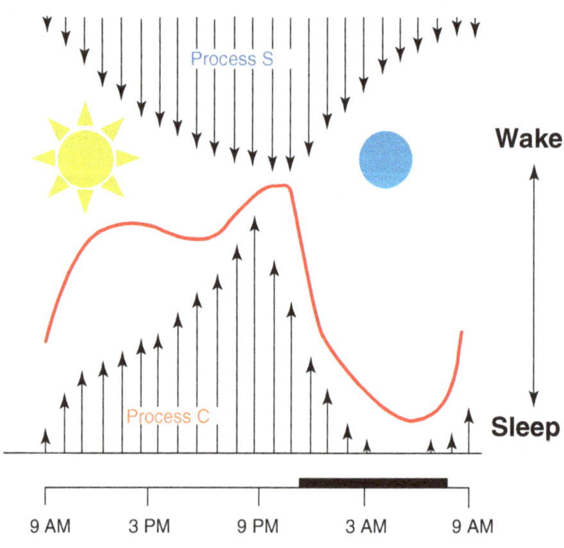

Fig. 3.2 Circadian oscillation (Process C) and homeostatic sleep drive (Process S) both influence the sleep-wake cycle. The red line indicates the wake propensity, with highest wakefulness seen at 9–10 PM and lowest wakefulness seen at approximately 6 AM. This illustrates that the circadian clock has a more powerful influence on wake propensity than homeostatic sleep drive. For instance, at 9 PM, although homeostatic sleep drive is strong, sleep does not readily occur because of the high level of alertness caused by the high circadian wakefulness signal. At approximately 6 AM, wake propensity is low despite the low homeostatic sleep drive, due to the trough in the circadian alerting signal. This figure appears in Chap. 11

The SCN receives external input primarily from the retina, which has specific photosensitive retinal ganglion cells that contain specialized photoreceptor cells with melanopsin. These cells, also known as intrinsically photosensitive retinal ganglion cells (ipRGCs), are distributed around the periphery of the retina. They are distinct from rods and cones, which are the visual cells. As a result, photosensitivity can be preserved even in conditions of visual loss [19]. Retinal physiology is described in greater detail in Chap. 14. These ipRGCs are most sensitive to light at a "short" wavelength of approximately 460 nm (450–480 nm (blue)) and least sensitive to "long" wavelengths of 595–660 nm (red/amber) [4, 19–21]. These cells convey photic input to the SCN via the retinohypothalamic tract, as well as to the pineal gland through the superior cervical ganglion [7].

The SCN also receives melatonin input from the pineal gland, which regulates its output. Efferent action on melatonin 1 (MT-1) and melatonin 2 (MT-2) receptors serve as darkness signals, sharply decreasing SCN activity. One of melatonin's functions, then, is to create a sleep permissive state during a limited time range. The timing of melatonin secretion, however, is regulated by output from the SCN and usually begins to rise approximately 2–3 hours before the natural sleep onset time, peaking in the middle of the sleep period [22]. Consequently, the onset of rise in melatonin levels under dim light conditions, also known as the dim light melatonin onset (DLMO), is one of several stable biological markers of circadian phase [22]. Operationally, this is defined when salivary melatonin rises above 2–3 pg/ml or plasma melatonin rises above 10 pg/ml (Fig. 3.3) [23, 24]. Melatonin production

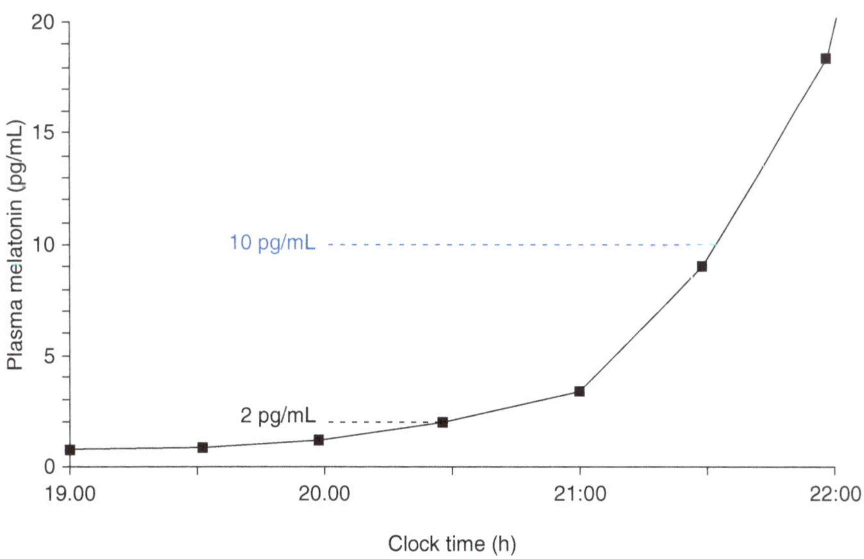

Fig. 3.3 The dim light melatonin onset (DLMO) in plasma is defined as the interpolated time when melatonin levels continuously rise above a threshold of either 2–3 pg/mL (3 pg/mL not shown) or 10 pg/mL (which usually occurs 1 hour later). In this figure, DLMO2 is approximately 2030 hours and DLMO10 is approximately 2130 hours. (Reprinted from Lewy et al. [23] (Open Access))

can be readily suppressed by light exposure through the retinal melanopsinergic system [25], beginning with light intensities under 200 lux, and more readily with light in the 460–480 nm wavelength [25]. Additionally, damage to the superior cervical ganglion, for instance, with cervical trauma, can also lead to lowered melatonin production and subsequent disruption of circadian rhythm [25].

Phase Response Curves of Light, Melatonin

Both light and melatonin can alter the circadian oscillator (SCN) and its subsequent output. Phase response curves (PRCs) of the circadian system for both light and melatonin have been derived. A PRC outlines the magnitude and direction of response (phase advance, phase delay, or neutral) of the circadian system to a zeitgeber for a given time. For a normally entrained person, light will advance the circadian rhythm if given in the early morning hours, while light will delay the rhythm if given in the evening hours [26]. Such findings have significant implications for accelerating entrainment to a new time zone or accommodating to shift work, for instance (see Chaps. 11 and 13 for more details; also see Fig. 3.4 for a diagram of typical phase relationships and Fig. 3.5 for an illustration of the PRC to light [27]).

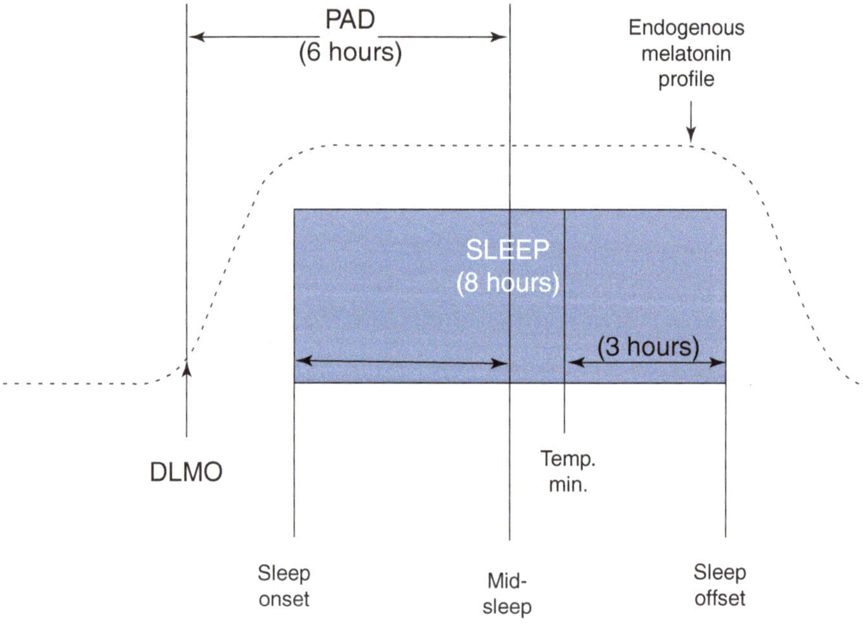

Fig. 3.4 Schematic diagram of normal phase relationships (rounded to the nearest integer) between sleep phase markers including dim light melatonin onset (DLMO), the endogenous melatonin profile, core body temperature minimum, and an 8-hour sleep time. The phase angle difference (PAD) is the hypothesized interval between the DLMO and mid sleep, shown as 6 hours in this figure, which is the average PAD for healthy controls. In patients with phase delay, for instance, the PAD would be ≤6 hours, while those who are advanced would have a PAD ≥6 hours. (Reprinted from Lewy et al. [23] (Open Access))

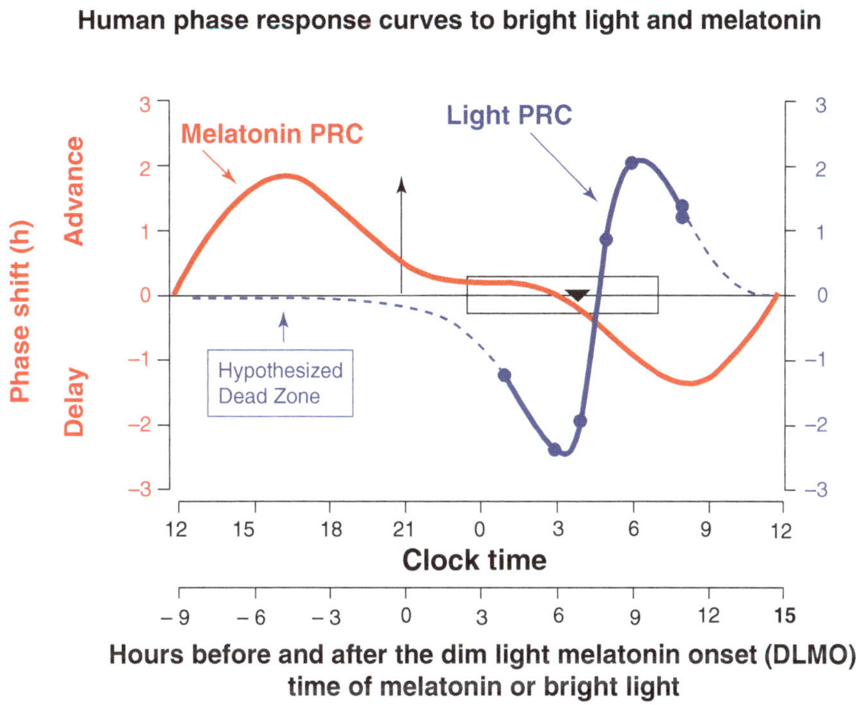

Fig. 3.5 Phase response curves for melatonin (shown in red) and light (shown in blue). The rectangle illustrates a hypothetical entrained sleep time of 7.5 hours, starting 2.5 hours after the DLMO (illustrated with ↑). The triangle shows the core body temperature minimum, 7 hours after the DLMO. These curves are derived from control subjects receiving melatonin doses of 3.0 mg per day or bright light pulses of 3500 lux at different times. Phase shifts are derived from circadian phase assessments conducted before and after 3 days of free running. (Reprinted from Eastman and Burgess [27], with permission from Elsevier)

The phase response curve of melatonin is approximately 180 degrees out of phase with the phase response curve to light, such that exposure in the late afternoon or early evenings will advance the sleep cycle with maximum advancement occurring when dosed ~5 hours before the DLMO [24, 28, 29]. Exposure in the mornings will delay the circadian cycle, though the amplitude of delay effect in the mornings is modest [5] (see Chap. 11 for more details, also see Fig. 3.5).

Chronotype (Circadian Typology) and Phase Tolerance

The term "chronotype," also known as circadian typology, is used to define individual preferences of sleep and wake times, with earlier chronotypes (sometimes called "larks") having preferences for earlier timing and later chronotypes ("owls") with preferences for later timing [30]. Approximately 40% of people display a morning or evening chronotype, while the remaining 60% have a neutral chronotype [31]. Variations in chronotype are determined partially by polymorphisms in circadian genes [7].

An individual's ability to tolerate sleeping at an abnormal circadian phase is referred to as "phase tolerance." Phase tolerance will also determine to a limited extent the presence of a circadian rhythm sleep/wake disorder, since symptom presentation will vary depending on a person's ability to adapt to the light dark cycle, for which there is significant variation [32, 33]. Phase tolerance may decline with age according to some reports, but data are conflicting [26, 33]. These changes have many implications, particularly for shift workers (see Chap. 11), including health professionals (see Chap. 12).

Circadian Changes with Age

Mounting evidence indicates the circadian system is not static after infancy as was once believed. Increasing evidence suggests the endogenous circadian period and light sensitivity are altered in puberty, resulting in the development of a delayed sleep phase in adolescence [34]. Carskadon proposed that adolescents develop a resistance to sleep pressure and simultaneously develop a delay in the circadian phase, providing a drive to stay up later in the evenings and awaken later in the mornings [35]. This delay in circadian phase is correlated with secondary sex development, with girls showing a delay in timing 1 year earlier than boys, paralleling their earlier pubertal onset [34, 36, 37]. Men, however, show greater magnitude of changes in chronotype from adolescence to adulthood than women [37]. This delay in timing of sleep onset has been seen in adolescents in over 16 countries spanning six continents [37]. Moreover, animal studies across six different mammalian species have also shown a delay in circadian phase with puberty, suggesting this is a preserved mammalian developmental stage [34]. Further studies suggest that adolescents show a blunting of response to the phase advance effects of morning light and an exaggerated response to the phase delaying effects of evening light exposure [34]. Additionally, adolescents show a decreased accumulation of homeostatic sleep drive compared with prepubertal children, further enabling a delay in sleep onset [38]. These changes in the circadian system and homeostatic drive in adolescence have significant implications for interpreting sleep complaints of adolescent patients. For instance, delaying school start times for adolescents, in recognition of these biological circadian changes that occur with pubescence, has been shown to improve academic performance and reduce absenteeism, in addition to having other positive benefits for students (described in further detail in Chap. 7).

This delay in circadian timing reaches peak eveningness at approximately age 20 and then continues to gradually advance toward morningness in the ensuing decades [6, 39]. Morningness-eveningness scores (MEQ, see later in this chapter) have been shown to increase by 1 point every 3.8 years [40, 41]. Greater differences in chronotype are seen with gender in the second and third decade of life, with men more frequently having later chronotypes earlier in life, but becoming earlier chronotypes after ages 40–50 [37, 39]. Variability in chronotype decreases significantly with age.

Responsiveness to light declines with age through a variety of factors. Aging is often associated with yellowing of the lenses of the eyes, which can result in decreased transmission of blue and green wavelengths due to decreases in lenticular transmittance [42]. Furthermore, there is a decrease in the number of photoreceptors within the circadian system although responsiveness to light is relatively preserved with age [42]. These factors may contribute to age-related circadian changes and subsequent development of circadian rhythm sleep/wake disorders, particularly advanced sleep phase wake disorder, in the geriatric population.

Epidemiology of Circadian Rhythm Sleep/Wake Disorders

Up to 3% of individuals may suffer from circadian rhythm sleep/wake disorders, but this number may be as high as 10% in adults and 16% in adolescents due to missed diagnoses or misdiagnoses according to some studies [43]. Other populations particularly vulnerable to the development of circadian rhythm disorders include the blind, as one in five experiences total absence of light perception [7]. In this patient population, prevalence estimates of circadian rhythm sleep-wake disorders are as high as 60–80% [7]. The International Classification of Sleep Disorders third edition (ICSD-3) identifies six major types of circadian rhythm sleep/wake disorders: advanced sleep phase type, delayed sleep phase type, irregular sleep/wake type, free-running type, jet lag type, and shift work type. Another common manual used for diagnosis is the Diagnostic and Statistical Manual of Mental Disorders, fifth edition (DSM-5), which has similar criteria and diagnostic categories as the ICSD-3 [44]. The key criteria that all these disorders share is an inability to fall asleep and awaken at desired times, leading to functional impairment. All arise due to the misalignment between the circadian system and the 24-hour external environment.

Diagnosis of Circadian Rhythm Sleep/Wake Disorders

The International Classification of Sleep Disorders third edition (ICSD-3) recommends that clinicians should use caregiver reports, sleep logs, and/or actigraphy to assist with the longitudinal evaluation of such disorders, for at least 7 days, but ideally for 14 days or more [5, 44]. The addition of chronotype questionnaires such as the Morningness-Eveningness Questionnaire (MEQ), the Composite Scale of Morningness (CSM), or the Munich Chronotype Questionnaire (MCTQ) described below can be useful. If possible, the use of circadian phase markers such as salivary DLMO, core body temperature and/or other objective markers for circadian rhythm is encouraged, although their role as diagnostic tools remains uncertain. These measures are described in detail in Chaps. 4 and 6. Such measures may also provide guidance regarding proper timing for administration of light and/or melatonin as treatment options [44].

Questionnaires

Several questionnaires are described in Chap. 5. Some of the most common chronotype questionnaires are the MEQ, developed by Horne and Ostberg in 1976, the CSM by Smith et al. in 1989, and the MCTQ, developed by Roenneberg et al. in 2003 [6, 45, 46]. The MEQ consists of 19 questions asking about the preferred timing of different daily activities, producing a composite score ranging from 16 to 86. Lower scores are suggestive of an evening chronotype, while higher scores suggest a morning chronotype. These scores will place subjects into one of five chronotypes: definite evening type (16–30), moderate evening type (31–41), intermediate (42–58), moderate morning type (59–69), and definite morning type (70–86) [43]. The CSM is a 13-item questionnaire derived partially from the MEQ as well as two other questionnaires, the diurnal type scale, and the circadian type scale [45, 46]. The MCTQ evaluates sleep timing on both work and free days to assess a person's chronotype. Another major difference with respect to the MEQ is that the MCTQ asks about actual sleep and wake times in addition to preferred timing of daily activities [43].

Actigraphy

Circadian rhythm disorders are routinely evaluated with actigraphy, a procedure utilizing the presence of body and/or limb movements to provide estimates of the sleep/wake schedule. These devices have been endorsed by the American Academy of Sleep Medicine (AASM) [47]. Data are typically obtained from devices worn on the wrist, but can also be worn on the ankle or waist to record movements through a piezoelectric or micromechanical accelerometer. Prolonged periods of data, usually weeks to months, can be recorded and later downloaded to an appropriate interface that can apply mathematical algorithms that produce temporal raster plots that ultimately provide sleep/wake estimates [47]. This technology is described in greater detail in Chap. 4. Commonly obtained sleep parameters include sleep onset latency (SOL), total sleep time (TST), wakefulness after sleep onset (WASO), and sleep efficiency (SE = TST/time in bed) [47]. Many actigraphy devices also have buttons that subjects can push to define certain events (e.g., bedtime), as well as light sensors that provide estimates of ambient exposure [47]. These data correlate well with sleep logs and caregiver reports, but actigraphy is completed more reliably [48, 49].

Treatment Options

Using the GRADE approach (Grading of Recommendations Assessment, Development and Evaluation), the American Academy of Sleep Medicine published recommendation statements for the treatment of circadian rhythm sleep-wake disorders in 2015 [5]. Interventions fall broadly into four main categories: (1) prescribed timing of sleep/wake activity and/or physical or social activities during the day; (2) light therapy and/or avoidance of light; (3) medications with chronobiotic effects and/or those that promote sleep or wakefulness; and (4) somatic interventions that alter body functions to ameliorate sleep/wake symptoms. The use of these modalities (if applicable) is discussed in each specific circadian rhythm sleep/wake disorder chapter, and select treatments are described below.

Light Therapy

Light therapy was first used in the 1980s when it was discovered that it could suppress melatonin production and alter circadian rhythms [50, 51]. Its efficacy on the treatment of circadian rhythm sleep/wake disorders can vary considerably depending on a wide variety of factors, including the luminosity of the light source, distance to light source, and the wavelength, timing, and duration of exposure. The brightness or luminosity of a light source is measured in lux (lx), a standard unit of light flow and measure of photopic illuminance. Higher lux is generally associated with higher efficacy for shifting circadian rhythms, but may impact compliance, particularly among older individuals [52]. Sunlight produces in excess of 100,000 lux, while indoor light typically is about 100 lux and rarely over 500 lux [53]. The threshold for suppression of melatonin had been previously assumed to be 2500 lux [53], but further work has described significantly lower lux as effective for melatonin suppression with longer exposure (e.g., <200 lux for 8 hours) [54] (see Chap. 15). Light therapy at >2500 lux administered before the core body temperature minimum (CBT_{min}) has been definitively demonstrated to delay the sleep/wake circadian rhythm, while light therapy after this will promote advances [55]. Consecutive days of exposure also affect the phase responses to light, as circadian resetting effects of light exposure can even be seen at 50 lux or lower [56]. As a result, chronic exposure to room light may have a greater impact than a few minutes of exposure to intense light.

The majority of light therapy devices emit fluorescent white light at 10,000 lux and contain ultraviolet light filters [57]. Additional monochromatic green or blue light-emitting diode (LED) light boxes are marketed with added benefits of portability and decreased intensity of white light, while purportedly preserving associated benefits, but evidence of equivalent efficacy has been inconsistent [58–60]. A common benchmark developed in the 1990s for therapeutic use of light therapy is 10,000 lux for 30 minutes a day, the timing of which depends on the nature of the disorder being treated [61]. For instance, timing will vary significantly for the treatment of jet lag disorder (see Chap. 13). Common side effects of light therapy include insomnia, headaches, eye strain, nausea, irritability, and agitation [62]. Additional potential light therapy adverse effects are described in Chap. 6.

Melatonin

A 0.1–0.3 mg dosage of melatonin will produce plasma concentrations between 100 and 200 pg/ml, considered a physiologic level, while 1.0 mg would be predicted to produce supraphysiologic levels of 500–600 pg/ml [25]. Maximum plasma concentrations are typically reached 45 minutes after administration [25]. Melatonin's half-life is short, with immediate release melatonin ceasing its action within 90 minutes [63]. Considering that it is metabolized by cytochrome P4501A2, medications that affect 1A2 activity could alter the available blood concentration [25]. As of 2018, phase response curves for doses above 5 mg have not been published [5] and, as such, are not routinely recommended for the treatment of circadian rhythm sleep/wake disorders [25]. Most data suggest that timing of melatonin administration is more important than dosage [5]. In many countries, it can be sold as an over-the-counter supplement and is not considered to be a pharmaceutical product. As a result, it is not subject to the same degree of regulation as a prescription medication, creating questions of quality and purity [63].

Summary

Circadian rhythms play a pivotal role in multiple physiologic functions. As a result, dysregulation may lead to the development of circadian rhythm sleep/wake disorders, which can have significant impacts on physical and mental health. These conditions are frequently misdiagnosed or overlooked. Gathering an appropriate history should include a detailed discussion of the timing of sleep/wake patterns and social/work schedules. Utilizing additional diagnostic tools such as chronotype questionnaires and actigraphy can lead to enhanced recognition of these disorders. Understanding the regulation and control of the sleep/wake cycle will lead to the development of appropriate treatment plans, which incorporate both chronobiotic strategies and behavioral modifications.

References

1. Martinez D, Lenz Mdo C. Circadian rhythm sleep disorders. Indian J Med Res. 2010;131:141–9.
2. Kanathur N, Harrington J, Lee-Chiong T Jr. Circadian rhythm sleep disorders. Clin Chest Med. 2010;31(2):319–25.

3. Pavlova M. Circadian rhythm sleep-wake disorders. Continuum (Minneap Minn). 2017;23(4, Sleep Neurology):1051–63.
4. Reid KJ, Zee PC. Circadian rhythm disorders. Semin Neurol. 2009;29(4):393–405.
5. Auger RR, Burgess HJ, Emens JS, Deriy LV, Thomas SM, Sharkey KM. Clinical practice guideline for the treatment of intrinsic circadian rhythm sleep-wake disorders: advanced sleep-wake phase disorder (ASWPD), delayed sleep-wake phase disorder (DSWPD), non-24-hour sleep-wake rhythm disorder (N24SWD), and irregular sleep-wake rhythm disorder (ISWRD). An update for 2015: an American Academy of Sleep Medicine clinical practice guideline. J Clin Sleep Med. 2015;11(10):1199–236.
6. von Schantz M. Natural variation in human clocks. Adv Genet. 2017;99:73–96.
7. Hartley S, Dauvilliers Y, Quera-Salva M-A. Circadian rhythm disturbances in the blind. Curr Neurol Neurosci Rep. 2018;18(10):65.
8. Czeisler CA, Duffy JF, Shanahan TL, Brown EN, Mitchell JF, Rimmer DW, et al. Stability, precision, and near-24-hour period of the human circadian pacemaker. Science. 1999;284(5423):2177–81.
9. Lewy AJ. Circadian misalignment in mood disturbances. Curr Psychiatry Rep. 2009;11(6):459–65.
10. Duffy JF, Cain SW, Chang AM, Phillips AJ, Münch MY, Gronfier C, et al. Sex difference in the near-24-hour intrinsic period of the human circadian timing system. Proc Natl Acad Sci U S A. 2011;108(Suppl 3):15602–8.
11. Dijk DJ, Archer SN. PERIOD3, circadian phenotypes, and sleep homeostasis. Sleep Med Rev. 2010;14(3):151–60.
12. Richardson GS. The human circadian system in normal and disordered sleep. J Clin Psychiatry. 2005;66(Suppl 9):3–9; quiz 42–3.
13. Archer SN, Schmidt C, Vandewalle G, Dijk DJ. Phenotyping of PER3 variants reveals widespread effects on circadian preference, sleep regulation, and health. Sleep Med Rev. 2018;40:109–26.
14. Borbély AA. A two process model of sleep regulation. Hum Neurobiol. 1982;1(3):195–204.
15. Wyatt JK. Circadian rhythm sleep disorders. Pediatr Clin North Am. 2011;58(3):621–35.
16. Bjorvatn B, Pallesen S. A practical approach to circadian rhythm sleep disorders. Sleep Med Rev. 2009;13(1):47–60.
17. Basheer R, Strecker RE, Thakkar MM, McCarley RW. Adenosine and sleep-wake regulation. Prog Neurobiol. 2004;73(6):379–96.
18. Dijk DJ, Czeisler CA. Contribution of the circadian pacemaker and the sleep homeostat to sleep propensity, sleep structure, electroencephalographic slow waves, and sleep spindle activity in humans. J Neurosci. 1995;15(5 Pt 1):3526–38.
19. Lucas RJ, Peirson S, Berson DM, Brown TM, Cooper HM, Czeisler CA, et al. Measuring and using light in the melanopsin age. Trends Neurosci. 2014;37(1):1–9.
20. Wright HR, Lack LC. Effect of light wavelength on suppression and phase delay of the melatonin rhythm. Chronobiol Int. 2001;18(5):801–8.
21. Brainard GC, Hanifin JP, Greeson JM, Byrne B, Glickman G, Gerner E, et al. Action spectrum for melatonin regulation in humans: evidence for a novel circadian photoreceptor. J Neurosci. 2001;21(16):6405–12.
22. Zee PC, Attarian H, Videnovic A. Circadian rhythm abnormalities. Continuum (Minneap Minn). 2013;19(1 Sleep Disorders):132–47.
23. Lewy AJ, Rough JN, Songer JB, Mishra N, Yuhas K, Emens JS. The phase shift hypothesis for the circadian component of winter depression. Dialogues Clin Neurosci. 2007;9(3): 291–300.
24. Pandi-Perumal SR, Smits M, Spence W, Srinivasan V, Cardinali DP, Lowe AD, et al. Dim light melatonin onset (DLMO): a tool for the analysis of circadian phase in human sleep and chronobiological disorders. Prog Neuropsychopharmacol Biol Psychiatry. 2007;31(1):1–11.
25. Cipolla-Neto J, Amaral FGD. Melatonin as a hormone: new physiological and clinical insights. Endocr Rev. 2018;39(6):990–1028.

26. Sack RL, Auckley D, Auger RR, Carskadon MA, Wright KP Jr, Vitiello MV, et al. Circadian rhythm sleep disorders: part I, basic principles, shift work and jet lag disorders. An American Academy of Sleep Medicine review. Sleep. 2007;30(11):1460–83.
27. Eastman CI, Burgess HJ. How to travel the world without jet lag. Sleep Med Clin. 2009;4(2):241–55.
28. Keijzer H, Smits MG, Duffy JF, Curfs LM. Why the dim light melatonin onset (DLMO) should be measured before treatment of patients with circadian rhythm sleep disorders. Sleep Med Rev. 2014;18(4):333–9.
29. Lewy AJ, Sack RA, Singer CL. Assessment and treatment of chronobiologic disorders using plasma melatonin levels and bright light exposure: the clock-gate model and the phase response curve. Psychopharmacol Bull. 1984;20(3):561–5.
30. Figueiro MG, Plitnick B, Rea MS. The effects of chronotype, sleep schedule and light/dark pattern exposures on circadian phase. Sleep Med. 2014;15(12):1554–64.
31. Adan A, Archer SN, Hidalgo MP, Di Milia L, Natale V, Randler C. Circadian typology: a comprehensive review. Chronobiol Int. 2012;29(9):1153–75.
32. Dawson D, Campbell SS. Timed exposure to bright light improves sleep and alertness during simulated night shifts. Sleep. 1991;14(6):511–6.
33. Moline ML, Pollak CP, Monk TH, Lester LS, Wagner DR, Zendell SM, et al. Age-related differences in recovery from simulated jet lag. Sleep. 1992;15(1):28–40.
34. Hagenauer MH, Perryman JI, Lee TM, Carskadon MA. Adolescent changes in the homeostatic and circadian regulation of sleep. Dev Neurosci. 2009;31(4):276–84.
35. Carskadon MA. Maturation of processes regulating sleep in adolescents. In: Marcus C, Carroll JL, Donnelly D, Loughlin GM, editors. Sleep in children: developmental changes in sleep patterns. Boca Raton: CRC Press; 2008. p. 95–109.
36. Carskadon MA, Vieira C, Acebo C. Association between puberty and delayed phase preference. Sleep. 1993;16(3):258–62.
37. Roenneberg T, Kuehnle T, Pramstaller PP, Ricken J, Havel M, Guth A, et al. A marker for the end of adolescence. Curr Biol. 2004;14(24):R1038–9.
38. Jenni OG, Achermann P, Carskadon MA. Homeostatic sleep regulation in adolescents. Sleep. 2005;28(11):1446–54.
39. Fischer D, Lombardi DA, Marucci-Wellman H, Roenneberg T. Chronotypes in the US – influence of age and sex. PLoS One. 2017;12(6):e0178782.
40. Robilliard DL, Archer SN, Arendt J, Lockley SW, Hack LM, English J, et al. The 3111 Clock gene polymorphism is not associated with sleep and circadian rhythmicity in phenotypically characterized human subjects. J Sleep Res. 2002;11(4):305–12.
41. von Schantz M, Taporoski TP, Horimoto ARVR, Duarte NE, Vallada H, Krieger JE, et al. Distribution and heritability of diurnal preference (chronotype) in a rural Brazilian family-based cohort, the Baependi study. Sci Rep. 2015;5:9214.
42. Kim SJ, Benloucif S, Reid KJ, Weintraub S, Kennedy N, Wolfe LF, et al. Phase-shifting response to light in older adults. J Physiol. 2014;592(1):189–202.
43. Kim MJ, Lee JH, Duffy JF. Circadian rhythm sleep disorders. J Clin Outcomes Manag. 2013;20(11):513–28.
44. Abbott SM, Reid KJ, Zee PC. Circadian rhythm sleep-wake disorders. Psychiatr Clin North Am. 2015;38(4):805–23.
45. Smith CS, Reilly C, Midkiff K. Evaluation of three circadian rhythm questionnaires with suggestions for an improved measure of morningness. J Appl Psychol. 1989;74(5):728–38.
46. Jankowski KS. Composite Scale of Morningness: psychometric properties, validity with Munich ChronoType Questionnaire and age/sex differences in Poland. Eur Psychiatry. 2015;30(1):166–71.
47. Smith MT, McCrae CS, Cheung J, Martin JL, Harrod CG, Heald JL, et al. Use of actigraphy for the evaluation of sleep disorders and circadian rhythm sleep-wake disorders: an American Academy of Sleep Medicine systematic review, meta-analysis, and GRADE assessment. J Clin Sleep Med. 2018;14(7):1209–30.

48. Bradshaw DA, Yanagi MA, Pak ES, Peery TS, Ruff GA. Nightly sleep duration in the 2-week period preceding multiple sleep latency testing. J Clin Sleep Med. 2007;3(6):613–9.
49. Auger RR, Varghese R, Silber MH, Slocumb NL. Total sleep time obtained from actigraphy versus sleep logs in an academic sleep center and impact on further sleep testing. Nat Sci Sleep. 2013;5:125–31.
50. Czeisler CA, Richardson GS, Coleman RM, Zimmerman JC, Moore-Ede MC, Dement WC, et al. Chronotherapy: resetting the circadian clocks of patients with delayed sleep phase insomnia. Sleep. 1981;4(1):1–21.
51. Lewy AJ, Wehr TA, Goodwin FK, Newsome DA, Markey SP. Light suppresses melatonin secretion in humans. Science. 1980;210(4475):1267–9.
52. Suhner AG, Murphy PJ, Campbell SS. Failure of timed bright light exposure to alleviate age-related sleep maintenance insomnia. J Am Geriatr Soc. 2002;50(4):617–23.
53. Eastman CI. Squashing versus nudging circadian rhythms with artificial bright light: solutions for shift work? Perspect Biol Med. 1991;34(2):181–95.
54. Gooley JJ, Chamberlain K, Smith KA, Khalsa SB, Rajaratnam SM, Van Reen E, et al. Exposure to room light before bedtime suppresses melatonin onset and shortens melatonin duration in humans. J Clin Endocrinol Metab. 2011;96(3):E463–72.
55. Auger RR. Advance-related sleep complaints and advanced sleep phase disorder. Sleep Med Clin. 2009;4(2):219–27.
56. Glickman G, Levin R, Brainard GC. Ocular input for human melatonin regulation: relevance to breast cancer. Neuro Endocrinol Lett. 2002;23(Suppl 2):17–22.
57. Brouwer A, Nguyen HT, Snoek FJ, van Raalte DH, Beekman ATF, Moll AC, et al. Light therapy: is it safe for the eyes? Acta Psychiatr Scand. 2017;136(6):534–48.
58. Meesters Y, Dekker V, Schlangen LJ, Bos EH, Ruiter MJ. Low-intensity blue-enriched white light (750 lux) and standard bright light (10,000 lux) are equally effective in treating SAD. A randomized controlled study. BMC Psychiatry. 2011;11:17.
59. Gordijn MCM, 't Mannetje D, Meesters Y. The effects of blue-enriched light treatment compared to standard light treatment in seasonal affective disorder. J Affect Disord. 2012;136(1–2):72–80.
60. Anderson JL, Hilaire MA, Auger RR, Glod CA, Crow SJ, Rivera AN, et al. Are short (blue) wavelengths necessary for light treatment of seasonal affective disorder? Chronobiol Int. 2016;33(9):1267–79.
61. Anderson JL, Glod CA, Dai J, Cao Y, Lockley SW. Lux vs. wavelength in light treatment of seasonal affective disorder. Acta Psychiatr Scand. 2009;120(3):203–12.
62. Terman M, Terman JS. Bright light therapy: side effects and benefits across the symptom spectrum. J Clin Psychiatry. 1999;60(11):799–808; quiz 809.
63. Golombek DA, Pandi-Perumal SR, Brown GM, Cardinali DP. Some implications of melatonin use in chronopharmacology of insomnia. Eur J Pharmacol. 2015;762:42–8.

Chapter 4
Physiologic Methods of Assessment Relevant to Circadian Rhythm Sleep-Wake Disorders

Vincent A. LaBarbera and Katherine M. Sharkey

Introduction

Various physiologic tools are available to aid in the assessment of circadian rhythms. Some measures are available primarily for use in research protocols, but are increasingly becoming available to sleep clinicians. The measures at our disposal are devised to objectively estimate variables such as circadian phase position, periodicity, and circadian amplitude, which when taken in conjunction with a careful history and physical exam allow for the assessment of essential aspects of circadian biology and behavior by the sleep clinician or researcher.

The physiologic methods that are described hereafter are wrist or body actigraphy, ambulatory light monitoring, melatonin monitoring, ambulatory core body temperature monitoring, polysomnography, multichannel ambulatory monitoring, and circadian gene expression in peripheral cells. Each of these methods has advantages and caveats; the type of circadian measure should be individualized to optimize cost-effectiveness, precision, level of acceptable invasiveness, and labor/time intensiveness. Confounding variables, such as narcotic/hypnotic or activating medication use, comorbid sleep disorders, comorbid psychiatric or medical conditions, preceding sleep duration and timing, and even the measuring tools themselves, can also influence, and be influenced by, these various techniques. As such, the astute clinician or researcher must select the measurement method with discretion. Those planning to assess circadian rhythms and/or circadian rhythm sleep-wake disorders

V. A. LaBarbera
Department of Neurology, Rhode Island Hospital/Warren Alpert Medical School of Brown University, Providence, RI, USA

K. M. Sharkey (✉)
Department of Medicine and Psychiatry and Human Behavior, Alpert Medical School of Brown University, Providence, RI, USA
e-mail: katherine_sharkey@brown.edu

R. R. Auger (ed.), *Circadian Rhythm Sleep-Wake Disorders*,
https://doi.org/10.1007/978-3-030-43803-6_4

may benefit from consulting with a sleep specialist and reviewing the latest guidelines in the International Classification of Sleep Disorders [1].

There are other factors, both intrinsic and extrinsic to the individual, that can affect the precision of circadian measurements. This phenomenon is called "masking," because the true endogenous circadian rhythm can be "masked" by physiologic processes unrelated to the output of the central circadian pacemaker. Examples of this, which will be discussed in more detail below, are the "masking" effects of sleep, food, movement, medications, and the menstrual cycle on the circadian rhythm of core body temperature. Masking effects can be mitigated using the constant routine (CR) protocol, as described in Chap. 2. Examples of CR are constant dim and/or long wavelength light, constant wakefulness in a recumbent posture, and feeding with frequent, regular, and identical small meals and liquid portions.

A final consideration in circadian rhythm measurement is the number of times that the parameter of interest is assessed. Measurement at a single time point provides a snapshot of internal circadian rhythms in the context of sleep timing, light-dark exposure, and other contributing factors close to the time of assessment. It is often of greater interest, however, to obtain serial measurements. In this scenario, the first measure is used to establish the baseline status and in some cases to determine if and when a circadian intervention should be initiated. Subsequent evaluations are used to track changes from baseline to measure any shift in the circadian rhythms.

Actigraphy

The most commonly used method of assessment is actigraphy, which is used for research, clinical, and, with the advent of wearables, personal home monitoring. An actigraph is a noninvasive ambulatory monitor – typically worn on the nondominant wrist and resembling a watch – that provides information about rest and activity patterns. Actigraphically measured estimates of rest and activity correlate well with gold-standard measures of sleep and wakefulness (i.e., polysomnography). Because it is not physically burdensome and time-consuming and is relatively inexpensive, actigraphy data are typically collected for multiple days and nights. The major drawback of wrist actigraphy is its inability to distinguish between true sleep and absence of motion; hence, in disorders of insomnia, where motionless wakefulness is not uncommon, actigraphy may overestimate sleep. Similarly, in disorders of abnormal sleep movement, actigraphy can overestimate wakefulness. Those considering use of wrist actigraphy should consider whether the data may be less reliable in their population of interest.

Actigraphs usually measure nondirectional acceleration. Oftentimes, these may be in arbitrary units (e.g., counts that are specific to the particular actigraph device), as opposed to established measurements of acceleration (meters per second squared or "g"). Common variables measured by actigraphs include duration of movement above a certain threshold ("time above threshold"), timestamps which cross an arbitrary zero-point ("zero-crossing"), or intensity of movement ("periodic integration") [2]. More recently developed actigraphs have been enhanced to include microelectrical mechanical sensors for three-dimensional acceleration measurement, body temperature measurement, user-input subjective measurements, light level detection, and decibel measurements [2].

Ancoli-Israel et al. described circadian rhythm analysis using actigraphy in a comprehensive review [3]. Actigraphy can identify patterns of sleep and wake that correlate with circadian rhythms of melatonin and core body temperature, particularly in individuals whose circadian rhythms are entrained to typical clock times. Actigraphy is also useful for documenting delayed or advanced sleep phase, disturbed sleep in shift workers, and non-24-hour sleep-wake schedules, especially if the individual being assessed is sleeping at his/her preferred times rather than according to the expected societal schedule [4–7]. Multiple methods to detect and analyze rhythmicity of activity have been proposed, including several variations on the periodogram, an analysis technique that examines time-series data to detect periodic signals. Twenty-four-hour circadian rhythms can be detected in raw actigraphy data using methods such as Fourier spectral analysis, Lomb and Scargle periodogram, Enright periodogram, ANOVA periodogram, and the Cosinor and cosine fit [2, 8–12].

When compared to the gold standard for sleep evaluation, polysomnography (PSG), wrist actigraphy has high accuracy and sensitivity, but relatively low specificity (86%, 96%, and 33%, respectively), at least in the laboratory setting [13]. According to the American Academy of Sleep Medicine (AASM) practice parameters, actigraphy is useful in characterizing and monitoring circadian rhythms not only in healthy adult populations, but also in children and infants; elderly adults and those in nursing facilities, with or without dementia; depressed or schizophrenic patients; and those in inaccessible locations, for example, pilots in spaceflights [14]. Particularly germane to clinicians, actigraphic measurement of sleep patterns is a recommended "guideline" for the evaluation of circadian rhythm sleep-wake phase disorders [15].

Ambulatory Light Monitoring

Photic stimulation is the strongest zeitgeber and provides important information to the internal biological clock. Light entering the eye excites retinal photoreceptors called intrinsically photosensitive retinal ganglion cells (ipRGCs), which then send action potentials via the retinohypothalamic tract to the suprachiasmatic nuclei (SCN) of the hypothalamus. The ipRGCs are specialized to transmit light information rather than to detect and transmit visual information like retinal rod and cone cells. As such, some people with blindness have intact light-dark input through the ipRGCs despite their visual impairment. The SCN function as the internal master clock of the body, and secondary efferents project from the SCN throughout the brain to synchronize physiology and behavior. The SCN govern the timing of melatonin release from the pineal gland, and melatonin secretion is closely linked to the timing of sleep and darkness. Light exposure patterns are therefore an important circadian measure, and knowledge about daily light-dark exposure can inform measurement of melatonin and other circadian rhythms.

Measured light levels can be used as a proxy measure for zeitgeber strength and provide the researcher with information about an individual's periodicity and circadian phase position. For example, if the amplitude of light exposure is low, it is more difficult for an individual's circadian rhythms to maintain entrainment. Although in rare situations of complete, continuous darkness, the need for a light zeitgeber may be overcome by exposure to social cues and interactions [16], in most settings, cycling periods of light and dark are needed to maintain entrainment. The zeitgeber of light exposure is now more significant as ever in sleep health, as light exposure from smartphones and other electronic devices has been shown to suppress melatonin release by up to 11% in a darkened room and up to 36% when used in a bright room [17].

The effects of disrupted sleep, both among inpatients and ambulatory subjects, can be monitored via actigraphy, as described above, but the amount of light to which a person is exposed has been a bit more elusive. In the research setting, various light meters have shown promise. Both wearable and stand-alone light meters can reliably capture light fluctuations in inpatient settings, and there is high congruence between a light meter that is set up in a room apart from a patient and one that is worn at wrist level [18]. Future work in this area could ultimately lead to home and workplace interventions regarding the timing of light exposure and personalized light-dark prescriptions that enhance circadian entrainment in individuals with circadian disorders [19].

Melatonin Rhythm Measurement

The hormone melatonin conveys information about the external light-dark cycle to the brain and the rest of the body. In entrained individuals, melatonin levels are low during the day, rise acutely in the evening (as long as environmental light levels are dim), remain high through the night, and decline to nearly undetectable levels in the morning. In the laboratory setting, the evening onset of melatonin secretion, termed the "dim light melatonin onset" (DLMO), is a well-studied circadian marker. The DLMO is considered the most accurate marker of circadian phase, as secretion of melatonin from the pineal gland is controlled directly by the suprachiasmatic nuclei. Although melatonin production can be masked by bright light, it is relatively robust in its resistance to the masking effects of other external stimulation, such as carbohydrate ingestion [20]. Given its reliability, robustness, and verifiability, it is often used as a "gold standard" measurement with which other assessments are compared [21].

The DLMO is of great utility to circadian and sleep researchers and clinicians as it can be measured noninvasively, most commonly using salivary or urinary samples [22, 23]. Typically, melatonin levels rise in the 2–3 hours preceding nocturnal sleep.

For this reason, sampling for DLMO determination requires a relatively short sampling window, which makes it an accessible tool [22]. Although most easily obtained via saliva, urinary and plasma analyses are also available. For all, the typical unit of measurement is picograms per milliliter.

Salivary sampling for DLMO takes place under dim light (fewer than 30 lux) every one-half to one hour. Samples should be collected at least 1 hour prior to and through the expected rise in melatonin. Although this method is portable and individuals can collect samples at home, this approach requires attention to several protocol considerations including the necessity to remain in dim light, the importance of accurate sample timing, and taking measures to avoid contaminating saliva samples with food or blood.

Urinary 6-sulfatoxymelatonin (aMT6s) is the primary urinary metabolite of melatonin. Collection of urine samples across 24 hours allows estimation of the rhythm of aMT6s excretion rate. This method does not require disruption of sleep as the overnight excretion rate can be measured using the first morning void. During the day, urine should be collected every 2–8 hours. The phase of the excretion rhythm is estimated from the peak of the cosine fitted curve, termed the "acrophase." Individual ability to void and compliance with collection are both limiting factors for this method of measurement.

Melatonin secretion can also be measured in plasma. In this method, blood is sampled at frequent intervals, most commonly via an intravenous catheter to avoid multiple phlebotomy sessions. The intravenous catheter should be inserted at least 2 hours prior to sampling to avoid alteration in melatonin levels associated with a painful stimulus and concurrent adrenergic surge. Midline intravenous access, central catheterization, or peripherally inserted central catheterization (PICC) may allow frequent sampling without frequent disruption of sleep, but does come with the potential risk of an invasive catheterization, such as local injury/discomfort, infection, or cardiac arrhythmia. Salivary acrophase correlates well with plasma acrophase of melatonin, albeit at a level of 30 percent that of the plasma level [24]. Plasma analysis may therefore allow for greater sensitivity and determination of circadian phase, due to the relatively higher melatonin levels present in plasma, and may be useful in individuals who secrete low levels of melatonin.

Absolute values, relative values (commonly a percentage of maximum or daily average) and curve fitting have all been used to determine circadian phase from the pattern of melatonin secretion. DLMO and its corollary, dim light melatonin offset (DLMOff), as well as the termination of melatonin synthesis ("Synoff"), are variables that can be measured as well to describe melatonin physiology. Synoff represents the transition from maximal nocturnal production to the morning decline, and DLMoff represents the return to daytime melatonin levels. Intraindividual melatonin rhythms, timing, and amplitudes are typically quite stable. However, interindividual differences can be quite large. Thus, measuring these markers of physiological transition as opposed to absolute values may be most reliable in group analyses.

DLMO and other melatonin measurements are not yet largely accessible to the sleep clinician, but there is limited evidence that DLMO measurement has utility in the diagnosis of circadian rhythm sleep disorders. For example, Rahman and colleagues demonstrated average DLMO was ~2.5 hours later in patients with delayed sleep-wake phase disorder (DSWPD) compared to non-DSWPD patients, whereas sleep latency was only ~40 minutes later. In this study, DLMO had a sensitivity and specificity of 90.3% and 84%, respectively, for diagnosis of DSWPD [25]. Although no melatonin assays are yet approved by the Food and Drug Administration in the United States for the diagnosis and management of circadian rhythm sleep disorders, several assays are commercially available in Europe [26].

Cost and frequency of sampling are the two biggest practical limitations of the melatonin assay [23]. Purchasing the reagents and monoclonal antibodies for each assay, which cost on the order of 12 dollars to 20 euros per sample, may be prohibitive to widespread adoption of these tools in the clinical realm [23, 26].

Core Body Temperature

Core body temperature variation is a well-known circadian rhythm, thought to be due to the interplay of heat production, derived from the metabolic activity of the viscera and the brain, which produces approximately 70% of the resting metabolic rate of the body, and heat loss, brought about by evaporation, conduction, convection, and radiation. This phenomenon was documented as early as 1842, with Gierse's study observing his own oral temperature reaching a maximum in the early evening and a minimum in the early morning hours [27]. More intricate studies under constant routine protocols revealed that distal skin temperature rises in the evening, but heat production, proximal skin temperature, and core body temperature decrease. In the morning hours, the pattern is reversed [28].

Core body temperature as a means of circadian monitoring has been studied in a wide variety of clinical research settings. A pitfall is that it can be masked by a variety of exogenous and endogenous factors. For example, in reproductive-age women, the temperature mesor (mean value) is higher in the luteal phase and lower in the pre-ovulatory phase and during menses when compared to the follicular phase. Consequently, the amplitude of the temperature circadian rhythm is reduced in the luteal phase. The period of circadian temperature variations is not affected by menstruation in the ambulatory population [29]. In stroke patients, body temperature demonstrates infradian rhythms, which is thought to reflect impaired consciousness post-stroke as well as decreased mobility and ambulation [30]. Patients with affective disorders exhibit a smaller core temperature rhythm amplitude than controls without a mood disorder, which is driven by higher nocturnal core temperature among those with affective disorders [31].

Core body temperature can be gathered via invasive methods, such as catheter insertion via the esophagus, rectum, or urinary bladder [32]. Esophageal and bladder temperature monitoring typically occur only in critical care settings and have been shown to be most similar to core body temperature measured via a pulmonary artery catheter [33]. For obvious reasons, these interventions are not pragmatic for the ambulatory setting, and peripheral body temperature readings are preferable. Wrist skin temperature has been shown to correlate well with DLMO; however, the sensitivity of peripheral thermometry is limited compared to core body temperature measures. Indeed, a recent systematic review concluded that peripheral thermometers do not have clinically acceptable accuracy when precise measurements of body temperature are required [34]. With that said, peripheral thermometry may be acceptable for assessing circadian changes as opposed to absolute values. Various technologies exist for ambulatory core body temperature monitoring. Wearable thermometry can be simple to install and use. Forehead dual heat flux methods [35] and iButton ® skin temperature monitors [36] have been studied as effective means of ambulatory monitoring. Other studies have shown utility with ingestible temperature monitors [29].

Although the constant routine is ideal for measuring the circadian rhythm of core body temperature, CRs are impractical for day-to-day or long-term recording. Thus, algorithms and filters associated with other vital sign changes, such as heart rate, have been devised to "unmask" the effects of physical activity and environmental factors and increase the precision of circadian temperature measurement [37].

Circadian Gene Expression in Peripheral Cells

Although advances in the basic science of molecular genetics in circadian rhythm sleep disorders continue to excite sleep researchers and clinicians, there are still limited applications in clinical practice. Early researchers in circadian rhythms observed the presence of an endogenous timekeeper. Later studies in model organisms such as *Neurospora*, *Drosophila*, *Caenorhabditis elegans*, and rodents revealed homologs and analogs of timekeeping genes, such as *per*, *tim*, *clock*, and *cry*. The first characterization of a Mendelian circadian trait in humans was in the landmark phenotypic study of a family with early sleep onset and offset, thus identifying an autosomal dominant trait with high penetrance for familial advanced sleep-wake phase disorder (FASP) [38, 39]. Ultimately, this was found to be a missense mutation in the human cry2 gene [40]. Since the discovery of FASP, other human circadian phenotypes have been identified and attributed to heredity, including familial natural short sleep (FNSS) [41, 42].

In the research realm, the noninvasive collection and measurement of expression of clock genes has been shown to be feasible. In a study by Akashi et al., circadian rhythmicity of hair follicle cells of the human scalp was measured via real-time polymerase chain reaction (PCR) with [43]. Although this technology has yet to reach clinical medicine, ongoing research may prove these strategies useful in evaluating circadian disorders.

Polysomnography

Polysomnography (PSG) is the gold standard for recording sleep and diagnosing many sleep disorders, e.g., obstructive sleep apnea or narcolepsy. Its utility in circadian rhythm sleep disorders is limited, however, because PSG is usually performed during a single sleep-wake cycle.

PSG utilizes electroencephalogram (EEG), electrooculogram, electromyogram, electrocardiogram, air flow, breathing effort, limb and body movements, and pulse oximetry. EEG frequency and other electrical phenomena recorded with EEG define the various stages of sleep. Sleep is divided into rapid-eye-movement (REM) and non-rapid-eye-movement (NREM) sleep, with stages N1 and N2 considered "light" sleep, N3 as deep or slow wave sleep, and REM sleep as a separate state of consciousness characterized by rapid eye movements, decreased muscle tone, and dream mentation. In general, deeper sleep is characterized by slower EEG frequencies.

In its comprehensive practice parameters for the evaluation of circadian rhythms, the AASM noted that polysomnography is not routinely indicated for the diagnosis of circadian rhythm sleep-wake disorders, but may be indicated to rule out another primary sleep disorder [15]. Polysomnography does have utility in some circadian research, however. For example, in forced desynchrony protocols (see Chap. 2), PSG measurements provide important information about sleep propensity at various points in the circadian cycle.

Multiple Channel Ambulatory Monitoring

Reliable detection, measurement, and interpretation of circadian rhythms and phase shifts using noninvasive ambulatory measurements, especially among individuals living at home and conducting their day-to-day activities, have proven difficult. Nevertheless, there are promising research protocols that may prove useful in clinical settings in the future. One such method is multichannel ambulatory monitoring, which combines several measures described above without relying on measurements of core body temperature or melatonin levels or requiring constant routine laboratory conditions. In one study of this approach [44], research subjects wore multichannel ambulatory monitors for 1 week without alteration to their daily routine, followed by a 32-hour constant routine procedure in the laboratory. Multiple regression techniques were applied to reduce the confounding effects of ambulatory measurements and revealed that the multiple channel approach had statistically significant improvement of variance of prediction error when compared to single predictors (actigraphy or a sleep diary). Compared to core body temperature, the multiple channel method also improved the range of prediction errors and showed a nonsignificant reduction in variance [44]. Multiple channel ambulatory monitoring has shown good accuracy with regard to temporal association with DLMO [21] and may be a promising new tool for making circadian phase estimates in an ambulatory setting, given its simplicity, applicability to clinical situations, and good reliability in early studies. With the advent of wearables and physiologic monitoring devices marketed directly to consumers for self-assessment, multichannel ambulatory monitors are likely to be used more commonly in future research and clinical assessments.

Heart Rate and Electrocardiographic Monitoring

Electrocardiogram (ECG or EKG) is a useful measurement in sleep medicine and sleep research and provides multiple parameters, e.g., heart rate variability, that have been used for everything from detecting sleep apnea to characterizing sleep stages. ECG is less commonly used to assess circadian rhythms, but this may change in the future. For example, core body temperature can be estimated using heart rate, and cardiac telemetry could be a practical and effective means to estimate core body temperature. R-R interval and mean heart rate have a strong correlation with core body temperature [45]. Further work is needed to fully understand phase differences between ECG and other circadian rhythms. For instance, Krauchi and colleagues showed that heart rate, among other variables, is phase advanced with regard to rectal temperature [46]. Nevertheless, with ambulatory ECG recordings that can run for several weeks, ECG may prove to be an important circadian measure in the coming years.

References

1. American Academy of Sleep Medicine, editor. International classification of sleep disorders. 3rd ed. Darien: American Academy of Sleep Medicine; 2014.
2. Roebuck A, Monasterio V, Gederi E, Osipov M, Behar J, Malhotra A, et al. A review of signals used in sleep analysis. Physiol Meas. 2014;35(1):R1–57.
3. Ancoli-Israel S, Cole R, Alessi C, Chambers M, Moorcroft W, Pollak CP. The role of actigraphy in the study of sleep and circadian rhythms. Sleep. 2003;26(3):342–92.
4. Youngstedt SD, Kripke DF, Elliott JA, Klauber MR. Circadian abnormalities in older adults. J Pineal Res. 2001;31:264–72.
5. Cole RJ, Smith JS, Alcala YC, Elliott JA, Kripke DF. Bright-light mask treatment of delayed sleep phase syndrome. J Biol Rhythms. 2002;17(1):89–101.
6. Carskadon MA, Acebo C, Richardson GS, Tate BA, Seifer R. An approach to studying circadian rhythms of adolescent humans. J Biol Rhythms. 1997;12(3):278–89.
7. Carskadon MA, Wolfson A, Acebo C, Tzischinsky O, Seifer R. Adolescent sleep patterns, circadian timing, and sleepiness at a transition to early school days. Sleep. 1998;21(8):871–81.
8. Scargle JD. Studies in astronomical time series analysis. II – statistical aspects of spectral analysis of unevenly spaced data. Astrophys J. 1982;263:835–53.
9. Refinetti R, Lissen GC, Halberg F. Procedures for numerical analysis of circadian rhythms. Biol Rhythm Res. 2007;38(4):275–325.
10. Enright J. The search for rhythmicity in biological time-series. J Theor Biol. 1965;8(3):426–68.
11. Sokolove PG, Bushell WN. The chi square periodogram: its utility for analysis of circadian rhythms. J Theor Biol. 1978;72(1):131–60.

12. Shono M, Shono H, Ito Y, Muro M, Maeda Y, Sugimori H. A new periodogram using one-way analysis of variance for circadian rhythms. Psychiatry Clin Neurosci. 2000;54(3):307–8.
13. Marino M, Li Y, Rueschman MN, Winkelman JW, Ellenbogen JM, Solet JM, Dulin H, et al. Measuring sleep: accuracy, sensitivity, and specificity of wrist actigraphy compared to polysomnography. Sleep. 2013;36(11):1747–55.
14. Littner M, Kushida CA, Anderson WM, Bailey D, Berry RB, Davila DG, et al. Standards of Practice Committee of the American Academy of Sleep Medicine. Practice parameters for the role of actigraphy in the study of sleep and circadian rhythms: an update for 2002. Sleep. 2003;26(3):337–41.
15. Morgenthaler TI, Lee-Chiong T, Alessi C, Friedman L, Aurora N, Boehlecke B, et al. Standards of Practice Committee of the AASM. Practice parameters for the clinical evaluation and treatment of circadian rhythm sleep disorders. Sleep. 2007;30(11):1445–59.
16. Aschoff J, Fatranská M, Giedke H, Doerr P, Stamm D, Wisser H. Human circadian rhythms in continuous darkness: entrainment by social cues. Science. 1971;171(3967):213–5.
17. Oh JH, Yoo H, Park HK, Do YR. Analysis of circadian properties and healthy levels of blue light from smartphones at night. Sci Rep. 2015;5:11325.
18. Higgins PA, Winkelman C, Lipson AR, Guo SE, Rodgers J. Light measurement in the hospital: a comparison of two methods. Res Nurs Health. 2007;30(1):120–8.
19. Mason IC, Boubekri M, Figueiro MG, Hasler BP, Hattar S, Hill SM, et al. Circadian health and light: a report on the National Heart, Lung, and Blood Institute's workshop. J Biol Rhythms. 2018;33(5):451–7.
20. Kräuchi K, Cajochen C, Werth E, Wirz-Justice A. Alteration of internal circadian phase relationships after morning versus evening carbohydrate-rich meals in humans. J Biol Rhythms. 2002;17(4):364–76.
21. Bonmati-Carrion MA, Middleton B, Revell V, Skene DJ, Rol MA, Madrid JA. Circadian phase assessment by ambulatory monitoring in humans: correlation with dim light melatonin onset. Chronobiol Int. 2014;31(1):37–51.
22. Benloucif S, Burgess HJ, Klerman EB, Lewy AJ, Middleton B, Murphy PJ, et al. Measuring melatonin in humans. J Clin Sleep Med. 2008;4(1):66–9.
23. Molina TA, Burgess HJ. Calculating the dim light melatonin onset: the impact of threshold and sampling rate. Chronobiol Int. 2011;28(8):714–8.
24. Voultsios A, Kennaway DJ, Dawson D. Salivary melatonin as a circadian phase marker: validation and comparison to plasma melatonin. J Biol Rhythms. 1997;12(5):457–66.
25. Rahman SA, Kayumov L, Tchmoutina EA, Shapiro CM. Clinical efficacy of dim light melatonin onset testing in diagnosing delayed sleep phase syndrome. Sleep Med. 2009;10(5):549–55.
26. Keijzer H, Smits MG, Duffy JF, Curfs LM. Why the dim light melatonin onset (DLMO) should be measured before treatment of patients with circadian rhythm sleep disorders. Sleep Med Rev. 2014;18(4):333–9.
27. Gierse A. Quaeniam sit ratio caloris organici, M. D. Thesis. 1842. Halle.
28. Kräuchi K. How is the circadian rhythm of core body temperature regulated? Clin Auton Res. 2002;12(3):147–9.
29. Coyne MD, Kesick CM, Doherty TJ, Kolka MA, Stephenson LA. Circadian rhythm changes in core temperature over the menstrual cycle: method for noninvasive monitoring. Am J Physiol Regul Integr Comp Physiol. 2000;279(4):R1316–20.
30. Takekawa H, Miyamoto M, Miyamoto T, Yokota N, Hirata K. Alteration of circadian periodicity in core body temperatures of patients with acute stroke. Psychiatry Clin Neurosci. 2002;56(3):221–2.
31. Carpenter JS, Robillard R, Hermens DF, Naismith SL, Gordon C, Scott EM, et al. Sleep-wake profiles and circadian rhythms of core temperature and melatonin in young people with affective disorders. J Psychiatr Res. 2017;94:131–8.
32. Lilly JK, Boland JP, Zekan S. Urinary bladder temperature monitoring: a new index of body core temperature. Crit Care Med. 1980;8(12):742–4.
33. Lefrant JY, Muller L, de La Coussaye JE, Benbabaali M, Lebris C, Zeitoun N, et al. Temperature measurement in intensive care patients: comparison of urinary bladder, oesophageal, rectal,

axillary, and inguinal methods versus pulmonary artery core method. Intensive Care Med. 2003;29(3):414–8.
34. Niven DJ, Gaudet JE, Laupland KB, Mrklas KJ, Roberts DJ, Stelfox HT. Accuracy of peripheral thermometers for estimating temperature: a systematic review and meta-analysis. Ann Intern Med. 2015;163(10):768–77.
35. Huang M, Tamura T, Tang Z, Chen W, Kanaya S. A wearable thermometry for core body temperature measurement and its experimental verification. IEEE J Biomed Health Inform. 2017;21(3):708–14.
36. Hasselberg MJ, McMahon J, Parker K. The validity, reliability, and utility of the iButton® for measurement of body temperature circadian rhythms in sleep/wake research. Sleep Med. 2013;14(1):5–11.
37. Nakano T, Koyama E, Imai T, Hagiwara H. Circadian rhythm estimation by core body temperature filtered with simultaneously recorded physiological data. Methods Inf Med. 1997;36(4–5):306–10.
38. Jones CR, Campbell SS, Zone SE, Cooper F, DeSano A, Murphy PJ, et al. Familial advanced sleep-phase syndrome: a short-period circadian rhythm variant in humans. Nat Med. 1999;5(9):1062–5.
39. Jones CR, Huang AL, Ptáček LJ, Fu YH. Genetic basis of human circadian rhythm disorders. Exp Neurol. 2013;243:28–33.
40. Hirano A, Shi G, Jones CR, Lipzen A, Pennacchio LA, Xu Y, et al. A Cryptochrome 2 mutation yields advanced sleep phase in humans. Elife. 2016;5:pii:e16695.
41. He Y, Jones CR, Fujiki N, Xu Y, Guo B, Holder JL Jr, et al. The transcriptional repressor DEC2 regulates sleep length in mammals. Science. 2009;325(5942):866–70.
42. Zhang L, Jones CR, Ptacek LJ, Fu YH. The genetics of the human circadian clock. Adv Genet. 2011;74:231–47.
43. Akashi M, Soma H, Yamamoto T, et al. Noninvasive method for assessing the human circadian clock using hair follicle cells. Proc Natl Acad Sci U S A. 2010;107:15643–8.
44. Kolodyazhniy V, Späti J, Frey S, Götz T, Wirz-Justice A, Kräuchi K, et al. Estimation of human circadian phase via a multi-channel ambulatory monitoring system and a multiple regression model. J Biol Rhythms. 2011;26(1):55–67.
45. Sim SY, Joo KM, Kim HB, Jang S, Kim B, Hong S, et al. Estimation of circadian body temperature rhythm based on heart rate in healthy, ambulatory subjects. IEEE J Biomed Health Inform. 2017;21(2):407–15.
46. Kräuchi K, Wirz-Justice A. Circadian rhythm of heat production, heart rate, and skin and core temperature under unmasking conditions in men. Am J Physiol. 1994;267(3 Pt 2):R819–29.

Chapter 5
Non-physiologic Methods of Assessment Relevant to Circadian Rhythm Sleep-Wake Disorders

Vincent A. LaBarbera and Katherine M. Sharkey

Introduction

Sleep clinicians and researchers have various tools at their disposal to characterize and study circadian rhythms and their related sleep-wake disorders. Assessments can range from subjective assessments of patients' experiences and ambulatory measures, which are more clinically accessible, to constant routines, forced desynchrony protocols, and measurements of circadian gene expression and other biomarkers (see Chap. 4). This chapter focuses on non-physiologic methods of assessment. The methods noted below can range from individual recollection of preferential time of sleep onset and waking, mood and daytime sleepiness questionnaires, and records of internet usage and other metadata to analyze circadian and/or chronobiologic behaviors.

Sleep Diaries

Sleep diaries – also referred to as sleep logs – are used routinely in clinical settings to assess sleep-wake disorders. Indeed, the International Classification of Sleep Disorders – Third Edition (hereafter ICSD-3) – published in 2014 [1] mandates sleep logs as essential diagnostic instruments for circadian-rhythm sleep-wake

V. A. LaBarbera
Department of Neurology, Rhode Island Hospital/Warren Alpert Medical School of Brown University, Providence, RI, USA

K. M. Sharkey (✉)
Department of Medicine and Psychiatry and Human Behavior, Alpert Medical School of Brown University, Providence, RI, USA
e-mail: katherine_sharkey@brown.edu

R. R. Auger (ed.), *Circadian Rhythm Sleep-Wake Disorders*,
https://doi.org/10.1007/978-3-030-43803-6_5

disorders. Guidelines state that a sleep diary should be completed for at minimum 7 days, with preference for 14 days, to capture both work and non-work days [2]. This type of paper-and-pencil monitoring and data gathering is fiscally prudent; however, it is dependent on the compliance of the patient or research participant and is subject to inherent confounders, such as recall bias, observer bias, and recency bias.

The determination of the utility and reliability of sleep diaries represent active areas of research, and recommendations are available for different populations. For example, in the adolescent demographic, a study including samples taken across four continents recommends at least five weekday nights of sleep diary completion to demonstrate good inter-cohort reliability. In a smaller subgroup of the same adolescent participants, sleep patterns were also stable across 12 consecutive school weeks [3]. A meta-analysis comparing sleep diaries and actigraphy for evaluation of insomnia showed that patients report 35–38 fewer minutes of sleep and ~24 minutes longer sleep onset latency on sleep diaries than was estimated with actigraphy [4]. The same analysis concluded that in insomnia treatment studies, actigraphy and sleep diaries performed equally well for determining clinically significant treatment response in terms of both shortened sleep latency and increased sleep duration after treatment.

Sleep diaries are easy to use and only take a few minutes each day to complete. When compared to the sleep clinician's or researcher's verbal history, where an individual may be vague or have inaccurate time estimates, a sleep diary attempts to mitigate those factors of recollection bias. Sleep diaries have been used in the research realm for at least 50 years [5] and have also been confirmed to be reliable when compared to wrist actigraphy [4, 6] or ambulatory electroencephalography [7].

Perhaps the first widely distributed sleep diary is the Pittsburgh Sleep Diary [8]. A prototypical sleep diary, the Pittsburgh Sleep Diary (and others that followed) includes information not only about sleep onset and wake time, but also about whether it was a workday versus non-workday, time (and type) of food and drink, medication use, consumption of alcohol and caffeine, and exercise. The time of initiation of bedtime is often marked to calculate the sleep onset latency, an estimate of how long it takes to fall asleep. Total sleep hours and number, time, and duration of awakenings are documented. Description of subjective mood or alertness before and after sleep can be monitored as well [9]. It is important to note that subjects should be asked about their preferences with regard to sleep times in addition to the actual times of sleep, as there may be large discrepancies between the two. Similarly, the patient's intended wake time may differ drastically from the actual time of waking, which may in turn cause symptoms such as sleep inertia or sleep drunkenness if these times are very disparate [10].

Although sleep diaries are relatively innocuous for the individual and the assessor, the drawbacks of daily commitment and the need to provide the subject with logs weeks in advance of an in-person interaction have led to a push to streamline into a reliable single-administration instrument. The Sleep Timing Questionnaire (STQ), a one-time questionnaire assessing the same measures as a sleep diary, has been shown to be an acceptable alternative to a daily sleep diary, with good test-retest reliability even when administered more than 100 days apart. The STQ has also been shown to be valid compared to the Pittsburgh Sleep Diary and to wrist actigraphy [11].

Once a patient or participant returns a sleep diary, the clinician or researcher can utilize data points to assess timing of sleep, regularity of sleep patterns, degree of sleep fragmentation, adequacy of sleep duration, and other factors integral to sleep health. The sleep diary can serve as a starting point for discussion about behaviors that may impact patients' sleep or wake functioning without his/her awareness. Practitioners of cognitive-behavioral therapy for insomnia (CBT-I) utilize sleep diary data to recommend prescribed sleep and wake times. For instance, in patients undergoing sleep restriction, prescribed time in bed is increased when the sleep diary "sleep efficiency" (defined as estimated total sleep time/time bed × 100) is consistently at or above 85% [12].

Since the inception of the Pittsburgh Sleep Diary, many iterations of sleep diaries have been created, estimated to be in the hundreds [9]. To resolve the problem of comparability, the Consensus Sleep Diary Working group aimed to devise a standardized consensus diary [13]. The nine components considered most crucial to a standardized sleep diary are as follows: "the time of getting into bed; the time at which the individual attempted to fall asleep; sleep onset latency; number of awakenings; duration of awakenings; time of final awakening; final rise time; perceived sleep quality (rated via Likert scale); and an additional space for open-ended comments from the respondent."

Questionnaires

Another important tool in the circadian clinician's or researcher's toolbox is a standardized chronotype questionnaire. Chronotype questionnaires aim to characterize the subject's preferences for timing of sleep and wakefulness, while also describing the subject's behavior in accordance with these preferences. More than a dozen chronotype questionnaires exist, and although they include slightly different questions and/or have been devised for and tested in different populations, they all measure related constructs [14].

Horne-Östberg Morningness-Eveningness Questionnaire (MEQ)

The Horne-Östberg Morningness-Eveningness Questionnaire (MEQ) is one of the first and most widely used chronotype assessments [15]. First implemented in research in 1976, the MEQ is a self-reported survey of 19 items, delving into the timing preference for sleep and other activities. The respondent is asked to indicate at what time of day he or she would prefer to do a specific activity, such as wake up or start sleep, take a test, or exercise. Lower MEQ scores indicate a preference for eveningness, and higher scores mean that the respondent tends toward morningness. The MEQ has a roughly normal distribution around neither type [16]. Scores ranging from 70 to 86 are deemed "Definitely Morning," scores of 42–58 are "Neither," and scores from 16 to 30 are "Definitely Evening." The coefficient alpha of the MEQ is reported at 0.82, indicating good inter-item consistency [15].

The MEQ scale has been externally validated with physiologic measures and shows significant correlations with endogenous circadian phase [16, 17] and period length [18]. The MEQ has since been used widely to link circadian preference to other measures and outcomes; for instance, morningness is associated with longer sleep duration, better subjective sleep quality, shorter sleep latency, and more regular bed times and rise times [19]. A study of spectral electroencephalogram (EEG) waveforms showed that a preference for morningness is also related to higher levels and faster decay rates of slow wave activity [20]. At least one study demonstrated a gender difference in the MEQ, with more women showing a preference for morningness [21]. On the other hand, eveningness has been associated with a host of negative outcomes including greater reports of fatigue, alcohol and caffeine use, worse academic performance in college students [21], more depressive symptoms [22], and higher incidence of Seasonal Affective Disorder (SAD) and Attention Deficit-Hyperactivity Disorder (ADHD) [23].

A criticism of the MEQ scale is its length and perception that the questions assess more than diurnal preferences [24]. In addition, the MEQ assesses respondents' stated preferences, rather than their actual behaviors. A foreshortened scale, the Reduced Morningness-Eveningness Questionnaire (rMEQ) [25–27], only contains five questions. The rMEQ cutoff scores define three chronotype groups: evening <12; neither = 12–17; and morning >17.

Torsvall and Akerstedt Diurnal Scale

The Torsvall and Akerstedt Diurnal scale [28] is a circadian preference questionnaire comprised of seven items, requesting information regarding one's preference of timing of his or her daily activities in circumstances with freedom to choose the schedule. The scale also asks the subject to rate how he or she would feel in certain scenarios. For example, question four reads "If you always had to rise at 0600, what do you think it would be like?" and question five reads, "When do you usually begin to feel the first signs of sleepiness and need for sleep?" While the brevity of this test is preferable logistically, compared to the MEQ, the coefficient alpha is 0.65, indicating marginal internal consistency.

The Torsvall and Akerstedt Diurnal Scale does not have as robust uptake in circadian research as the MEQ, apart from a few Japanese studies that have used it as a marker of morningness in relation to mental health. One such study showed that patients with post-tramautic stress disorder (PTSD) after a natural disaster had better sleep quality if they were morning types rather than evening types [29].

Composite Scale of Morningness

The Composite Scale of Morningness (CSM) [24] by Smith et al. was created in an attempt to mitigate the weaknesses of the MEQ and Torsvall and Akerstedt scales (and a third scale by Folkard et al. [30], which had relatively poor inter-item psychometric properties). This 13-item scale was used to characterize morningness and was employed to select and place shift workers to match their preferred circadian patterns. The CSM has excellent internal consistency of 0.87, which has held up across various cultures, nationalities, and stages in life (e.g., college students, working adults), but has not been validated against physiological markers of the biological clock [31]. The MEQ and CSM both show high reliability, whereas the reliability of the rMEQ is not as strong [14].

Preferences Scale

Smith and colleagues developed another scale, the Preferences Scale (PS), using data on university students in four countries. This scale attempted to unyoke certain activities from the typical 24-hour clock cycle and, in doing so, increase its suitability for "people with alternate sleep-wake schedules and lifestyles," a strength not seen in other diurnal-centric scales [32]. Later, this scale was also used in six countries, from both temperate and nontemperate climates, and in subjects from many different cultural backgrounds [33]. The PS is comprised of 12 items and asks respondents to consider their behavior in relation to other people in a five-point Likert format from "much later than most people" to "much earlier than most people." The alpha coefficient ranged from 0.80 to 0.90 depending on country of origin. The scale's reliability, cross-cultural applicability, and questions that are disentangled from the "typical" diurnal schedule make it a useful tool. The PS is highly correlated with the CSM and has been validated using subjective, but not objective, measures [14].

Munich Chronotype Questionnaire

The Munich Chronotype Questionnaire (MCTQ) has been utilized successfully in a large cohort of subjects in Europe, consists of 16 self-reported items, and was developed by Roenneberg, Wirz-Justice, and Merrow [34]. The MCTQ aims to elucidate actual sleep behavior of a subject over the most recent four-week period. Specifically, it aims to report on specific times during which a certain behavior is performed, and differentiates between work days and free days. It has been administered to thousands of participants since its inception and data collected using this instrument has contributed to landmark circadian concepts, including the phenomenon of the delay in chronotype that accompanies the transition to adolescence and early adulthood

[35] and the construct of social jet lag, characterized by marked difference in sleep behavior between workdays and free days [36]. Survey participants also self-assess as to which chronotype they fit, from "extreme early" to "extreme late." The MCTQ Shift [37] is a validated iteration of the MCTQ, from which the chronotypes of shift workers can be described.

Children's Chronotype Questionnaire

The Children's Chronotype Questionnaire (CCTQ) was designed specifically for a parent or caretaker to answer questions about his or her child's sleep habits [38]. This 27-item questionnaire records information about the days and nights of "scheduled" and "free" days. The parent or caregiver completes the CCTQ by answering a final question regarding the type of person the child in question most resembles, a morning type, an evening type, or whether he or she falls on the spectrum in between. There are also a series of questions within the survey that specifically ask about behaviors of the child during specific times of the day, such as preference of time to start school, energy levels in relation to time of awakening, and others. The CCTQ assesses the midsleep point on free days (MSF) and provides a morningness/eveningness scale score. This questionnaire has been validated in prepubertal children, aged 4–11, using actigraphy, with strong test-retest reliability.

Internet/Twitter/Apps

In recent years, technology has become a pervasive component of our daily lives. With ease of access to smart phones, tablets, video games, and the internet, newly recognized psychological and sleep pathologies have emerged [39–41]. Social media use has been employed to monitor circadian rhythms [42, 43]. Using a novel approach to characterize "social jet lag," researchers tabulated Twitter ® use in approximately 246,000 users from 2012 to 2013 and categorized, with a resolution down to the county level, varied circadian changes in sleep patterns.

Specific mobile telephone and computer applications are now becoming more popular as avenues to collect data on circadian rhythms. Many are available for personal use and are available for purchase on the Apple ® App Store with the search of "circadian rhythm." Research protocols, such as the My CircadianClock study from the Salk institute [44], are attempting to utilize crowdsourced data from these apps to analyze how various behaviors affect circadian rhythms. Circadian/sleep apps have been shown to perform similarly to paper sleep diaries [45]. Circadian rhythms as measured by social media platforms may also be useful tools to identify emotional states, such as anger, fatigue, sadness, and positive mood [46].

Non-physiologic methods of assessing circadian/sleep behavior and circadian preferences are useful tools for rapid, inexpensive assessment of factors relevant to circadian rhythm sleep-wake disorders, and other behaviors. They are also practical for measuring circadian behavior in large samples or in settings where physiologic methods are not available. Researchers and clinicians interested in circadian rhythms should be familiar with various sleep and circadian diaries and questionnaires so that they may select the most appropriate measure for their specific sample or area of interest.

References

1. American Academy of Sleep Medicine, editor. International classification of sleep disorders. 3rd ed. Darien: American Academy of Sleep Medicine; 2014.
2. Morgenthaler TI, Lee-Chiong T, Alessi C, et al. Practice parameters for the clinical evaluation and treatment of circadian rhythm sleep disorders: an American Academy of Sleep Medicine report. Sleep. 2007;30(11):1445–9.
3. Short MA, Arora T, Gradisar M, Taheri S, Carskadon MA. How many sleep diary entries are needed to reliably estimate adolescent sleep? Sleep. 2017;40(3):1–10.
4. Smith MT, McCrae CS, Cheung J, Martin JL, Harrod CG, Heald JL, Carden KA. Use of actigraphy for the evaluation of sleep disorders and circadian rhythm sleep-wake disorders: an American Academy of Sleep Medicine systematic review, meta-analysis, and GRADE assessment. J Clin Sleep Med. 2018;14(7):1209–30.
5. McGhie A, Russell SM. The subjective assessment of normal sleep patterns. J Ment Sci. 1962;108:642–54.
6. Wilson KG, Watson ST, Currie SR. Daily diary and ambulatory activity monitoring of sleep in patients with insomnia associated with chronic musculoskeletal pain. Pain. 1998;75(1):75–84.
7. Rogers AE, Caruso CC, Aldrich MS. Reliability of sleep diaries for assessment of sleep/wake patterns. Nurs Res. 1993;42(6):368–72.
8. Monk TH, Reynolds CF 3rd, Kupfer DJ, Buysse DJ, Coble PA, Hayes AJ, et al. The Pittsburgh sleep diary. J Sleep Res. 1994;3:111–20.
9. Ibáñez V, Silva J, Cauli O. A survey on sleep questionnaires and diaries. Sleep Med. 2018;42:90–6.
10. Trotti LM. Waking up is the hardest thing I do all day: sleep inertia and sleep drunkenness. Sleep Med Rev. 2017;35:76–84.
11. Monk TH, Buysse DJ, Kennedy KS, et al. Measuring sleep habits without using a diary: the sleep timing questionnaire (STQ). Sleep. 2003;2:208–12.
12. Kyle SD, Aquino MR, Miller CB, Henry AL, Crawford MR, Espie CA, Spielman AJ. Towards standardisation and improved understanding of sleep restriction therapy for insomnia disorder: a systematic examination of CBT-I trial content. Sleep Med Rev. 2015;23:83–8.
13. Carney CE, Buysse DJ, Ancoli-Israel S, Edinger JD, Krystal AD, Lichstein KL, et al. The consensus sleep diary: standardizing prospective sleep self-monitoring. Sleep. 2012;35(2):287–302.
14. Di Milia L, Adan A, Natale V, Randler C. Reviewing the psychometric properties of contemporary circadian typology measures. Chronobiol Int. 2013;30(10):1261–71.
15. Horne J, Ostberg O. A self-assessment questionnaire to determine morningness-eveningness in human circadian rhythms. Int J Chronobiol. 1976;4(2):97–110.
16. Baehr EK, Revelle W, Eastman CI. Individual differences in the phase and amplitude of the human circadian temperature rhythm: with an emphasis on morningness–eveningness. J Sleep Res. 2000;9:117–27.
17. Duffy JF, Dijk DJ, Hall EF, Czeisler CA. Relationship of endogenous circadian melatonin and temperature rhythms to self-reported preference for morning or evening activity in young and older people. J Invest Med. 1999;47(3):141–50.

18. Duffy JF, Rimmer DW, Czeisler CA. Association of intrinsic circadian period with morningness-eveningness, usual wake time, and circadian phase. Behav Neurosci. 2001;115(4):895–9.
19. Soehner AM, Kennedy KS, Monk TH. Circadian preference and sleep-wake regularity: associations with self-report sleep parameters in daytime-working adults. Chronobiol Int. 2011;28(9):802–9.
20. Mongrain V, Carrier J, Dumont M. Circadian and homeostatic sleep regulation in morningness-eveningness. J Sleep Res. 2006;15(2):162–6.
21. Taylor DJ, Clay KC, Bramoweth AD, Sethi K, Roane BM. Circadian phase preference in college students: relationships with psychological functioning and academics. Chronobiol Int. 2011;28(6):541–7.
22. Abe T, Inoue Y, Komada Y, Nakamura M, Asaoka S, Kanno M, et al. Relation between morningness-eveningness score and depressive symptoms among patients with delayed sleep phase syndrome. Sleep Med. 2011;12(7):680–4.
23. Rybak YE, McNeely HE, Mackenzie BE, Jain UR, Levitan RD. Seasonality and circadian preference in adult attention-deficit/hyperactivity disorder: clinical and neuropsychological correlates. Compr Psychiatry. 2007;48(6):562–71.
24. Smith CS, Reilly C, Midkiff K. Evaluation of three circadian rhythm questionnaires with suggestions for an improved measure of morningness. J Appl Psychol. 1989;74(5):728–38.
25. Caci H, Deschaux O, Adan A, Natale V. Comparing three morningness scales: age and gender effects, structure and cut-off criteria. Sleep Med. 2009;10(2):240–5.
26. Adan A, Almirall H. Horne & Östberg morningness-eveningness questionnaire: a reduced scale. Personal Individ Differ. 1991;12(3):241–53.
27. Natale V, Esposito MJ, Martoni M, Fabbri M. Validity of the reduced version of the morningness–eveningness questionnaire. Sleep Biol Rhythms. 2006;4:72–4.
28. Torsvall L, Akerstedt T. A diurnal type scale. Construction, consistency and validation in shift work. Scand J Work Environ Health. 1980;6(4):283–90.
29. Wada K, Kuroda H, Nakade M, Takeuchi H, Harada T. Epidemiological studies on the relationship between PTSD symptoms and circadian typology and mental/sleep health of young people who suffered a natural disaster, Great Hanshin-Awaji Earthquake. Nat Sci. 2014;6(5):338–50.
30. Folkard S, Monk TH, Lobban MC. Towards a predictive test of adjustment to shift work. Ergonomics. 1979;22:79–91.
31. Levandovski R, Sasso E, Hidalgo M. Chronotype: a review of the advances, limits and applicability of the main instruments used in the literature to assess human phenotype. Trends Psychiatry Psychother. 2013;35(1):3–11.
32. Smith CS, Folkard S, Schmieder RA, Parra LF, Spelten E, Almirall H. The preferences scale: multinational assessment of a new measure of morningness. Proc Hum Factors Ergon Soc Annu Meet. 1993;37(13):925–9.
33. Smith CS, Folkard S, Schmieder RA, Parra LF, Spelten E, Almirall H, et al. Investigation of morning-evening orientation in six countries using the preferences scale. Personal Individ Differ. 2002;32(6):949–68.
34. Roenneberg T, Wirz-Justice A, Merrow M. Life between clocks: daily temporal patterns of human chronotypes. J Biol Rhythms. 2003;18(1):80–90.
35. Roenneberg T, Kuehnle T, Pramstaller PP, Ricken J, Havel M, Guth A, Merrow M. A marker for the end of adolescence. Curr Biol. 2004;14(24):R1038–9.
36. Wittmann M, Dinich J, Merrow M, Roenneberg T. Social jetlag: misalignment of biological and social time. Chronobiol Int. 2006;23(1–2):497–509.
37. Juda M, Vetter C, Roenneberg T. The Munich ChronoType Questionnaire for Shift-Workers (MCTQShift). J Biol Rhythms. 2013;28(2):130–40.
38. Werner H, LeBourgeois MK, Geiger A, Jenni OG. Assessment of chronotype in four- to eleven-year-old children: reliability and validity of the Children's Chronotype Questionnaire (CCTQ). Chronobiol Int. 2009;26(5):992–1014.
39. Harbard E, Allen NB, Trinder J, Bei B. What's keeping teenagers up? Prebedtime behaviors and actigraphy-assessed sleep over school and vacation. J Adolesc Health. 2016;58(4):426–32.

40. Lam LT. Internet gaming addiction, problematic use of the internet, and sleep problems: a systematic review. Curr Psychiatry Rep. 2014;16(4):444.
41. Touitou Y, Touitou D, Reinberg A. Disruption of adolescents' circadian clock: the vicious circle of media use, exposure to light at night, sleep loss and risk behaviors. J Physiol Paris. 2016;110(4 Pt B):467–79.
42. Leypunskiy E, Kıcıman E, Shah M, Walch OJ, Rzhetsky A, Dinner AR, et al. Geographically resolved rhythms in twitter use reveal social pressures on daily activity patterns. Curr Biol. 2018;28(23):3763–75.
43. Roenneberg T. Twitter as a means to study temporal behaviour. Curr Biol. 2017;27(17):R830–2.
44. myCircadianClock. Salk Institute for Biological Studies in La Jolla, California. http://mycircadianclock.org/. Accessed 14 Sept 2019.
45. Tonetti L, Mingozzi R, Natale V. Comparison between paper and electronic sleep diary. Biol Rhythm Res. 2016;47(5):743–53.
46. Dzogang F, Lightman S, Cristianini N. Circadian mood variations in Twitter content. Brain Neurosci Adv. 2017;1:1–23.

Chapter 6
Delayed Sleep-Wake Phase Disorder

Gregory S. Carter and R. Robert Auger

Introduction

Delayed sleep-wake phase disorder (DSWPD) is the most common circadian rhythm sleep-wake disorder (CRSWD) observed among adolescents and is encountered less frequently among older demographics [1, 2]. Roenneberg et al. [3] identified a prevalence inflection point of *delayed circadian preference* which reaches its maximum around 20 years of age, followed by a reversion toward an advanced circadian preference.

Individuals with DSWPD exhibit an extreme "night owl" preference that results in an inability to initiate sleep and to arise at conventional "early bird" times [4]. Resultant chronic sleep deprivation and sleep inertia (an inability to achieve full alertness upon arising in the morning) result in impaired daytime function [5]. Difficulties conforming to traditional school, social, and employment schedules may cause considerable distress for patients and those within their orbits. Mood disorders, maladaptive sleep-related behaviors, and other comorbidities complicate the individuals' attempts to self-correct and, by relation, clinicians' attempts to manage symptomatology [6–9]. The precise etiology of DSWPD is unclear, but it appears to relate to both endogenous and exogenous factors. Evidence-based treatment options are limited, but one can employ rational therapeutics utilizing a combination of scientific knowledge and accumulated clinical experience.

G. S. Carter (✉)
Department of Neurology and Neurotherapeutics, The University of Texas Southwestern Medical Center at Dallas, Dallas, TX, USA
e-mail: Gregory.Carter@UTSouthwestern.edu

R. R. Auger
Mayo Center for Sleep Medicine and Department of Psychiatry and Psychology, Mayo Clinic College of Medicine, Rochester, MN, USA

R. R. Auger (ed.), *Circadian Rhythm Sleep-Wake Disorders*,
https://doi.org/10.1007/978-3-030-43803-6_6

Clinical Features

The International Classification of Sleep Disorders, Third Edition (ICSD-3) [4], provides diagnostic criteria for DSWPD (Table 6.1) and describes typical associated symptoms. The manual also refers to useful complementary assessment tools, including chronotype questionnaires [10] and physiologic measurements of circadian timing, such as the dim light melatonin onset (DLMO) [11]. Use of the latter as a diagnostic tool has demonstrated controversial findings.

Murray et al. [12] reported on DSWPD subjects recruited from the communities of Melbourne, Sidney, and Adelaide, Australia. The authors looked at the timing of DLMO as related to the participant's desired or required bedtime (DBT). Subjects maintained their usual sleep-wake patterns (documented with sleep diaries and wrist actigraphy) for at least 7 days prior to coming into the laboratory. If DLMO occurred at least 30 minutes prior to the participant's DBT, the subjects' sleep difficulties were deemed non-circadian in nature. Forty-three percent of DSWPD participants ($n = 79$, age 28.7 ± 9.8 years) exhibited this finding, with a mean DLMO 1.48 ± 0.78 hours before their work or school night DBT (10:49 PM ± 49 minutes), which was 3.20 ± 1.06 hours prior to their habitual sleep time (11:52 PM ± 1:16 hours). While this would appear to delineate physiologic versus behavioral DSWPD, the significance of this distinction (e.g., from a treatment perspective) is indeterminate at this juncture.

Sonheim et al. [13] highlighted DSWPD symptoms and impairments in an objective fashion, including the disabling complaint of sleep inertia that commonly occurs subsequent to early rising related to early school or work start times. Their study included 9 rigorously diagnosed Norwegian patients aged 22.5 ± 2.2 years (four males) and a matched healthy control group. Not unexpectedly with a polysomnogram (PSG) start time of 12:00 AM, sleep latency was greater in patients than in controls [41 ± 37 vs. 7 ± 7 minutes, respectively ($p = 0.003$)] and total sleep time was reduced [333 ± 49 minutes vs. 379 ± 22 minutes, respectively ($p = 0.01$)].

Table 6.1 International Classification of Sleep Disorders, Third Edition, diagnostic criteria for delayed sleep-wake phase disorder

A. There is a significant delay in the phase of the major sleep episode in relation to the desired or required sleep time and wake-up time, as evidenced by a chronic or recurrent complaint by the patient or a caregiver of inability to fall asleep and difficulty awakening at a desired or required clock time.
B. The symptoms are present for at least 3 months
C. When patients are allowed to choose their ad libitum schedule, they will exhibit improved sleep quality and duration for age and maintain a delayed phase of the 24-hour sleep-wake pattern
D. Sleep log and, whenever possible, actigraphy monitoring for at least 7 days (preferably 14 days) demonstrate a delay in the timing of the habitual sleep period. Both workdays/school days and free days must be included within this monitoring.
E. The sleep disturbance is not better explained by another current sleep disorder, medical or neurological disorder, mental disorder, medication use, or substance use disorder.

Source: American Academy of Sleep Medicine (2014)

At 7:00 AM an alarm clock was activated that started at 72 decibels (dB) for a duration of 4.4 seconds, with increases in 2 dB increments at 5-second intervals until the subject was awake or a maximum of 104 dB was reached. If this maximal stimulus failed to produce an arousal from sleep, the subjects were awakened manually. Three (33%) of the DSWPD patients (but none of the controls) failed to awaken to the alarm and remained in rapid-eye-movement (REM) sleep. The awakening thresholds of the other six patients were not significantly different than those of controls. The subgroup that did not arouse to the alarm had less REM sleep percentage than their arousable counterparts (14 ± 5% vs. 19 ± 2%, respectively $p = 0.04$) and more stage N1sleep (35 ± 9 minutes vs. 17 ± 12 minutes, respectively $p = 0.04$).

Immediately upon awakening all participants were given a continuous performance test [14], each serving as his/her own control, with comparison of scores from their baseline evening assessment. Healthy controls showed a statistically insignificant improvement of 3.4% in their morning response times compared to the previous evening (347.8 ± 59.6 milliseconds [ms] vs. 335.9 ± 56.9 ms). In comparison, the nine DSWPD patients showed an 8.6% statistically significant worsening in response times (300.6 ± 25.9 ms to 326.7 ± 44.3 ms, $p = 0.013$).

Differential Diagnosis

The clinical differential of DSWPD includes an evening circadian preference without significant distress or impaired functioning, inadequate sleep hygiene with resistance to going to bed at an appropriate time, use of sleep-disrupting drugs or substances, and/or chronic insomnia disorder. Epidemiological surveys often differentiate chronic insomnia from DSWPD via a paucity of sleep maintenance difficulties in the latter, as well as a resolution of subjective sleep quality and duration when DSWPD patients are allowed to sleep at desired times on unrestricted days [5, 8, 15–18].

Comorbidities

There are several conditions seen with increased frequency in patients with DSWPD. There is conjecture that this is at least partially related to the delayed timing of patients' circadian rhythms, either in a direct sense (e.g., genetic susceptibilities that increase the probability of both DSWPD and the comorbid phenotype) or in an indirect manner, i.e., as a consequence of occupational, academic, and/or social impairments. The condition seldom presents as a circadian misalignment in isolation, and associated comorbidities may be most challenging for clinicians.

Rajaratnam et al. [5] selected 295 participants between the ages of 18 and 65 deemed to be at high risk for DSWPD, based upon completed surveys culled from 13,844 US respondents. A comparison group of over 700 subjects was comprised of "low-risk" non-evening types. The high-risk group (aged 31.2 ± 12.2 years) was significantly younger ($p < 0.001$) than their counterparts (40.0 ± 15.0). The authors evaluated the presence of general impairments using an online version of the Sheehan Disability Scale [19]. The high-risk cohort exhibited comparatively increased daytime impairments, including moderate, marked, or extreme disruptions in work and/or school environments (32.9% vs. 18.0%) ($p < 0.001$). In addition, absenteeism was reported in 12.2% of the high-risk group vs. 6.5% of the low-risk group ($p = 0.025$).

Likely contributing in part to the findings described above, depression and personality vulnerabilities are frequently comorbid with DSWPD [20–23]. Shirayama et al. [20] recruited 22 ICSD-diagnosed DSWPD subjects and 20 matched controls through the National Center of Neurology and Psychiatry in Kodaira, Japan. The DSWPD patients consisted of 12 men (25.5 ± 5.5 years) and 10 women (29.2 ± 5.1 years) with symptom durations of 7.1 ± 4.8 and 8.9 ± 4.1 years, respectively. No prior treatment was reported by 16 (73%) subjects (10 men and 6 women).

The two groups received three types of psychological tests, including the Yatabe-Guilford test (Y-G test) [24], the Minnesota Multiphasic Personality Inventory (MMPI) [25], and the Rorschach test [26]. The DSWPD Y-G and MMPI assessments revealed statistically significant increases in depressive symptoms. The Rorschach test results reflected decreased self-awareness of the inclination toward immediate gratification, which the authors attributed to difficulties learning from experience, perhaps perpetuated by social withdrawal from peers and/or aforementioned depressive symptoms.

Abe et al. [27] more specifically examined the prevalence and characteristics of depression in 90 Japanese ICSD-defined DSWPD patients (aged 27.1 ± 9.2 years) recruited from an outpatient sleep clinic. Symptoms were assessed by the Zung self-rating depression scale [28], and raw scores were divided into four severity levels, as described by Barrett et al. [29]. Sixty-four percent of patients had depression scores in the moderate to severe range (much higher than the general population [30, 31]), with diurnal variation, sleep disturbance, fatigue, and psychomotor retardation reported most often as accompanying symptoms.

Wilhelmsen-Langeland et al. [21] used the NEO-PI-R (Neuroticism, Extroversion, Openness-Personality Inventory-Revised) [32] to assess personality traits among 40 Norwegian ICSD-diagnosed DSWPD patients (20.7 ± 3.1 years) compared to 21 age-matched controls, all recruited through local school advertisements. The DSWPD group scored higher in neuroticism (anxiety and depression) and lower in conscientiousness (dutifulness, achievement striving, and self-discipline). The authors' interpretation was that the requirement to arise from bed early in the morning, with attendant sleep deprivation, increased the risk for interpersonal conflicts. Internalization of these conflicts lowers self-esteem and sense of personal competence, thereby reducing the capacity to enforce self-discipline. The strongest area of deviation from normative values was demonstrated in conscientiousness, with 62.5% of DSWPD patients scoring low or very low. This NEO-PI-R facet is a known predictor of poor treatment compliance, which led the authors to conclude that effective treatments must incorporate motivational interventions.

Epidemiology

DSWPD prevalence reports vary considerably [33], but the condition clearly occurs most commonly among adolescents and young adults. United States DSWPD prevalence studies have only been published in abstract form and are therefore not included within.

Ohayon et al. [34] described an adolescent DSWPD prevalence of 0.4% in France, Great Britain, Germany, and Italy. The sample was composed of a total of 1125 adolescents aged 15–18 years of age. Each country conducted separate surveys using the same methodology from 1993 to 1997. The latest census figures were used to build representative samples according to age, gender, and geographic distribution. Diagnostic interviews were conducted by telephone using the Sleep-EVAL computer program, a level 2 expert system capable of formulating diagnostic hypotheses and validating through further questions. Importantly, the European schools documented later school start times than most in the United States, which would serve to decrease prevalence estimates in comparison to American counterparts [35].

Lovato et al. [8] reported on 374 Australian adolescents (aged 15.6 ± 1.0 years) recruited from eight schools in urban Adelaide. Subjects completed 7-day sleep diaries in association with wrist actigraphy. While 51.9% of their sample met one ICSD criterion and 14% met two criteria, only 1.1% met full DSWPD criteria. Hazama et al. [36] gave a self-administered questionnaire of their own composition to a total of 4971 Japanese junior high (1240), senior high (1205), and university students (2526) from rural areas. Questions were based on ICSD diagnostic criteria (1997) [15] but with a symptom duration of at least 6 months. Analyses provided an overall DSWPD prevalence estimate of 0.48%, which increased to 1.66% in the subgroup of university students.

Sivertsen et al. [16] surveyed 9338 Norwegian adolescents aged 16–18 (17.8 ± 0.8 years). Questions approximated ICSD criteria [33] and incorporated school attendance assessments. The overall DSWPD prevalence rate was 3.3% and those afflicted had significantly higher rates of school non-attendance than their non-afflicted counterparts ($p < 0.001$). Danielsson et al. [37] randomly sent questionnaires to 1000 young people in urban Sweden aged 16–26 years, of whom 671 (ages 21.8 ± 3.1 years) responded. The questionnaire approximated DSM-5 (*Diagnostic and Statistical Manual of Mental Disorder*, Fifth Edition) [38] rather than (very similar) ICSD criteria [4, 33]. Twenty-seven participants (4%) met DSWPD criteria.

The prevalence rates cited above range from 0.4% to 4.0% in adolescents and young adults. Differences may relate to numerous factors including the methodology of the surveys, school start times, and the geographical representation of the samples, including urban or rural environments.

Pathophysiology

A delay in circadian phase correlates with the beginning of puberty in both humans and other mammalian species [39–50], although in humans a combination of physiology and behavioral influences is readily apparent. There is also evidence [39, 42, 46] to support a decreased response to homeostatic sleep drive during human adolescence, as well as a decreased sensitivity to the phase advancing effects of morning light and an increased sensitivity to the phase delaying effects of evening light [47, 48]. This latter finding is also seen in pubertal mice [49, 50] and relies on the presence of gonadal hormones. All of these factors serve to facilitate adolescents' typically later times of sleep onset and offset.

Although one may envision the phenotype of DSWPD as an extreme manifestation of one or more of these phenomena, the precise etiology is unclear. Some studies [51, 52] describe longer circadian period lengths among DSWPD patients, making it harder to maintain non-delayed entrainment within the 24-hour light/dark cycle. Micic et al. [51] recruited participants through advertisements posted at universities across Adelaide, Australia. Six DSWPD participants (mean age 22.0 ± 3.3 years) and an age-matched comparison group of 7 "good sleepers" completed the protocol. Circadian period length (tau) was determined by an ingestible core body temperature capsule that transmitted readings every minute during a 78-hour experimental routine [53]. The DSWPD participants had a mean tau of 24 hours and 54 minutes (SD = 23 minutes), while the control group had a mean tau of 24 hours and 29 minutes (SD = 16 minutes) ($p = 0.04$).

Though not directly addressed in this chapter, it is important not to neglect the numerous external factors that may contribution to the induction, exacerbation, or perpetuation of DSWPD, including behavioral [20, 40, 54, 55] variables, reduced parental influence on bedtimes [56], the evening use of blue screens and social media [57], and part-time employment after school [58]. More direct maladaptive sleep-related behaviors are also prominent, such as markedly later rise times on weekends [59], which can lessen the homeostatic drive that favors an earlier sleep-onset time and mask the advancing effects of morning light. A key issue for practitioners is to disentangle the myriad internal and external factors contributing to sleep/wake delays.

Genetic Factors

Genetic screening for clock gene polymorphisms is not presently commercially available, but has the potential to aid our understanding of the pathophysiology of DSWPD [60–65]. The circadian rhythms of mammals involve a highly complex set of interactions with a transcriptional and translational feedback loop of clock component genes [66, 67]. What follows is a simplified model to allow a preliminary understanding of relevant genetic anomalies and their putative roles in DSWPD (Fig. 6.1).

The clock gene encodes a transcriptional regulatory protein, CLOCK, which forms a heterodimer in the cytoplasm with another protein called BMAL-1 (Brain and Muscle Aryl hydrocarbon Receptor Nuclear Translocator-Like-1 protein) [67]. The bmal-1 gene is rhythmically transcribed out of phase with period (per) and cryptochrome (cry) genes. The CLOCK/BMAL-1 heterodimer is able to pass into and accumulate in the nucleus of the cell until it is degraded by phosphorylation. In the nucleus, this heterodimer activates the transcription of the per and cry genes with slow production of their respective proteins. These proteins form CRY/PER heterodimers in the cytoplasm, which then translocate into the nucleus, causing inhibition of the CLOCK/BMAL-1-induced transcription of the per and cry genes, until phosphorylation/degradation prevents further translocation. The cycle then starts anew as fresh CLOCK/BMAL-1 heterodimer induces a new round of gene transcription. This transcription-translational feedback loop is completed in slightly over 24 hours.

There have been several studies [60–65] looking at polymorphisms or mutations in this system. Ebisawa et al. [63] reported an association of per3 polymorphisms with DSWPD in Japan. Archer et al. [64] reported an association of a length polymorphism in the same with evening preference and DSWPD. Most recently, Patke et al. [60] described a family of Turkish descent with familial DSWPD transmitted in

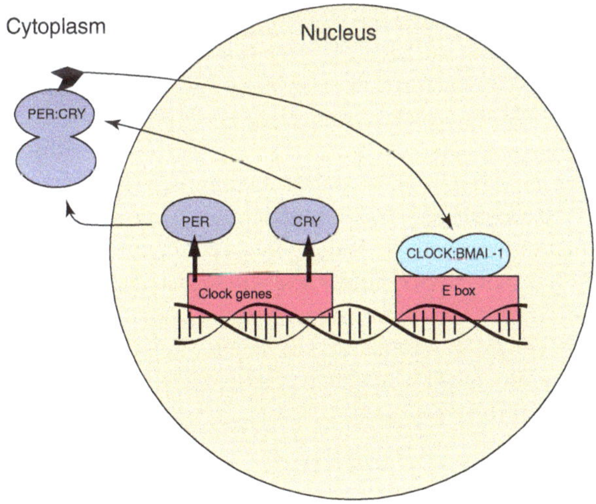

Fig. 6.1 A simplified model of the cryptochrome (cry) and period (per) genes and their role in the ~24-hour translational/transcriptional-based feedback oscillator that comprises the molecular circadian clock. The CLOCK/BMAL-1 heterodimer binds to the E-box promoter of the clock genes, activating (+) the production of CRY and PER proteins. These proteins leave the nucleus, combine into a heterodimer in the cytoplasm (if not degraded by phosphorylation), and return to the nucleus and accumulate, inhibiting (−) the CLOCK/BMAL-1 activation of cry and per gene transcription and completing the feedback loop

an autosomal dominant fashion. Afflicted individuals demonstrated an adenine-to-cytosine point mutation in the cry1 gene, which caused a 24-amino-acid skip in the production of altered CRY1 protein. Fibroblast cultures revealed a gain of function, leading to increased nuclear localization and enhanced inhibition of the CLOCK/BMAL-1 heterodimer's activation of gene transcription. This led to a delay in the re-initialization of the transcriptional and translational feedback loop and lengthening of the circadian period. Indeed, the measured circadian period of one subject was prolonged at 24.5 hours [60], compared to a normal control value of 24.2 hours [68]. Further genetic screening of the family showed no other changes in clock-related genes. The authors looked at available human databases of human genetic variations and found the frequency of the cry1 mutation to be up to 0.6%, consistent with the overall reported frequency of DSWPD. This finding is also of potential interest from a comorbidity standpoint, as both Soria [69] and Hua [70] reported over representation of cry1 polymorphisms in patients with major depression.

Treatment

Therapeutic circadian-based interventions for DSWPD consist primarily of post-awakening light therapy and/or strategically timed melatonin. Cognitive behavioral therapy for comorbid insomnia (CBT-I) commonly accompanies these treatments. Large-scale randomized controlled trials for DSWPD are lacking, as illustrated by the American Academy of Sleep Medicine's (AASM) most recently published circadian rhythm sleep-wake disorder practice guidelines [71]. Affirmative findings are described below, with notation of those studies that were not included in the AASM's systematic evidence review. Negative findings were plentiful due to the rigorous GRADE (Grading of Recommendations, Assessment, Development, and Evaluations) system employed. As a case in point, there was insufficient evidence to endorse post-awakening light treatment as monotherapy. While the use of GRADE should ultimately propel greater quality clinical research, clinicians will invariably need to employ less rigorously studied interventions and to extrapolate well-described scientific data into clinical practice to satisfactorily address patients' needs. Patient frustrations leading to noncompliance are common, as responses are not rapid and effects are often modest. Clinicians should therefore be prepared to exhibit patience, to provide encouragement, and to adopt a long-term view required for management of chronic illnesses.

Combination Light Therapy Studies

Gradisar et al. [72] *combined* post-awakening light therapy with cognitive behavioral therapy for insomnia (CBT-I) over a period of 8 weeks. The study included 23 adolescent DSWPD patients (ages 14.7 ± 1.7 years) who were compared to an age-matched wait list group. Sleep diaries were used to ascertain subjective sleep parameters.

Upon arising from bed, the participants were exposed to either natural sunlight or a 1000 lux broad-spectrum lamp, for 30–120 minutes. In counterclockwise increments of 30 minutes per day, the participants advanced to a target time of 6:00 AM. At that point, light therapy was discontinued, and the participants were encouraged to maintain the earlier rise time. This occurred in conjunction with six 45–60-minute sessions of CBT-I with a clinical psychologist, with significant parental involvement. At the end of the trial, statistically and clinically significant differences (the latter predefined by the AASM practice parameter Task Force) were noted in the active treatment group on school nights versus the wait list group in regard to total sleep time (8.1 ± 0.6 hours vs. 6.9 ± 1.1 hours) and initial sleep latency (22.2 ± 12.8 minutes vs. 65.3 ± 42.0 minutes), respectively. Moreover, the dropout rate in the treatment group was more favorable than that of the wait list group (11.5% vs. 26.1%, respectively). Six-month follow-up assessments revealed durable responses.

A separate combination treatment study was published subsequent to the AASM practice parameter's availability and, as such, was not subject to the same evidence-based review. Danielsson et al. [73] conducted a two-phase study of light therapy (LT) and CBT-I in individuals aged 16–26 years. The 57 participants were randomized into two groups. In the initial 2-week phase, both groups were exposed daily to 30–45 minutes of 10,000 lux bright white light post-awakening. Eleven individuals (19%), roughly split between the two groups, were either noncompliant or intolerant of treatment, leaving 46 (81%) to proceed to the second 4-week phase of LT + CBT-I versus LT alone. Within the former group, 4 weekly 90–120-minute CBT-I sessions were conducted by a clinical psychologist.

Sleep diaries were used to measure sleep onset and offset. The first 2-week phase showed statistically significant improvements from baseline among all participants (advances in sleep onset from 3:10 AM ± 1:23 to 1:19 AM ± 1:26 and 2:56 AM ± 1:28 to 1:08 AM ± 1:00 and in sleep offset from 10:27 AM ± 1:57 to 8:09 AM ± 1:22 and 10:22 AM ± 2:05 to 8:14 AM ± 0:55 among pre-CBT-I + LT combined and LT-alone groups, respectively). During the subsequent 4 weeks of treatment, five out of the six dropouts originated from the combined CBT-I + LT group, and there were no further statistically significant sleep-onset or offset changes. The relatively higher dropout rate in the CBT-I + LT group is difficult to reconcile with that observed with the Gradisar study. The addition of CBT-I did not confer additional benefit versus LT alone overall, although participants were older than those in the Gradisar study (perhaps requiring less external influences). Neither the Gradisar nor the Danielsson study was designed in a manner that allowed one to extricate the individual benefits of LT and CBT-I.

Adverse Effects Associated with Light Therapy

No studies have been done that specifically address potential harms of light therapy in patients with DSWPD, but no serious adverse effects have been reported. In the Cochrane Systematic Review for the treatment of nonseasonal depression, Tuunainen and colleagues [74] found that hypomania was the sole adverse effect more common among patients receiving light therapy versus controls (Relative Risk 4.91 [CI 1.66–4.46]). Nevertheless, light therapy has been safely used for the treatment of bipolar depression, with careful monitoring [75]. Other minor, but commonly described adverse effects include eye strain, nausea, and agitation, which all tend to spontaneously remit. Treatment-emergent headaches also commonly remit [76], but light therapy can induce migraines in approximately one-third of those susceptible [77]. Although commercially available products do not emit ultraviolet light, patients with eye disease and/or those using photosensitizing medications should only use light therapy with periodic ophthalmological and/or dermatological monitoring of the underlying condition [76, 78, 79]. Reassuringly, one study reported no changes in extensive ophthalmologic examinations among seasonal affective disorder patients without preexisting conditions after up to 6 years of daily use in the fall and winter months [78].

Melatonin

Two adult studies of exogenous melatonin performed by the same investigators [80, 81] demonstrated substantial select polysomnography-measured benefits in sleep parameters (total sleep time, initial sleep latency), but an affiliated circadian phase marker was not employed. The Rahman study [81] ($n = 20$, randomized, double-blind, placebo-controlled, crossover design, mean age 30.8 ± 12.4 and 35.6 ± 14.0 years for females and males, respectively) utilized a 5 mg melatonin dose administered between 7:00 PM and 9:00 PM for a period of 28 days. The Kayumov study [80] ($n = 20$, same design/age distribution) used the same dose, but scheduled it at 7:00 PM the first week and between 7:00 PM and 9:00 PM the second and third weeks (according to subjects' preferences) with a consistent time during the fourth week (average chosen time 9:00 PM). The most definitive results were obtained from an analysis of a subset of patients without comorbid depression from the Rahman study ($n = 12$, increased total sleep time = 56.00 minutes [CI 48.51–63.49]). Initial sleep latency was assessed with the same subcategorization. Among the subgroup with comorbid depression ($n = 28$), sleep latency decreased by 43.52 minutes [CI −34.45 to −52.60]. Among the nondepressed subjects ($n = 12$), sleep latency decreased by 37.70 minutes [CI −31.75 to −43.65].

One AASM-reviewed randomized, placebo-controlled double-blind study contained solely pediatric DSWPD patients, ranging in age from 6 to 12 years [82]. The three active treatment groups received melatonin at dosages of 0.05 mg/kg, 0.1 mg/kg, and 0.15 mg/kg, with respective mean doses of 1.6 ± 0.4 mg ($n = 16$), 2.9 ± 0.9 mg ($n = 19$), and 4.4 ± 1.0 mg ($n = 18$). The placebo-controlled group consisted of 17 patients. The duration of treatment was six nights, with instructions to consistently take melatonin 1.5–2.0 hours prior to habitual bedtime (unclear if equivalent to habitual sleep-onset time), with consistent nightly timing. The data of 64 participants were utilized for actigraphy/sleep-related analyses. With respect to AASM Task Force-defined critical outcomes, sleep-onset time favorably advanced in comparison to placebo among the 0.15 mg/kg group only (mean difference -42.77 minutes [CI −21.77 to −63.78]). Nevertheless, sleep latency improved among all three groups (statistical and Task Force-defined clinically significant differences) in comparison to placebo in order of increasing dosage (mean difference −38.39 minutes [CI −18.24 to −58.53], −44.24 minutes [CI −24.04 to −64.44], and −43.80 minutes [CI −24.06 to −63.54], respectively). A positive relationship between DLMO phase advances and an earlier circadian time of administration (TOA) was described with no differences between the various melatonin dosage groups. No advantages between clock determined TOA versus circadian determined TOA were demonstrated in relation to sleep-onset and initial sleep latency times.

Two randomized, placebo-controlled studies by another group of investigators examined the use of melatonin for DSWPD among children/adolescents with various psychiatric comorbidities (all were diagnosed with attention deficit hyperactivity disorder) [83, 84]. Participants aged 6–12 years received fast-release melatonin for 4 weeks at dosages of 3 or 5 mg at either 6:00 PM or 7:00 PM. The more recent study [83] based dosage on weight (3 mg if <40 kg; 5 mg if >40 kg at 7:00 PM) while the earlier protocol [84] uniformly provided 5 mg at 6:00 PM. Combined analyses ($n = 132$) revealed an advance in DLMO of nearly 1 hour in comparison to placebo (mean difference −54.22 minutes [CI −31.67 to −76.78]). Actigraphically assessed sleep-onset time ($n = 130$) also advanced (mean difference −36.57 minutes [CI −16.96 to −56.18]). Other actigraphically derived sleep parameters were obtained only in the more recent study ($n = 105$), but failed to detect statistical and/or Task Force-defined clinically significant changes. Subjective assessments in the earlier study failed to demonstrate significant differences in sleep parameters [84].

Subsequent to publication of the AASM Practice Parameters, Sletten et al. [85] tested the efficacy of 0.5 mg of immediate-release melatonin for ICSD-defined DSWPD in an Australian multicenter trial ($n = 116$). Inclusion criteria mandated that subjects have a physiologic delayed circadian phase, defined by occurrence of the DLMO within 30 minutes of their habitual bedtime (HBT) or any time thereafter. Participants were randomized to melatonin ($n = 54$; 29.94 ± 9.63 years) or placebo ($n = 50$; 28.88 ± 10.46 years), with instructions to take the capsule 1 hour before HBT. Sleep diaries and actigraphy were utilized throughout the 4-week treatment period. On average, the participants took the study drug on 21.11 ± 3.18 nights, and only 5% took drug every night. Posttreatment DLMO values did not reach statistical significance. Actigraphy data demonstrated statistically significant improvements in the melatonin group for sleep-onset latency (95% CI −9.68 [−16.69 to −2.67]minutes) ($p = 0.007$), sleep efficiency (95% CI 2.72 [0.59–4.86] %) ($p = 0.013$), and sleep-onset time (95% CI −0:29 [−0:54 to −0:04 hours:minutes) ($p = 0.023$). The sleep diaries revealed similar findings, albeit with additional improvements in sleep offset time (95% CI −0:35 [−1:00 to −0:11] hours:minutes ($p = 0.005$).

The astute reader will note demonstration of successful sleep-related outcomes without changes in the circadian phase marker and vice versa, both within the studies reviewed above and elsewhere [71]. Related, the use of chronobiotically timed melatonin was given a weak endorsement within the AASM practice guidelines [71]. Nonclinical scientific evidence among healthy individuals suggests that lower doses of melatonin than were used in the cited studies (with the exception of the Sletten study [85]) may actually be more effective than higher doses, however, possibly because the latter exerts effects on both phase advance and phase delay portions of the phase response curve.

Moreover, since phase response curves for a melatonin dose of 0.5 mg (healthy adult populations) show maximal phase advances when taken approximately 10–12 hours prior to the mean midpoint of sleep on "free" or unrestricted sleep nights (or 6–8 hours prior to the mean sleep-onset time on free sleep nights) [86], timing may have been problematic in these studies, which in some instances may have resulted in melatonin functioning as an hypnotic rather than a chronobiotic.

When taken correctly and consistently, low-dose melatonin can reliably affect a 90 minute (±30 minutes) phase advance. Once a new steady-state circadian phase has been achieved (after 1–2 weeks), the midpoint of sleep on free days can be recalculated, and the dosing time adjusted. Additional stepwise advances might be necessary until satisfactory results are achieved [87]. While maintenance protocols have not been established, clinical practice supports subsequent use of melatonin at a fixed time, with earlier dosing if sleep consistently occurs later than desired and later dosing if sleep consistently occurs earlier than desired [88]. Similarly, the timing of post-awakening light therapy can be adjusted with the same logic. If spontaneous awakenings occur earlier, the timing of light should change accordingly, until the desired time of awakening is achieved. One can then continue to administer light at a fixed time. A trial off of melatonin and/or light can be pursued with rigid adherence to the desired sleep/wake schedule, with a low threshold to return to treatment if symptoms recur. At least three laboratory-based studies among healthy adults describe a synergistic effect (with respect to circadian phase advances) when strategically timed light and melatonin are used together [89–91], but there are no clinical studies to definitively substantiate this practice [92, 93]. Further investigations are required to determine which timing and dosage of therapy (or therapies) result in the best outcomes for those with DSWPD.

Adverse Effects Associated with Melatonin

Melatonin is considered a dietary supplement and is therefore not subject to the scrutiny afforded to the United States Food and Drug Administration (FDA)-approved medications. Concerns have been raised about the purity of available formulations, as well as the reliability of stated doses per tablet. Formulations that are United States Pharmacopeia Convention Verified can be considered the most reliable in this regard. In general, melatonin is associated with a lack of reported serious adverse effects [94–98]. A review by the National Academy of Sciences stated that short-term use of ≤10 mg/daily (higher than typical chronobiotic doses) appears to be safe in healthy adults but recommended caution in children/adolescents and women of reproductive age (see further below). Adverse effects such as headaches, somnolence, hypotension, hypertension, gastrointestinal upset, and exacerbation of *alopecia areata* have been reported at higher melatonin doses in healthy adults, and the same effects have been reported at lower doses among those with relevant pre-existing conditions. Melatonin has also been associated with an increase in depressive symptoms [99], and caution is advised when prescribing to patients taking warfarin and to patients with epilepsy, as a result of various case reports submitted to the World Health Organization [95]. A recent publication described impairment in glucose tolerance among healthy women [100] subsequent to acute melatonin administration.

Studies that address long-term effects are scarce, as are studies that specifically involve pediatric/adolescent populations. A randomized, placebo-controlled trial that investigated the toxicology of a 28-day treatment with 10 mg melatonin (solely comprised of healthy male adult participants) revealed no group differences with respect to adverse effects on polysomnography, subjective sleepiness, numerous clinical laboratory examinations, or other subjectively recorded events [101]. Similarly, in a meta-analysis that reviewed controlled trials with melatonin ($n = 10$ studies, >200 subjects) with use for ≤3 months, there were few reports of adverse events [99]. A long-term follow-up study of pediatric patients with DSWPD + attention deficit hyperactivity disorder who utilized melatonin doses up to 10 mg (mean follow-up time of approximately 4 years) detected no serious adverse events in serial interviews with the children's parents, and 65% of participants continued to use the medication daily [102]. A follow-up open-label prospective study of subjects with neurodevelopmental disabilities comorbid with DSWPD who received controlled-release melatonin (max dosage 15 mg) up to 3.8 years similarly described no adverse events [103, 104]. Patients and caregivers are nevertheless frequently wary to use this supplement, due to concerns related to potential adverse effects on growth hormone regulation (10 mg dose) [105] and on reproductive function/development (3 mg dose) [106]. Possibly relevant to the latter concern, Tanner (pubertal) stages [107, 108] were assessed serially in a questionnaire-based study involving children/adolescents (mean duration ~3 years), in an effort to compare pubertal development among those using melatonin (mean dose ~3 mg) during prepuberty to non-melatonin users in the general Dutch population (controls) [109]. No significant group differences were detected.

Less Explored Interventions

Creative and off-label interventions are often required for DSWPD. Esaki et al. [110] reported a pilot study of blue light-blocking glasses for evening use in DSWPD. They conducted an open-label design of seven patients (five males, mean age 18.11 ± 3.18 years) who completed the 4-week trial. The participants completed baseline actigraphy and DLMO measurements during Week 1 and at the end of the trial. During Weeks 2–4 the participants wore glasses that blocked wavelengths below 530 nm from 9:00 PM until bedtime and removed the glasses only in the dark. The DLMO showed an advance of 78 minutes (95% CI: −34 to 183 minutes) though this was not statistically significant ($p = 0.145$). Actigraphy data showed a significant mean advance in sleep-onset time of 132 minutes (95% CI: 13–252 minutes) ($p = 0.034$). This small study provides some preliminary evidence that "blue-blocker" glasses (which are already commercially available) may be beneficial for DSWPD patients.

There are isolated reports regarding the use of hypnotics in DSWPD (typically as an adjunctive treatment with chronotherapy), but there is insufficient rigor in methodology for purposes of evidence analysis [111, 112]. Two reports describe DSWPD patients' resistance to the effects of traditional hypnotics [113, 114]. Nevertheless, a laboratory-based study that imposed a 4-hour phase advance on healthy subjects described sleep-related benefits (polysomnographic and subjective measures) with zolpidem [115]. The off-label use of neuroleptic medications has also been reported. Omori et al. [116] described an open-label, flexible-dose, 4-week trial of "low dose" aripiprazole in 12 adult subjects, the majority of whom suffered from comorbid depression and were on complex medication regimens, including combinations of antidepressants and mood stabilizers. Evaluations included 1 week of prescreening and assessments every 1–2 weeks. Sleep diaries revealed significant advances in sleep onset [1.1 hours ± 1.2 ($p = 0.021$)] and sleep offset [2.5 hours ± 1.3 ($p = 0.001$)] and a decrease in total sleep time [1.3 hours ± 1.4 hours ($p = 0.018$)]. The latter was viewed as favorable in this particular study as a countermeasure for sleep inertia. None of these trials employed circadian phase markers.

Finally, modafinil and armodafinil are widely used as wake-promoting agents for numerous FDA-approved indications, including daytime sleepiness from shift-work disorder [117]. It may be used off-label for occasional adjunctive use in patients with DSWPD, although there are no studies that can support or refute this indication.

Conclusions and Future Directions

This chapter presents DSWPD clinical features, diagnostics, comorbidities, epidemiology, pathophysiology, genetics, and available treatment modalities. Circadian-based basic science developments continue to outpace clinical research pertaining to CRSWDs. The recently published (2015) AASM practice guidelines [71] point out these deficiencies and may serve as a roadmap for future studies that will propel higher-quality, more sophisticated therapies.

Generally speaking, larger more rigorously designed studies (randomized placebo-controlled trials) with ICSD-3 defined DSWPD [4] are required, and replication of results from separate centers is essential. More specifically, future studies could advance the field by including detailed therapeutic information, such as the method and means of treatment delivery (e.g., protective eyewear vs. volitional avoidance of light, light therapy intensity/wavelength/proximity, continuous vs. pulsed light administration [including gradually vs. abruptly changing illumination] [118], melatonin formulation, relationship of treatment timing with respect to a defined physiologic circadian phase marker or other sleep parameter, inclusion/exclusion of prescribed sleep/wake schedules or other behavioral interventions, and study environment [laboratory vs. non-laboratory]). Future research should also address the "dose" of light utilized including lux level, duration [119], as well as season [120] and other environmental factors that affect overall light exposure history [121].

Field-based studies are sorely needed. While it is necessary to extrapolate information gleaned from healthy subjects in simulated settings (due to the paucity of clinical research), one must also be cautious not to let tightly controlled bench research prematurely dictate clinical treatment. As a prime example, there are currently no data to support devices that solely deliver blue short-wavelength light in the treatment of DSWPD, and two laboratory-based studies that describe no additional benefit with blue-enriched bright light [122, 123], despite the fact that these wavelengths have been identified as especially important for circadian phase resetting in nonclinical experiments [124].

From the standpoint of outcomes, similar *clinically relevant sleep-related* measures will be required for inter-study comparative purposes (polysomnography vs. actigraphy vs. subjective reports, and physiologic or non-physiologic circadian markers). Systematic measures of treatment compliance are also required to accurately inform clinical practice. In the melatonin study performed by Sletten and colleagues [85], completers took the study drug on only 21.11 ± 3.18 nights of the 28-day trial, with an average of 5.30 ± 1.00 tablets per week. Only 5% of the participants were fully compliant/adherent and took capsules all 28 days. Reporting of such should be uniform and as objective as possible in circadian-based studies, so that readers can better interpret results (and understand their limitations). A separate strategy that has been investigated for post-awakening light therapy (for which compliance is also poor) [125] examines the bypassing of this barrier via delivery of treatment through closed eyelids (i.e., during sleep) [126–129].

Inter-study medication (e.g., melatonin) comparisons will require equivalent dosing, timing (with respect to clock time, typical sleep-onset time, or other physiologic/non-physiologic circadian marker), and treatment durations, to accurately gauge benefit. The issue of formulation may also be relevant in melatonin studies (regular vs. sustained release vs. sublingual, etc.), and one group suggested that slow exogenous melatonin metabolism could be responsible for a lack of sustained effect in select instances [130].

Taking into account melatonin safety concerns (particularly among children and those of reproductive age), future properly powered studies should be performed to identify the lowest effective melatonin dosage and duration of treatment (acute and maintenance). Long-term physiologic studies are needed to accurately ascertain any serious chronic risks, particularly as melatonin supplements are not subject to FDA oversight [131]. At least two other investigations (involving ramelteon) also suggest a potential future DSWPD treatment role for melatonin agonists [132, 133].

Related to long-term risks of circadian-based interventions in general, research is needed to determine the minimum required duration of specific treatments (or if they are required indefinitely) and/or to determine maintenance treatment schedules.

Further studies that investigate multimodal or combination therapies are needed to determine whether combinations may prove to be synergistic and to extricate independent effects of treatment modalities, so that relative successes and failures can be exploited for differing clinical scenarios. With respect to the latter point, in the previously cited Gradisar study [72] (involving adolescents with DSWPD), light therapy was discontinued (and apparently not required) once a target wake time was reached, at which time solely behavioral interventions ensued. It is not clear to what degree this treatment could be generalized to all DSWPD populations.

Demonstration of superiority (or lack thereof) of circadian versus clock-hour timing for interventions should engender studies that aim to explore demonstrable benefits of phase assessments in the clinical setting, which in turn could serve to delineate relative chronobiotic versus hypnotic effects of medications or supplements. Some of the reviewed interventions demonstrated successful sleep-related outcomes without changes in the circadian phase marker and vice versa. If the importance of circadian phase timing of administration is demonstrated, it will be necessary to determine light and melatonin phase response curves for adult and pediatric populations afflicted with DSWPD (as they may differ from normal populations [86, 134]). Complicating matters, alterations in phase relationships between the circadian timing system and the timing of sleep among those with DSWPD could impact the ability of interventions to exert benefits, even with knowledge of the pertinent phase response curves. For example, longer intervals from various endogenous melatonin parameters [135] and core body temperature minima [136–138] to sleep offset have frequently been described among adult patients with DSWPD as compared to controls. However, this finding has not been demonstrated among protocols in which subjects are forced to maintain a more conventional sleep/wake schedule [139–141], suggesting that this observation may simply be a consequence of longer habitual total sleep time. Greater elucidation is required. On a separate note, effective treatments may need to address concomitant impairment of homeostatic sleep processes in DSWPD and among adolescents in general [46, 142]. Whether hypnotics have a role in this setting deserves to be further explored [115].

This chapter reflects the biological underpinnings associated with DSWPD. Studies are needed to investigate and understand predominant exogenous and endogenous contributors to the development and perpetuation of this condition, so that different subtypes (and possibly different treatment/prophylactic regimens) can be identified. In the case of adolescents/young adults and, to a lesser degree, other adults, numerous exogenous factors, such as increased autonomy with respect to sleep time, employment, and involvement in extracurricular activities, have been identified as variables contributing to the generally observed delay in sleep/wake patterns [34], but have not been studied among adolescent DSWPD cohorts specifically [143, 144]. Additionally, repeated exposure to frustrations at not being able to fall asleep at a desired time can lead to the development of a concomitant conditioned insomnia, which can perpetuate sleep difficulties. Exposure to indoor lighting during evening hours [145–148] and/or delays in weekend wake times [59, 149, 150] have also been implicated as contributors to persistently delayed sleep/wake times, but have not been specifically implicated in adolescent DSWPD [151]. Some have urged that school lighting environments be optimized for maximal circadian benefits [152]. Identification and manipulation of exogenous variables in trials of DSWPD may prove fruitful.

The associated development of clinical profiles would enable clinicians to better ascertain which patients might respond to suggested treatments, and related research is encouraged. In the Gradisar study [72] of adolescents with DSWPD, barriers to successful outcomes with light therapy included school nonattendance, unrestricted sleep during vacation periods, and (not surprisingly) amotivation. Patients fitting this profile are perhaps better suited to less complex interventions. In a separate study involving young adult subjects with DSWPD and non-24-hour sleep-wake rhythm disorder receiving melatonin, a higher response rate correlated indirectly with a shorter habitual total sleep time, as well as a later age of onset [153]. Information such as this may eventually allow clinicians to optimally tailor treatment.

In select cases, accommodation to a DSWPD patient's circadian preference may be most practical, and further studies examining implementation of such schedules are desirable. Believing that some DSWPD cases are refractory to treatment, Dagan and Abadi [154] recommended foregoing therapy and instead urged implementation of rehabilitation and accommodation to the preferred sleep/wake schedule in select instances, including support for disability from duties that require strict sleep/wake schedules, and encouragement to pursue endeavors with more flexible scheduling. The benefits of such accommodation were demonstrated in a separate military-based study, with evidence of superior performance and mood among those enabled to adapt a relatively delayed sleep/wake schedule, which correlated with increased total sleep time [155]. A later school start time may be sought for adolescents, if practical and available. This intervention alone can significantly increase total sleep time and mitigate associated impairments [156–161]. This topic is reviewed authoritatively in a separate chapter.

In sum, although much work remains, significant progress has been made in the recognition/treatment of DSWPD since the inception of Sleep Medicine as a distinct medical discipline. The aim of this chapter is to provide a framework for clinicians to make properly informed treatment decisions, with the additional hope that it will serve as an impetus to address clinical research deficiencies, and promote novel inquiries for treatments of this challenging and interesting condition.

References

1. Schrader H, Bovim G, Sand T. The prevalence of delayed and advanced sleep phase syndromes. J Sleep Res. 1993;2:51–5.
2. Regestein QR, Monk TH. Delayed sleep phase syndrome: a review of its clinical aspects. Am J Psychiatry. 1995;152(4):602–8.
3. Roenneberg T, Kuehnle T, Pramstaller PP, Ricken J, Havel M, Guth A, et al. A marker for the end of adolescence. Curr Biol. 2004;14(24):R1038–9.
4. American Academy of Sleep Medicine. Delayed sleep-wake phase disorder. In: American Academy of Sleep Medicine, editor. International classification of sleep disorders. 3rd ed. Darien: American Academy of Sleep Medicine; 2014. p. 189–224.

5. Rajaratnam SMW, Licamele L, Birznieks MS. Delayed sleep phase disorder risk is associated with absenteeism and impaired functioning. Sleep Health. 2015;1(2):121–7.
6. Ando K, Hayakawa T, Ohta T, Kayukawa Y, Ito A, Iwata T, et al. Long-term follow-up study of 10 adolescent patients with sleep-wake schedule disorders. Jpn J Psychiatry Neurol. 1994;48(1):37–41.
7. Alvarez B, Dahlitz MJ, Vignau J, Parker JD. The delayed sleep phase syndrome. Clinical and investigative findings in 14 subjects. J Neurol Neurosurg Psychiatry. 1992;55(8):665–70.
8. Lovato N, Gradisar M, Short M, Dohnt H, Micic G. Delayed sleep phase disorder in an Australian school-based sample of adolescents. J Clin Sleep Med. 2013;15(9):939–44.
9. Richardson CE, Gradisar M, Barbero SC. Are cognitive "insomnia" processes involved in the development and maintenance of delayed sleep wake phase disorder. Sleep Med Rev. 2016;26:1–8.
10. Adan A, Almirall H. Horne and Ostberg morningness-eveningness questionnaire: a reduced scale. Personal Individ Differ. 1991;12:241–53.
11. Deacon S, Arendt J. Posture influences melatonin concentrations in plasma and saliva in humans. Neurosci Lett. 1994;167(2):191–4.
12. Murray JM, Sletten TL, Magee M, Gordon C, Lovato N, Bartlett DJ, et al. Prevalence of circadian misalignment and its association with depressive symptoms in delayed sleep phase disorder. Sleep. 2017;40(1):1–10.
13. Solheim B, Langsrud K, Kallestad H, Olsen A, Bjorvatn B, Sand T. Difficult morning awakening from rapid eye movement sleep and impaired cognitive function in delayed sleep phase disorder patients. Sleep Med. 2014;15(10):1264–8.
14. Conners K, MHS Staff. Conners' continuous performance test (CPT II): version 5 for windows. Technical guide and software manual. Toronto: Multi-Health Systems Inc.; 2004.
15. American Academy of Sleep Medicine. International classification of sleep disorders, revised. Westchester: American Academy of Sleep Medicine; 1997.
16. Siversen B, Pallesen S, Stormark KM, Boe T, Lundervold AJ, Hysing M. Delayed sleep phase syndrome in adolescents: prevalence and correlates in a large population-based study. BMC Public Health. 2013;13:1163–72.
17. Trenkwalder C, Hening WA, Walters AS, Campbell SS, Rahman K, Chkroverty S. Circadian rhythm of periodic limb movements and sensory symptoms of restless legs syndrome. Mov Disord. 1999;14(1):102–10.
18. Bogan RK. Effects of restless legs syndrome (RLS) on sleep. Neuropsychiatr Dis Treat. 2006;2(4):513–9.
19. Sheehan DV. The anxiety disease. New York: Scribner; 1983.
20. Shirayama M, Shirayama Y, Iida H, Kato M, Kajimura N, Watanabe T, et al. The psychological aspects of patients with delayed sleep phase syndrome (DSPS). Sleep Med. 2003;4:427–33.
21. Wilhelmsen-Langeland A, Saxvig IW, Pallesen S, Nordhus IH, Vedaa O, Sorensen E, et al. The personality profile of young adults with delayed sleep phase disorder. Behav Sleep Med. 2014;12(6):481–92.
22. Dagan Y, Sela H, Omer H, Hallis D, Dar R. High prevalence of personality disorders among circadian rhythm sleep disorders (CRSD) patients. J Psychosom Res. 1996;41:357–63.
23. Micic G, Lovato N, Gradisar M, Lack LC. Personality differences in patients with delayed sleep-wake phase disorder and non-24-h sleep wake rhythm disorder relative to healthy sleepers. Sleep Med. 2016;30:128–35.
24. Tsujioka B, Sonohura T, Yatabe T. A factorial study of the temperament of Japanese college male students by the Yatabe-Guilford personality inventory. Psychologia. 1957;1:110–9.
25. Hathaway SR, McKinley JC. A multiphasic personality schedule (Minnesota): I. Construction of the schedule. J Psychol. 1940;10:249–54.
26. Ainsworth MD, Klopfer WG, Klopfer B. Part two: interpretation. In: Klopfer B, Ainsworth MD, Klopfer WG, Holt RR, editors. Development of the Rorschach technique. San Diego: Harcourt Brace Jovanovich; 1954. p. 249–402.
27. Abe T, Inoue Y, Komada Y, Nakamura M, Asaoka S, Kanno M, et al. Relation between morningness-eveningness core and depressive symptoms among patients with delayed sleep phase syndrome. Sleep Med. 2011;12(7):680–4.

28. Zung WW. A self-rating depression scale. Arch Gen Psychiatry. 1965;12:63–70.
29. Barrett J, Hurst MW, DiScala C, Rose RM. Prevalence of depression over a 12-month period in a nonpatient population. Arch Gen Psychiatry. 1978;35:741–4.
30. Childa F, Okayama A, Nishi N, Sakai A. Factor analysis of Zung scale scores in a Japanese general population. Psychiatry Clin Neurosci. 2004;58:420–6.
31. Yamazaki S, Fukuhara S, Green J. Usefulness of five-item and three-item mental health inventories to screen for depressive symptoms in the general population of Japan. Health Qual Life Outcomes. 2005;3:48–54.
32. Costa PT, McCrae RR. NEO-PI-R. Professional manual. Revised NEO personality inventory (NEO-PI-R) and NEO five-factor inventory (NEO-FFI). Lutz: Psychological Assessment Resources; 1992.
33. American Academy of Sleep Medicine. The international classification of sleep disorders: diagnostic and coding manual. 2nd ed. Westchester: American Academy of Sleep Medicine; 2005.
34. Ohayon MM, Roberts RE, Zulley J, Smirne S, Priest RG. Prevalence and patterns of problematic sleep among older adolescents. J Am Acad Child Adolesc Psychiatry. 2000;39(12):1549–56.
35. Crowley SJ, Acebo C, Carskadon MA. Sleep, circadian rhythms, and delayed phase in adolescence. Sleep Med. 2007;8:602–12.
36. Hazama GI, Inoue Y, Kojima K, Ueta T, Nakagome K. The prevalence of probable delayed-sleep-phase syndrome in students from junior high school to university in Tottori, Japan. Tohoku J Exp Med. 2008;216(1):95–8.
37. Danielsson K, Markstrom A, Broman JE, von Knorring L, Jansson-Froejmark M. Delayed sleep phase disorder in a Swedish cohort of adolescents and young adults: prevalence and associated factors. Chronobiol Int. 2016;33(10):1331–9.
38. American Psychiatric Association. Diagnostic and coding manual of mental disorder. 5th ed. Arlington: American Psychiatric Publishing; 2013.
39. Carskadon MA, Harvey K, Duke P, Anders TF, Litt IF, Dement WC. Pubertal changes in daytime sleepiness. Sleep. 1980;2:453–60.
40. Wolfson AR, Carskadon MA. Sleep schedules and daytime functioning in adolescents. Child Dev. 1998;69(4):875–87.
41. Carskadon MA, Vieira C, Acebo C. Association between puberty and delayed phase preference. Sleep. 1993;16(3):258–62.
42. Hagenauer MH, Perryman JI, Lee TM, Carskadon MA. Adolescent changes in the homeostatic and circadian regulation of sleep. Dev Neurosci. 2009;31:276–84.
43. Carskadon MA, Acebo C, Richardson GS, Tate BA, Seifer R. An approach to studying circadian rhythms in adolescents. J Biol Rhythm. 1997;12:278–89.
44. Carskadon MA, Acebo C, Jenni OG. Regulation of adolescent sleep: implications for behavior. Ann N Y Acad Sci. 2004;1021:276–91.
45. Taylor DJ, Jenni OG, Acebo C, Carskadon MA. Sleep tendency during extended wakefulness: insights into adolescent sleep regulation and behavior. J Sleep Res. 2005;14:239–44.
46. Jenni OG, Achermann P, Carskadon MA. Homeostatic sleep regulation in adolescents. Sleep. 2005;28:1446–54.
47. Carskadon MA, Acebo C, Arnedt JT. Failure to identify pubertally-mediated melatonin sensitivity to light in adolescents. Sleep. 2002;25:A191.
48. Aoki H, Ozeki Y, Yamada N. Hypersensitivity of melatonin suppression in response to light in patients with delayed sleep phase syndrome. Chronobiol Int. 2001;18(2):263–71.
49. Weinert D, Eimert H, Erkert HG, Schneyer U. Resynchronization of the circadian corticosterone rhythm after a light/dark shift in juvenile and adult mice. Chronobiol Int. 1994;11:222–31.
50. Weinert D, Kompauerova V. Light induced phase and period responses of circadian activity rhythms in laboratory mice of different age. Zoology. 1998;101:45–52.
51. Micic G, De Bruyn A, Lovato N, Wright H, Gradisar M, Ferguson S, et al. The endogenous circadian temperature period length (tau) in delayed sleep phase disorder compared to good sleepers. J Sleep Res. 2013;22(6):617–24.

52. Micic G, Lovato N, Gradisar M, Burgess HJ, Ferguson SA, Lack L. Circadian melatonin and temperature taus in delayed sleep-wake phase disorder and non-24-hour sleep wake rhythm disorder patients: an ultradian constant routine study. J Biol Rhythm. 2016;31(4):387–405.
53. Brown EN, Czeisler CA. The statistical analysis of circadian phase and amplitude in constant-routine core-temperature data. J Biol Rhythm. 1992;7:177–202.
54. Takahashi Y, Hohjoh H, Matsuura K. Predisposing factors in delayed sleep phase syndrome. Psychiatry Clin Neurosci. 2000;54:356–8.
55. Jabeen S. Delayed sleep phase syndrome: a forerunner of psychiatric distress. J Coll Physicians Surg Pak. 2013;23(12):874–7.
56. Carskadon MA. Patterns of sleep and sleepiness in adolescences. Pediatrician. 1990;17(1):5–12.
57. Van den Bulck J. Television viewing, computer game playing, and internet use and self-reported time in bed and time out of bed in secondary-school children. Sleep. 2004;27(1):101–4.
58. Carskadon MA, Mancuso J, Rosekind MR. Impact of part-time employment on adolescent sleep patterns. Sleep Res. 1989;18:114.
59. Taylor A, Wright HR, Lack LC. Sleeping-in on the weekend delays circadian phase and increases sleepiness the following week. Sleep Biol Rhythms. 2008;6:172–9.
60. Patke A, Murphy PJ, Onat OE, Krieger AC, Ozcelik T, Campbell SS, et al. Mutation of the human circadian clock gene cry1 in familial delayed sleep phase disorder. Cell. 2017;169(2):203–15.
61. Katzenberg D, Young T, Finn L, et al. A CLOCK polymorphism associated with human diurnal preference. Sleep. 1998;21:569–76.
62. Katzenberg D, Young T, Lin L, Finn L, Mignot E. A human period gene (hper1) polymorphism is associated with diurnal preference in normal adults. Psychiatr Genet. 1999;9:107–9.
63. Ebisawa T, Ushiyama M, Kajimura N, Mishima K, Kamei Y, Katoh M, et al. Association of structural polymorphisms in the human period3 gene with delayed sleep phase syndrome. EMBO Rep. 2001;21:342–6.
64. Archer SN, Robilliard DL, Skene DJ, Smits M, Williams A, Arendt J, et al. A length polymorphism in the circadian clock gene per3 is linked to delayed sleep phase syndrome and extreme diurnal preference. Sleep. 2003;26:413–5.
65. Pereira DS, Tufik S, Louzada FM, Benedito-Silva AA, Lopez AR, Lemos NA, et al. Association of the length polymorphism in the human per3 gene with the delayed sleep-phase syndrome: does latitude have an influence upon it? Sleep. 2005;28(1):29–32.
66. Vitaterna MH, Turek FW, Jiang P. Genetics and genomics of circadian clocks. In: Kryger M, Roth T, Dement WC, editors. Principles and practice of sleep medicine. 6th ed. Philadelphia: Elsevier; 2017. p. 272–80.
67. Buhr ED, Takahashi JS. Molecular components of the mammalian circadian clock. In: Kramer A, Merrow M, editors. Circadian clocks. Handbook of experimental pharmacology. Berlin: Springer-Verlag; 2013. p. 217.
68. Czeisler CA, Duffy JF, Shanahan TL, Brown EN, Mitchell JF, Rimmer DW, et al. Stability, precision, and near-24-hour period of the human circadian pacemaker. Science. 1999;284:2177–81.
69. Soria V, Martinez-Amoros E, Escaramis G, Valero J, Perez-Egea R, Garcia C, et al. Differential association of circadian genes with mood disorders: CRY1 and NPAS2 are associated with unipolar major depression and CLOCK and VIP with bipolar disorder. Neuropsychopharmacology. 2010;35:1279–89.
70. Hua P, Liu W, Chen D, Zhao Y, Chen L, Zhang N, et al. Cry1 and Tef gene polymorphisms are associated with major depressive disorder in the Chinese population. J Affect Disord. 2014;157:100–3.
71. Auger RR, Burgess JH, Emens SJ, Deriy VL, Thomas MS, Sharkey MK. Clinical practice guidelines for the treatment of intrinsic circadian rhythm sleep-wake disorders: advanced sleep-wake phase disorder (ASWPD), delayed sleep-wake phase disorder (DSWPD), non-24-hour sleep-wake rhythm disorder (N24SWD), and irregular sleep-wake rhythm disorder (ISWRD). An update for 2015. J Clin Sleep Med. 2015;11(10):1199–236.

72. Gradisar M, Dohnt H, Gardner G, Paine S, Starkey K, Menne A, et al. A randomized controlled trial of cognitive-behavior therapy plus bright light therapy for adolescent delayed sleep phase disorder. Sleep. 2011;34(12):1671–80.
73. Danielsson K, Jansson-Frojmark M, Broman J, Markstrom A. Cognitive behavioral therapy as an adjunct treatment to light therapy for delayed sleep phase disorder in young adults: a randomized controlled feasibility study. Behav Sleep Med. 2016;14(2):212–32.
74. Tuunainen A, Kripke DF, Endo T. Light therapy for non-seasonal depression. Cochrane Database Syst Rev. 2004;(2):CD004050.
75. Dauphinais DR, Rosenthal JZ, Terman M, DiFebo HM, Tuggle C, Rosenthal NE. Controlled trial of safety and efficacy of bright light therapy vs. negative air ions in patients with bipolar depression. Psychiatry Res. 2012;196:57–61.
76. Terman M, Terman JS. Light therapy for seasonal and nonseasonal depression: efficacy, protocol, safety, and side effects. CNS Spectr. 2005;10:647–63.
77. Ulrich V, Olesen J, Gervil M, Russell MB. Possible risk factors and precipitants for migraine with aura in discordant twin-pairs: a population-based study. Cephalalgia. 2000;20:821–5.
78. Gallin PF, Terman M, Reme CE, Rafferly B, Terman JS, Burde RM. Ophthalmologic exam in patients with seasonal affective disorder, before and after bright light therapy. Am J Ophthalmol. 1995;119:202–10.
79. Reme CE, Rol P, Grothmann K, Kasse H, Terman M. Bright light therapy in focus: lamp emission spectra and ocular safety. Technol Health Care. 1996;4:403–13.
80. Kayumov L, Brown G, Jindal R, Butto K, Shapiro CM. A randomized, double-blind, placebo-controlled crossover study of the effect of exogenous melatonin on delayed sleep phase syndrome. Psychosom Med. 2001;63:40–8.
81. Rahman SA, Kayumov L, Shapiro CM. Antidepressant action of melatonin in the treatment of delayed sleep phase syndrome. Sleep Med. 2010;11:131–6.
82. van Geijiswijk IM, van der Heijden KB, Egberts AC, Korzilius HP, Smits MG. Dose finding of melatonin for chronic idiopathic childhood sleep onset insomnia: an RCT. Psychopharmacology. 2010;212:379–91.
83. Van der Heijden KB, Smits MG, Van Someren EJ, Ridderinkhof KR, Gunning WB. Effect of melatonin on sleep, behavior, and cognition in ADHD and chronic sleep-onset insomnia. J Am Acad Child Adolesc Psychiatry. 2007;46:233–41.
84. Smits MG, Nagtegaal EE, van der Heijden J, Coenen AM, Kerkhof GA. Melatonin for chronic sleep onset insomnia in children: a randomized placebo-controlled trial. J Child Neurol. 2001;16:86–92.
85. Sletten LT, Magee M, Murray MJ, Gordon JC, Lovato N, Kennaway JD, et al. Efficacy of melatonin with behavioral sleep-wake scheduling for delayed sleep-wake phase disorder: a double-blind, randomized clinical trial. PLoS Med. 2018;15(6):e1002587.
86. Burgess HJ, Revell VL, Molina TA, Eastman CI. Human phase response curves to three days of daily melatonin: 0.5 mg versus 3.0 mg. J Clin Endocrinol Metab. 2010;95:3325–31.
87. Nesbitt AD. Delayed sleep-wake phase disorder. J Thoracic Dis. 2018;10(Supplement 1):S103–11.
88. Emens JS, Eastman CL. Diagnosis and treatment of non-24-h sleep-wake disorder in the blind. Drugs. 2017;77(6):637–50.
89. Revell VL, Burgess HJ, Gazda CJ, Smith MR, Fogg LF, Eastman CI. Advancing human circadian rhythms with afternoon melatonin and morning intermittent bright light. J Clin Endocrinol Metab. 2006;91:54–9.
90. Burke TM, Markwald RR, Chinoy ED, Snider JA, Bessman SC, Jung CM, et al. Combination of light and melatonin time cues for phase advancing the human circadian clock. Sleep. 2013;36:1617–24.
91. Paul MA, Gray GW, Lieberman HR, Love RJ, Miller JC, Trouborst M, et al. Phase advance with separate and combined melatonin and light treatment. Psychopharmacology. 2011;214:515–23.
92. Wilhelmsen-Langeland A, Saxvig WI, Pallesen S, Nordhus HI, Vedaa O, Lundervold JA, Bjorvatn B. A randomized controlled trial with bright light and melatonin for the treatment

of delayed sleep phase disorder: effects on subjective and objective sleepiness and cognitive function. J Biol Rhythm. 2013;28(5):306–21.
93. Saxvig WI, Wilheimsen-Langeland A, Pallesen S, Vedaa O, Nordhus HI, Bjorvatn B. A randomized controlled trial with bright light and melatonin for delayed sleep phase disorder: effects on subjective and objective sleep. Chronobiol Int. 2014;31(1):72–86.
94. Ferracioli-Oda E, Qawasmi A, Bloch MH. Meta-analysis: melatonin for the treatment of primary sleep disorders. PLoS One. 2013;8:e63773.
95. Herxheimer A, Petrie KJ. Melatonin for the prevention and treatment of jet lag. Cochrane Database Syst Rev. 2002;(2):CD001520.
96. Spadoni G, Bedini A, Rivara S, Mor M. Melatonin receptor agonists: new options for insomnia and depression treatment. CNS Neurosci Ther. 2011;17:733–41.
97. Board on Life Sciences, Institute of Medicine and National Research Council of the National Academies. Dietary supplements: a framework for evaluating safety. Washington, D.C.: The National Academies Press; 2005.
98. Buscemi N, Vandermeer B, Hooton N, Pandya R, Tjosvold L, Hartling L, et al. The efficacy and safety of exogenous melatonin for primary sleep disorders. A meta-analysis. J Gen Intern Med. 2005;20:1151–8.
99. Werneke U, Turner T, Priebe S. Complementary medicines in psychiatry: review of effectiveness and safety. Br J Psychiatry. 2006;188:109–21.
100. Rubio-Sastre P, Scheer FA, Gomez-Abellan P, Madrid JA, Garaulet M. Acute melatonin administration in humans impairs glucose tolerance in both the morning and evening. Sleep. 2014;37:1715–9.
101. Seabra ML, Bignotto M, Pinto LR Jr, Tufik S. Randomized, double-blind clinical trial, controlled with placebo, of the toxicology of chronic melatonin treatment. J Pineal Res. 2000;29:193–200.
102. Hoebert M, van der Heijden KB, van Geijlswijk IM, Smits MG. Long-term follow-up of melatonin treatment in children with ADHD and chronic sleep onset insomnia. J Pineal Res. 2009;47:1–7.
103. Wasdell MB, Jan JE, Bomben MM, Freeman RD, Rietveld WJ, Tai J, et al. A randomized, placebo-controlled trial of controlled release melatonin treatment of delayed sleep phase syndrome and impaired sleep maintenance in children with neurodevelopmental disabilities. J Pineal Res. 2008;44:57–64.
104. Carr R, Wasdell MB, Hamilton D, Weiss MD, Freeman RD, Tai J, et al. Long-term effectiveness outcome of melatonin therapy in children with treatment-resistant circadian rhythm sleep disorders. J Pineal Res. 2007;43:351–9.
105. Valcavi R, Zini M, Maestroni GJ, Conti A, Portioli I. Melatonin stimulates growth hormone secretion through pathways other than the growth hormone-releasing hormone. Clin Endocrinol. 1993;39:193–9.
106. Luboshitzky R, Shen-Orr Z, Nave R, Lavi S, Lavie P. Melatonin administration alters semen quality in healthy men. J Androl. 2002;23:572–8.
107. Marshall WA, Tanner JM. Variations in the pattern of pubertal changes in girls. Arch Dis Child. 1969;44:291–303.
108. Marshal WA, Tanner JM. Variations in the pattern of pubertal changes in boys. Arch Dis Child. 1970;45:13–23.
109. van Geijlswijk IM, Mol RH, Egberts TC, Smits MG. Evaluation of sleep, puberty and mental health in children with long-term melatonin treatment for chronic idiopathic childhood sleep onset insomnia. Psychopharmacology. 2011;216:111–20.
110. Esaki Y, Kitajima T, Ito Y, Koike S, Nakao Y, Tsuchiya A, et al. Wearing blue light-blocking glasses in the evening advances circadian rhythms in patients with delayed sleep phase disorder. Chronobiol Int. 2016;33(8):1037–44.
111. Ito A, Ando K, Hayakawa T, Iwata T, Kayukawa Y, Ohta T, et al. Long-term course of adult patients with delayed sleep phase syndrome. Jpn J Psychiatry Neurol. 1993;47:563–7.
112. Ohta T, Iwata T, Kayukawa Y, Okada T. Daily activity and persistent sleep-wake schedule disorders. Prog Neuropsychopharmacol Biol Psychiatry. 1992;16:529–37.
113. Mizuma H, Miyahara Y, Sakamoto T, Kotorii T, Nakazawa Y. Two cases of delayed sleep phase syndrome (DSPS). Jpn J Psychiatry Neurol. 1991;45:163–4.

114. Yamadera H, Takahashi K, Okawa M. A multicenter study of sleep-wake rhythm disorders: therapeutic effects of vitamin B12, bright light therapy, chronotherapy and hypnotics. Psychiatry Clin Neurosci. 1996;50:203–9.
115. Walsh JK, Deacon S, Dijk DJ, Lundahl J. The selective extrasynaptic GABAA agonist, gaboxadol, improves traditional hypnotic efficacy measures and enhances slow wave activity in a model of transient insomnia. Sleep. 2007;30:593–602.
116. Omori Y, Kanbayashi T, Sagawa Y, Imanishi A, Tsutsui K, Takahashi Y, et al. Low dose of aripiprazole advanced sleep rhythm and reduced nocturnal sleep time in the patients with delayed sleep phase syndrome: an open-labeled clinical observation. Neuropsychiatr Dis Treat. 2018;14:1281–6.
117. Czeisler AC, Walsh KJ, Roth T, Hughes JR, Wright PK, Kingsbury L, et al. Modafinil for excessive sleepiness associated with shift-work sleep disorder. NEJM. 2005;353(5): 476–86.
118. Kondo M, Tokura H, Wakamura T, Hyun KJ, Tamotsu S, Morita T, et al. Influences of twilight on diurnal variation of core temperature, its nadir, and urinary 6-hydroxymelatonin sulfate during nocturnal sleep and morning drowsiness. Coll Antropol. 2009;33(1):193–9.
119. Chang AM, Santhi N, St Hilaire M, Gronfier C, Bradstreet DS, Duffy JF, et al. Human responses to bright light of different durations. J Physiol. 2012;590:3103–12.
120. Higuchi S, Motohashi Y, Ishibashi K, Maeda T. Less exposure to daily ambient light in winter increases sensitivity of melatonin to light suppression. Chronobiol Int. 2007;24:31–43.
121. Chang AM, Scheer FA, Czeisler CA. The human circadian system adapts to prior photic history. J Physiol. 2011;589:1095–102.
122. Smith MR, Eastman CI. Phase delaying the human circadian clock with blue-enriched polychromatic light. Chronobiol Int. 2009;26:709–25.
123. Smith MR, Revell VL, Eastman CI. Phase advancing the human circadian clock with blue-enriched polychromatic light. Sleep Med. 2009;10:287–94.
124. Sack RL, Auckley D, Auger RR, Carskadon MA, Wright KP Jr, Vitiello MV, et al. Circadian rhythm sleep disorders: part I, basic principles, shift work and jet lag disorders. An American Academy of Sleep Medicine review. Sleep. 2007;30:1460–83.
125. Bjorvatn B, Pallesen S. A practical approach to circadian rhythm sleep disorders. Sleep Med Rev. 2009;13:47–60.
126. Cole RJ, Smith JS, Alcala YC, Elliott JA, Kripke DF. Bright-light mask treatment of delayed sleep phase syndrome. J Biol Rhythm. 2002;17:89–101.
127. Figueiro MG, Rea MS. Preliminary evidence that light through the eyelids can suppress melatonin and phase shift dim light melatonin onset. BMC Res Notes. 2012;5:221.
128. Fromm E, Horlebein C, Meergans A, Niesner M, Randler C. Evaluation of a dawn simulator in children and adolescents. Biol Rhythm Res. 2011;42:417–25.
129. Terman M, Jiuan ST. Circadian rhythm phase advance with dawn simulation treatment for winter depression. J Biol Rhythm. 2010;25:297–301.
130. Braam W, van Geijlswijk I, Keijzer H, Smits MG, Didden R, Curfs LM. Loss of response to melatonin treatment is associated with slow melatonin metabolism. J Intellect Disabil Res. 2010;54:547–55.
131. Kennaway DJ. Potential safety issues in the use of the hormone melatonin in paediatrics. J Paediatr Child Health. 2015;51:584–9.
132. Richardson GS, Zee PC, Wang-Weigand S, Rodriguez L, Peng X. Circadian phase-shifting effects of repeated ramelteon administration in healthy adults. J Clin Sleep Med. 2008;4:456–61.
133. Rajaratnam SM, Polymeropoulos MH, Fisher DM, et al. Melatonin agonist tasimelteon (VEC-162) for transient insomnia after sleep-time shift: two randomized controlled multicentre trials. Lancet. 2009;373:482–91.
134. St Hilaire MA, Gooley JJ, Khalsa SB, Kronauer RE, Czeisler CA, Lockley SW. Human phase response curve to a 1 h pulse of bright white light. J Physiol. 2012;590:3035–45.
135. Shibui K, Uchiyama M, Okawa M. Melatonin rhythms in delayed sleep phase syndrome. J Biol Rhythm. 1999;14:72–6.
136. Ozaki S, Uchiyama M, Shirakawa S, Okawa M. Prolonged interval from body temperature nadir to sleep offset in patients with delayed sleep phase syndrome. Sleep. 1996;19:36–40.

137. Watanabe T, Kajimura N, Kato M, Sekimoto M, Nakajima T, Hori T, et al. Sleep and circadian rhythm disturbances in patients with delayed sleep phase syndrome. Sleep. 2003;26:657–61.
138. Uchiyama M, Okawa M, Shibui K, Kim K, Tagaya H, Kudo Y, et al. Altered phase relation between sleep timing and core body temperature rhythm in delayed sleep phase syndrome and non-24-hour sleep-wake syndrome in humans. Neurosci Lett. 2000;294:101–4.
139. Mundey K, Benloucif S, Dubocovich ML, Ze PC. Phase-dependent treatment of delayed sleep phase syndrome with melatonin. Sleep. 2005;28:1271–8.
140. Wyatt JK, Stepanski EJ, Kirkby J. Circadian phase in delayed sleep phase syndrome: predictors and temporal stability across multiple assessments. Sleep. 2006;29:1075–80.
141. Chang AM, Reid KJ, Gourineni R, Zee PC. Sleep timing and circadian phase in delayed sleep phase syndrome. J Biol Rhythm. 2009;24:313–21.
142. Uchiyama M, Okawa M, Shibui K, Liu X, Hayakawa T, Kamei Y, et al. Poor compensatory function for sleep loss as a pathogenic factor in patients with delayed sleep phase syndrome. Sleep. 2000;23:553–8.
143. Carskadon MA. Patterns of sleep and sleepiness in adolescents. Pediatrician. 1990;17:5–12.
144. Wyatt JK. Delayed sleep phase syndrome: pathophysiology and treatment options. Sleep. 2004;27:1195–203.
145. Peixoto CA, da Silva AG, Carskadon MA, Louzada FM. Adolescents living in homes without electric lighting have earlier sleep times. Behav Sleep Med. 2009;7:73–80.
146. Vollmer C, Michel U, Randler C. Outdoor light at night (LAN) is correlated with eveningness in adolescents. Chronobiol Int. 2012;29:502–8.
147. Cajochen C, Frey S, Anders D, Späti J, Bues M, Pross A, et al. Evening exposure to a light-emitting diodes (LED)-backlit computer screen affects circadian physiology and cognitive performance. J Appl Physiol. 2011;110:1432–8.
148. Figueiro MG, Rea MS. Evening daylight may cause adolescents to sleep less in spring than in winter. Chronobiol Int. 2010;27:1242–58.
149. Burgess HJ, Eastman CI. A late wake time phase delays the human dim light melatonin rhythm. Neurosci Lett. 2006;395:191–5.
150. Crowley SJ, Carskadon MA. Modifications to weekend recovery sleep delay circadian phase in older adolescents. Chronobiol Int. 2010;27:1469–92.
151. Auger RR, Burgess HJ, Dierkhising RA, Sharma RG, Slocumb NL. Light exposure among adolescents with delayed sleep phase disorder: a prospective cohort study. Chronobiol Int. 2011;28:911–20.
152. Figueiro MG, Rea MS. Lack of short-wavelength light during the school day delays dim light melatonin onset (DLMO) in middle school students. Neuroendocrinol Lett. 2010;31:92–6.
153. Kamei Y, Hayakawa T, Urata J, Uchiyama M, Shibui K, Kim K, et al. Melatonin treatment for circadian rhythm sleep disorders. Psychiatry Clin Neurosci. 2000;54:381–2.
154. Dagan Y, Abadi J. Sleep-wake schedule disorder disability: a lifelong untreatable pathology of the circadian time structure. Chronobiol Int. 2001;18:1019–27.
155. Miller NL, Tvaryanas AP, Shattuck LG. Accommodating adolescent sleep-wake patterns: the effects of shifting the timing of sleep on training effectiveness. Sleep. 2012;35:1123–36.
156. Danner F, Phillips B. Adolescent sleep, school start times, and teen motor vehicle crashes. J Clin Sleep Med. 2008;4(6):533–5.
157. Thacher VP, Onyper VS. Longitudinal outcomes of start time delay on sleep, behavior, and achievement in high school. Sleep. 2016;39(2):271–81.
158. National Sleep Foundation. Sleep in America poll summary findings. Washington, D.C.: National Sleep Foundation; 2006. p. 2006.
159. Wahlstrom K. Changing times: findings from the first longitudinal study of later school start times. NAASP Bull. 2002;86:3–21.
160. Owens JA, Belon K, Moss P. Impact of delaying school start time on adolescent sleep, mood, and behavior. Arch Pediatr Adolesc Med. 2010;164:608–14.
161. Wahlstrom K. The prickly politics of school starting times. Phi Delta Kappa. 1999;80:344–7.

Chapter 7
Circadian Rhythms and School Start Times: The Indivisible Link Between Medicine and Education

Kyla L. Wahlstrom

Introduction

National efforts to align the starting times of secondary schools with the circadian rhythms of adolescents have been more than 23 years in the making. Still, only a small percentage of the more than 45,000 high schools in the United States have starting times after 8:30 a.m., which is the time recommended by the American Academy of Pediatrics for all secondary schools [1]. This means that 87% of all teens in the United States are in schools with early starting times. This chapter will review the research findings over the years that have conclusively shown that early starting times are in direct conflict with teens' circadian phase delay, with resulting serious negative outcomes associated with teen sleep deprivation. This chapter will also explore the evolution of the later start time movement and concludes with the yet-unsolved issues that persist and impede the change in start times to occur nationwide.

An especially important aspect of this topic is the continued lack of integration in the literature between the medical and educational research communities. It is not often that medical research has immediate relevance to discussions among educators. However, the sleep/wake cycle of adolescents has been found to have a clear relationship to the timing of the school day, and the early starts are seen as a major impediment to teens obtaining the recommended amount of sleep per night. As the story of early start times unfolds in this chapter, the fields of medicine and educational policy become instant, and somewhat indivisible, partners.

K. L. Wahlstrom (✉)
Department of Organizational Leadership, Policy and Development, College of Education and Human Development, University of Minnesota, Minneapolis, MN, USA
e-mail: wahls001@umn.edu

R. R. Auger (ed.), *Circadian Rhythm Sleep-Wake Disorders*,
https://doi.org/10.1007/978-3-030-43803-6_7

Early Research

From the late 1980s through the 1990s, research began to accumulate that clearly identified the links between the circadian phase delay in teenagers and how it manifested in teens' day-to-day functioning.

Circadian Rhythms

A growing body of literature about circadian rhythms began to support the profound effects of various rhythmic patterns in living organisms. Weinert et al. [2] contended that circadian rhythms develop in early childhood, although at different rates and times. There was mounting evidence that circadian rhythms were generated endogenously (internally) and that these patterns persisted in the absence of any periodic environmental or social cues [3]. Redfern et al. [4] believed that the endogenous clock (maintained by the hypothalamus and, to a lesser extent, the pineal gland) and the exogenous system (affected by many factors including social situation, light/dark phases, and exercise) were not necessarily directly in relation with one another. Other findings regarding circadian rhythms included their relation to physiology [5], melatonin secretion [6–8], and depression symptomatology [7, 9]. Finally, it was noted that natural circadian patterns are very resistant to change [3]. In terms of how all of this pertains to adolescence, Dahl and Carskadon asserted in 1995 [10] that adolescents experience a natural circadian phase delay and, therefore, tend to stay up later and sleep later in the morning than in preadolescence.

Adolescent Sleep Patterns

Carskadon noted that as teenagers move through their teenage years, their slow wave sleep decreases by about 40% by the time they reach mid-puberty [11] and that daytime sleepiness increases concurrently, even without any change in the adolescent's nighttime sleep length [12]. Sleep researchers concur that biology and social pressures are major factors in determining adolescent sleep [13]. Contributing factors include the hormonal changes occurring during puberty that are associated with changes in their circadian rhythms, parental involvement in setting teens' bedtimes and awakenings, curfews, school schedules, part-time employment, and teens' use of alcohol, caffeine, and other drugs [14, 15]. For example, parents of even 12–13-year-olds frequently stop setting bedtimes and enforcing weekend wake-up times, while school schedules get earlier as students mature.

Similar studies on students were completed in other locations in the world, including Brazil [16], Italy [17, 18], and Israel [19, 20]. Those studies revealed that the sleep–wake cycle for students in those countries is nearly identical to that found among students in the United States. In other words, the circadian phase delay occurring in adolescents' neurological systems was not culturally based; it was, instead, a phenomenon of human development.

Consequences of Unmet Sleep Needs

As teens' sleep undergoes a circadian phase delay during adolescence and their sleep duration reduces due to early starting schools, identified risks include daytime sleepiness, vulnerability to catastrophic accidents, mood and behavior problems, increased susceptibility to drug and alcohol misuse, and development of major sleep disorders [21]. In a study conducted by Wolfson et al. [22], it was found that conduct/aggressive behaviors were highly associated with shorter sleep and earlier school start times. Decreased motivation and emotional regulation were also found to be significantly related to sleep deprivation in teens [23–25]. Noland et al. [26] found sleep-deprived teens to have increased stress levels, to be more likely to be overweight and to exhibit greater use of sleeping pills, cigarettes, and alcohol. These results signaled important relationships between sleep quantity and behavioral difficulties in adolescents.

Early Studies in Educational Settings

As early as 1989, Allen and Mirabile [27] examined self-reported wake patterns for students during the school year from two different senior high schools with different start times (8:00 a.m. and 7:30 a.m.). Both student groups reported short sleep time (average of 7 hours) with bedtimes at approximately 11:00 p.m. on weekdays. On weekends, sleep time averaged 9 hours and bedtime mean was around 1:00 a.m.

In 1991, Allen further examined the effects of varied school start times [28] by using a sleep–wake questionnaire with 11th–12th grade students at four high schools. Results indicated that during the school week, students from early, as compared to late, starting schools showed shorter average sleep duration and, during the weekend, a tendency for longer sleep time. He concluded that forced awakenings alone did not reset the circadian rhythm; similarly, the repeated forced awakenings for 5 out of 7 days in the week also did not appear to reset the adolescents' sleep phase delay.

Then, in 1992, Allen explored the relation between the amount of sleep phase delay and major academic outcomes [29], including grades, for 12th grade students in an early start school (7:40 a.m.). Allen found students with greater school-week sleep lag were at a significant disadvantage for academic achievement, owing to their reported decreased alertness during the school day. A later study [30] by Kowalski and Allen in 1995 compared 12th graders in a late start school (9:30 a.m.) to 11th–12th graders in an early start school (7:20 a.m.). Results indicated that total sleep time during the school week was significantly longer for the late start group (mean of 7.5 hours vs. 6.9 hours), although weekend bedtime did not significantly vary. In addition, poorer school grades were associated with increased school sleep lag and impaired alertness later in the day. The students' maximum sleepiness occurred during the early part of the school day. Similar findings were also reported by Wolfson and Carskadon [31] and Carskadon et al. [32]. Also at that time, a lone article appeared in an educational journal that expressed concern about how the sleep phase delay in adolescents may have a negative impact on students taking the Scholastic Aptitude Test (SAT) [33].

1996: The First Comprehensive Study of Later Start Times

Thus, by the mid-1990s, there were significant research findings that teens' health and performance were being severely compromised with early starting schools. However, the finding about the circadian phase delay in adolescents was unknown to school leaders and educational policy makers. This is because the research on teen sleep resided only within the medical literature and virtually none of it had appeared in the journals that educational leaders normally read. The significant event that opened the door to bring the medical findings on teens and sleep to the attention of educators was a resolution initiated by the president of the Minneapolis Psychiatric Society in September 1993 at the annual meeting of the Minnesota Medical Association (MMA) House of Delegates. The MMA contacted more than 450 school districts in the state asking superintendents to consider pushing back start times to at least 8 a.m. That letter explained the recent findings regarding the circadian phase delay in teens and the potential benefit that a later starting time might have for area high schools. One year later, in 1994, a follow-up letter from the MMA and the American Sleep Disorders Association (based in Rochester, MN, at that time) was sent to school superintendents, asking if any changes had been made to school start times. None were reported. Then, in 1995, one district, Edina (MN) Public Schools, decided to delay their high school start time effective for the 1996 school year.

In August 1996, this author received a phone call at the University of Minnesota Educational Research Center from the Edina district superintendent, Dr. Ken Dragseth. Edina Public Schools is a wealthy first-ring suburban district of Minneapolis, MN. The school board had recently approved a policy resolution to change the starting time of their one high school with 1800 students in grades 9–12 from 7:20 a.m. to 8:30 a.m. The plan involved shifting the bus transportation schedule to have the elementary schools start first, then the middle schools, and finally the high school. Dr. Dragseth was seeking an evaluation to determine if the action made a difference, such as in learning and in students' general health and well-being.

To be honest, I was skeptical that the suggestion that a later starting time might be helpful for teens, given my past 19 years' experience as a classroom teacher and school principal. Nevertheless, a comprehensive plan to evaluate the change over the course of the school year was developed that included obtaining survey and interview data from students, teachers, parents, and administrators. We also examined the effects on transportation, athletics, students' after-school employment, extracurricular activities, teacher contracts, and impacts on custodians and cafeteria workers. Prior to this study, no others had examined the later starting time of schools from a system-wide perspective.

The findings from the evaluation in the Edina high school [34] revealed numerous positive outcomes, plus a few challenges that needed to be addressed. The self-report student survey revealed that students were obtaining nearly 1 hour more of sleep every school night after the later start was implemented and that their weekend oversleep was significantly less. Both students and teachers alike reported that students were significantly more awake during the first hour of class and that their grades were better because of more alertness. Students noted that they were not as likely to fall asleep while doing their homework, which they thought might be a reason for improved academic performance.

School administrators noticed a more calm tone throughout the building and there were fewer discipline referrals to the office. Counselors had fewer students come to their offices with complaints of anxiety or depression, plus there were fewer relationship problems among the students being brought to the counselors. School nurses echoed similar positive outcomes, noting that fewer students were coming to them with somatic complaints, headaches, and feelings of general malaise. Custodians noted that there was less trash in the hallways from students who were eating packaged snacks and drinking soda between classes as a means to wake up. After-school employment by students was also not negatively affected. A survey of parents revealed 92% of the parents noted their children "were easier to live with" with the later start time.

A few issues also arose with the implementation of the later start. Sports practices after school began slightly later in the afternoon, and some parents were unhappy with the need to change to a later family dinner time. However, some coaches simply reduced the amount of daily practice time by 20–30 minutes. That lessened the problem somewhat and, surprisingly, less practice time did not adversely affect the winning records for the sports teams. After-school childcare also arose as an issue because the later dismissal time for the high school students affected their time to be available for babysitting.

Finally, a completely unexpected problem arose with the traffic patterns around the high school. With the high school starting and dismissing later, the traffic patterns at the nearby intersections dramatically changed—for the worse. The police and traffic control officers had to be called to facilitate the movement of traffic, especially related to the buses and the cars of students and parents compounding the already-heavy traffic on the local roadways, competing with the commuters who were on their way to work. Traffic control officers, semaphores, and turn lanes were put in place to address the problem.

1997: Later Starts for Minneapolis Public High Schools

In 1997, following the report of the successes experienced in the Edina High School, the Minneapolis School District decided to shift the starting times of their seven comprehensive high schools from 7:15 a.m. to 8:40 a.m. The Minneapolis District had an interim superintendent at that time, and she believed that the later starting time would be of benefit to their over 10,000 high school students, where over 67% were students of color and over 83% were on free/reduced price lunch support. Findings from a longitudinal study completed there [35] from 1997 to 2002 revealed many of the same findings discovered in Edina, plus Minneapolis schools reported a statistically significant improvement in attendance and a reduction in tardiness. In urban schools such as those in Minneapolis, when students miss their bus due to sleeping through their alarm, many do not have an alternative means of getting to school. This has also been found to be an outcome in lower income communities, where transportation to school can be a contributing factor in low attendance rates and greater rates of tardiness when students miss the bus. Finally, importantly, rates of students maintaining enrollment in their Minneapolis high schools (i.e., not dropping out) improved by up to 8% over the 3 years after the later starting time was implemented.

The Middle Years: 2002–2014

During the years from 2002 through 2014, the number of schools and districts across the United States that had implemented a later starting time for their high schools continually increased. At the same time, the number of studies examining the outcomes for students in later starting schools also increased. The outcomes studies were consistent in their findings—students obtained more sleep per night, they were more alert in classes and had less reported depressive feelings, and they had better attendance, less tardiness, and less risky behaviors.

Emerging National Interest

At the time that the studies were in progress in the Edina and Minneapolis Schools, the local news reporters were highlighting the positive findings on the front pages of the area newspapers. These stories got picked up by national news outlets, and the findings were being re-reported in more than 30 newspapers across the United States, including all major news venues, such as the *New York Times*, the *Washington Post*, the *Philadelphia Enquirer*, the *Dallas Morning News*, the *Los Angeles Times*, the *Seattle Times*, the *Chicago Tribune*, the *Miami Herald*, etc. It was this flurry of national news reporting that shifted the later school start time from being a local initiative to a national news event. Nationally recognized sleep researchers were being interviewed about the biology of adolescent sleep, while the local school superintendents in the areas served by those newspapers were being asked whether they, too, would consider changing their high school starting times to provide the same benefits to their own local high school students.

There is no doubt that the news coverage of the first studies done in Minneapolis [35] and Edina [34] was the catalyst to move the topic of the circadian phase delay in teens into the national spotlight. The prior research conducted by sleep researchers in the late 1980s and 1990s remained in the obscure realm of reports in medical journals. It was not until the research was viewed in the light of the day-to-day lives of thousands of teens in natural settings that the findings were viewed as being applicable to all teens in any locale.

The National Sleep Foundation (NSF) also had an essential role in the early years of presenting information to the general public. At several of their annual meetings in the early 2000s, the NSF included speakers whose work focused on sleep research on adolescents, and also had educational researchers present findings and strategies for districts considering making a change in their start times. Also, as early as 2005, the Associated Professional Sleep Societies (APSS) had a strategic national role in sharing information about the outcomes for students whose high schools started later. This was an important audience of sleep researchers with whom to engage, as they then brought the findings for a later start time back home to their local institutions/research labs and school districts.

As more and more research was conducted by sleep researchers, those findings continued to be reported in medical journals. The research was usually conducted in less than a year's time, most sample sizes were relatively small, and often the studies were conducted in controlled settings instead of natural settings. Journal articles reaching educators during the early 2000s were few in number. It was not until the article by this author in 2002 [35], published in an educational journal, which discussed longitudinal research findings and substantiated the long-term outcomes and benefits for students in later starting Minneapolis high schools, did the discussion increase among educational leaders across the United States.

The queries by superintendents steadily increased as the findings for later starting students were shared in an even-wider variety of media outlets. News shows such as *Good Morning America, CBS Morning News, ABC News Tonight,* and *All Things Considered* began to highlight the issue, leading to hundreds of inquiries. Most of the inquiries I received from superintendents and school board members consisted of one primary question: "Does the later start time improve students' grades?" I replied that the finding for academic grades was not fully answered as of yet, having only positive trend lines to report, but that the other benefits such as improved health and well-being were proven. With the response that improved academic performance was not substantiated, usually the person inquiring was no longer interested in having a further discussion. Most school leaders noted that they would be unable to initiate a local discussion to potentially shift to later high school start times without having proof that this action would improve students' grades. The nationwide push toward "accountability" for student performance and addressing the "achievement gap" at this time in the mid-2000s was shaping nearly all decisions that school districts made. Improved student health and well-being were apparently not sufficient to motivate a school leader to begin the difficult task of suggesting a change in school start times.

Additional Outcomes Studied

During this period of 2002–2014, research on outcomes for adolescents obtaining less-than-optimal sleep began to encompass many new variables. At this time, middle school students also began to be included in many studies [36, 37]. Caffeine use began to be studied as well [38, 39].

There were two studies that stand out during this time as having wide-ranging impact on discussions related to teen sleep deprivation. McKnight-Eily et al. in 2011 [40] used the data from the Youth Risk Behavior Survey (YRBS), a biannual national survey of adolescent behavior, to discover that teens who obtained less than 8 hours of sleep per night were at a statistically higher risk of using drugs, cigarettes, and alcohol than their peers who slept 8 or more hours per night. Furthermore, the degree of risk substantially increased with every hour less of sleep. The study's findings were instrumental in determining that 8 hours of sleep is considered a "turning point" in healthy vs. non-healthy behavior choices in adolescents. The study continues to be viewed as seminal.

The other topic that was eventually noticed by the local policy makers related to teen car crashes due to drowsy driving. Danner and Phillips [41] and Vorona et al. [42] found that the rates of teen car crashes were statistically related to the starting time of high school. In the Danner and Phillips study, the average crash rate for teenage drivers dropped 16.5% in one county in 2 years after the school start time was changed to about 1 hour later for those area high schools, while crash rates throughout the rest of the state *increased by* 8.9% during the same time period. The conclusion was that teens with sleep duration of >8 hours per night were statistically less likely to be drowsy drivers. This was an important finding because car crashes are the #1 cause of teen deaths in the United States.

The use of technology by teens from 2005 and beyond also began to emerge as a highly significant factor in teen sleep or the lack thereof. Studies of teens using computers late into the night or being engaged in social media on their cell phones at all hours of the night revealed serious negative outcomes [43]. Having a computer in the bedroom was related to later bedtimes, later wake-up times, and shortened sleep duration [25]. More than 50% of teens who texted or surfed the internet at bedtime were more likely to have problems falling asleep, plus having mood, behavior, and cognitive problems during the day [44].

Unfortunately, once again, these findings were not reported in educational journals. The disconnect or gap was growing wider with respect to the increasing abundance of evidence regarding the benefits of later start times for secondary schools and what educational leaders actually knew. Primarily due to the mass media coverage of some of the research at this time, from reporters covering national medical and safety conferences, educational leaders were eventually alerted to the key link between adolescent sleep and the safety, performance, and well-being of their students.

The CDC Call for a Comprehensive Study

In 2010, the Centers for Disease Control and Prevention awarded a research grant to this author to study the full range of outcomes for students whose high schools changed to a later starting time. The comprehensive study was the largest of its kind, collecting data in eight high schools in five districts in three states (Colorado, Wyoming, and Minnesota). Over 9000 students completed the *Teen Sleep Habits Survey* [45] which queried their sleep/wake habits and also asked questions about sleepiness, emotional status, activity levels, technology use, homework, and car crashes. The data collected in Jackson Hole, Wyoming, were particularly notable because they were collected both prior to and after the start time change. In addition, their time change from 7:35 a.m. to 8:55 a.m. was the largest amount of change and latest start time of all eight high schools in the study.

The findings from the study [46] included the following:

- Nearly 60% of the students in schools that started at 8:30 a.m. or later and 66% of the students whose school started at 8:55 obtained 8 or more hours of sleep per night.
- Teens reporting less than 8 hours of sleep reported significantly more depressive symptoms than their peers who got 8 or more hours per school night.
- Students who reported using tobacco or drugs in the last 2 weeks were also more likely to report high levels of depressive feelings as measured on the depression scale on the survey. (This finding is significant due to the strong correlation previously found between teen sleep deprivation and drug, cigarette, and alcohol use [40].)
- Caffeine use was also significantly associated with hours of sleep, as teens reporting less than 8 hours per school night reported significantly greater use of caffeinated drinks.
- The survey revealed that more than 88% of students have a cell phone in their bedroom and 41% have a computer. Furthermore, having a computer or phone in the bedroom was significantly related to obtaining less than 8 hours of sleep.
- There was a decrease in tardiness overall for students in grades 9–12, with a statistically significant reduction in tardiness in the schools that had greatest delays in school start times (e.g., 8:35 a.m. and 8:55 a.m.).
- Statistically significant increases in first period GPA in *all* core courses of English, math, social studies, and science were received by students in the high school with the latest starting time of 8:55.
- Two of the five schools saw a statistically positive increase in their overall performance on national standardized achievement tests, such as the SAT and ACT.
- In Jackson Hole, Wyoming, during the first year of implementation of a later start time 8:55 a.m., the rate of car crashes decreased 70%. The fact that over 66% of Jackson Hole High School students obtained more than 8 hours of sleep each school night was hypothesized as the reason for the significant drop.

The most salient action following the release of the findings from the 2014 CDC-funded study was the creation of the report and policy recommendation from the American Academy of Pediatrics (AAP) [1], which was a nationwide call to action to begin all secondary schools in the United States at 8:30 a.m. or later. The AAP report was the result of a lengthy review of all of the research, both medical and educational, and was written by a team of physicians and educational policy leaders. Together, they admonished all persons who cared about or cared for teenagers to learn more about the sleep patterns of adolescents, asserting that the severe lack of sleep for today's teens was a public health crisis. The AAP report has been seen as a critical step by both educational and health care professionals to raise the national awareness about the biological reason (i.e., the circadian phase delay) for starting schools later. Many of the national professional health associations, including the American Medical Association [47], the American Psychological Association [48], and the National School Nurses Association, have issued statements of support for the 8:30 a.m. start time.

Continuing Questions/Dilemmas: 2015 to Present

As of this writing, many issues related to later starting times remain unaddressed, and/or nuanced concerns have arisen due to their implementation. In addition, new areas of inquiry are being brought to the fore as the state of adolescent sleep research continues to reveal new insights.

Persistent Dilemmas

The #1 issue that has persisted for over 20 years and that continues to halt any discussion to implement a later start time is the political resistance that is present in many school districts, from members of the school board to the superintendent to the parents in the community. Interviews with these stakeholders over the years have revealed that the decision to alter a school's start time—in effect, to alter the community's rhythm—is considered to be extremely risky [49]. The school leaders have revealed a fear of being replaced in their roles if the debate about a later start time divides and polarizes the community. As a result, many districts have chosen not to change school starting times for any grade level. Reasons cited are often erroneous or misleading, such as the contention that a change would create an enormous increase in transportation costs, that students would participate less in sports and after-school activities, or that kids will use the extra time to stay up later—all of which have been shown in research reports to be unfounded [34, 35, 50]. Districts that have successfully made the change have dealt constructively with the legitimate concerns such as after-school childcare, concerns about elementary children waiting for the bus on dark winter mornings, and parent work schedules.

Another issue that has persisted over the years is the lack of education of the general public, and even many physicians, about adolescent sleep patterns. Persons who do not know about the circadian phase delay in teens usually view teens' late-to-bed late-to-rise behavior as being merely lazy. This lack of knowledge by many/most is the most common reason that the later school start time discussions get scuttled before any actions can occur. Parents need to be educated about their teenager's sleep patterns and good sleep hygiene, education which often has only occurred when a school district has formed a community committee to discuss the possibility of changing to a later start. Teachers and principals also need to learn more about the biology of adolescent sleep and how it affects student behavior and learning throughout the day.

Finally, an issue that has repeatedly emerged over the years in critiques and meta-analyses of the research studies [51–53] is the lack of randomized controlled trials (RCTs), the gold standard in experimental research. In the case of studying outcomes for students whose schools start early or late, the true research design would control for contextual variables that are present in every individual school building, such as differences among teachers, school culture and climate, peer and social structures, the physical setup of the school itself, etc. Thus, an RCT would randomly assign all students in that building to an early start or a late start and examine outcomes for them over a period of three or more years. Using students as research subjects is difficult to do in any study, but to allow random assignment for when the students go to school would disrupt a family, disrupt parents' work and family schedules, and create unfathomable logistical problems with the teachers' contracts and bus schedules. In other words, having true RCTs to examine outcomes for public school students is completely infeasible, yet the critics state that if the gold standard is not met, then conclusive results for the outcomes cannot be established.

There is, however, one singular RCT that assessed outcomes for early vs. late start. That study by Carrell et al. [54] was done at the United States Air Force Academy from 2004 to 2008 where, over the years, incoming freshman students were randomly assigned to early vs. late starting first hour classes. They found that the condition of early school start times negatively affected academic performance, with "early" students performing significantly worse in their first hour classes. Those same students also performed worse in their overall academic work throughout the day.

The Air Force Academy is a unique setting where the students' lives and schedules are fully controlled. A randomized controlled trial was thus uniquely possible to be done there. It is unimaginable that this would be possible in any public high school in the United States. Thus, the criterion of an RCT as the only way to substantiate the positive outcomes for later start times is untenable. Rather, the enormous body of positive evidence [55–60] that is correlational for the outcomes for hundreds of thousands of teens should be left to stand as sufficient evidence that later start times are beneficial. In fact, more than 1800 articles have already been written ([52], p. 15) about the outcomes and benefits of later starting times.

Current and Emerging Issues

There are issues related to the sleep patterns of teens that have emerged from more recent research. These are less directly related to the starting time of school per se and are more concerned with the nuances of adolescent sleep. These unknowns, if answered, may strengthen or shift the policy actions related to the starting time of all schools at all levels. Some of the issues are as follows.

Role of Circadian Preference in Assessing Benefits Several current articles [61–63] examined circadian preference in better understanding outcomes for adolescents, especially for those whose sleep/wake schedule is altered. For example, even as early as 1995, Callan's study [33] of morningness/eveningness suggested that administering the SAT only in the morning may discriminate against some students.

Sleep Duration Versus Sleep Timing There are reports now available [63] that distinguish between the *amount* of sleep an adolescent receives and *when* that sleep occurs. Sleep duration may be more strongly related to its timing than understood previously. Studies of sleep efficiency may further show that duration is an incomplete measure of the quality of the sleep obtained.

Role of Rapid-Eye-Movement (REM) Sleep in Teens' Development With REM sleep known to be related to memory consolidation for learning and cognition [64], current studies of neural development in teens may provide even stronger evidence that adolescents need to be allowed to sleep to the conclusion of their natural sleep cycle. Current forced awakenings for teens whose schools start early does not generally allow for all periods of their REM sleep to occur. Emotional regulation is also strongly associated with REM sleep [23], and the elimination of one or more periods of REM sleep due to early school start times may become even more clearly and directly linked to teen depression.

Role of Sleep in Teens' Health Conditions Obesity, diabetes, and cardiovascular issues are known to be strongly associated with sleep in adults. Current studies are now examining how the early starting times of high schools are potentially contributing to those conditions [65–67] in adolescents due to sleep deprivation. Studies of technology use during normal sleep time are also seen as affecting teen health [68]. Additional studies that would follow early start teens' health into adulthood would be a logical next area of inquiry.

Effect on Preteens of Starting Elementary Schools Earlier As the later high school start time efforts have been implemented, there have been many requests for research on elementary preadolescent children whose elementary schools were shifted to an early start time when the bus schedules were "flipped" to transport secondary students later. Very few studies have been done on the effect of an earlier start time on elementary-aged students [69] who are now arising earlier due to the changes made at the high school level. This is likely because the sleep of preteenage children is considered to be malleable (as compared with teen sleep), and putting them to bed earlier would not necessarily be an issue of reduced sleep. However, the thought that parents of preteens should just put their younger children to bed earlier to adjust to the early elementary start has received resistance by many parents. Working parents see this as a loss of time to spend with their child in the evening. Also, some young children are naturally "owls, which makes putting them to bed early a challenge in allowing them adequate sleep if the elementary schools are shifted to an earlier time.

Effect of Circadian Phase Delay and Its Persistence into Early College Years Postsecondary schools are beginning to examine [70, 71] the sleep patterns and habits of freshman and sophomore students and how they play a role in their overall functioning, including academic performance and course taking. As the research in circadian rhythms expands, it will likely inform academic counseling and recommendations for creating healthy choices and life experiences for students in postsecondary institutions.

Additional issues within the educational policy arena are also swirling around as educators discuss and debate the evidence related to later school start times. Some of these include the following.

The Continued Inability to Compare Academic Outcomes Between Districts and Across States Each state and district in the United States has control over the testing of their students. The test items and schedule of grade levels tested differ from state to state, plus districts can decide when the tests are administered and to whom. This type of "local control" makes comparisons among schools who are making changes in their start times an analytical nightmare. Matched sample studies within the same setting address these issues well. Thus, when a reviewer seeks to assess the validity of the findings in any given study, the editor and reader must look for a careful attempt made by the researcher to account for the differences across the settings. This would also include comparisons involving the performance on national tests, such as the American College Testing (ACT) and SAT. Some districts require all high school juniors/senior to take the ACT/SAT, while others make it discretionary. This lack of uniformity of who is taking the national exams creates potential bias in the data.

The Concept of "Local Control" for School Districts Local control is a founding principle of American schooling which allows local communities to make nearly all decisions about the education of its children. (Note: The accountability movement in 2003, known as "No Child Left Behind," was one of the first-ever national attempts to defy local control in order to standardize academic assessment and was resisted by some states for many years, even after it was passed as a national policy.) Administrators do not want any local advocacy group or state policymaker to interfere in any educational decision normally made at the district level.

Attempts to create legislation at the national level from 1999 to 2017 [72] to entice districts across the United States to shift to a later start time have historically been met with strong resistance by school boards and teachers' unions. However, in

2019, California's legislature approved a bill that mandated high schools across the state to start no earlier than 8:30 am and middle schools to start no earlier than 8:00 am, effective for the 2022 school year. Newly elected Governor Newsom signed the bill into law [73] on October 13, 2019. This is landmark legislation that may cause districts and other states to consider similar actions, particularly with the efforts to frame later start times as a public health policy initiative. Future assessments and outcomes findings for California students will help guide this issue as it continues to evolve and will reflect what we are learning about making this change on such a large scale.

Conclusion

Later start time is not a "magic bullet" to solve the issues confronting teens' health and well-being. However, it is a tool, among many, that can help our young people to go through their teenage years and enter adulthood in a better state. The current medical research studies on circadian timing, sleep duration, chronotype, REM sleep, etc., will provide increasing substance to understanding the role of sleep in teens' health and well-being. At the same time, further studies completed by educational researchers will add to the joint body of knowledge about the interaction between teen sleep and educational outcomes. As this chapter began, it will end with the renewed emphasis on the need to continue to jointly address adolescent sleep and education. Medical researchers need to examine and cite the scientific literature from educational journals, just as educational researchers must do the same from the medical literature.

It was over 15 years before the educational literature and the medical literature began to "interact" by the mid-2010s, yet still today, medical research articles cite few, if any, related findings from educational research that would place their findings into context for further application and continued study. By the same token, educational research studies rarely co-examine the political realities of school start time when assessing outcomes for later start times. School reform efforts toward implementing a later start time are often woefully lacking an in-depth view of the relationship between the medical facts and the political tensions of attempting a change. As researchers, all of us must persist in emphasizing the biology of teen sleep while we are supportive of changes to later start times in secondary schools.

The interrelated dynamics include the school board and their political relationship to the superintendent, the role of principals and their involvement in the decision, and the voices of teachers, students, and families and their perceived needs. All must be taken into account when we examine the data supporting and challenging the change. It is an extremely complicated process, but one that can bring lifelong benefits to the adolescents whose well-being depends on the actions of the adults in their lives.

References

1. Adolescent Sleep Working Group Committee on Adolescence and Council on School Health. School start times for adolescents. Pediatrics. 2014;134(3):642–9.
2. Weinert D, Sitka U, Minors DS, Waterhouse JM. The development of circadian rhythmicity in neonates. Early Hum Dev. 1994;36(2):117–26.
3. Kraft M, Martin RJ. Chronobiology and chronotherapy in medicine. Dis Mon. 1995;41(8):506–75.
4. Redfern P, Minors D, Waterhouse J. Circadian rhythms, jet lag, and chronobiotics: an overview. Chronobiol Int. 1994;11(4):253–65.
5. Tolstoi LG. A review of chronobiology and chronopharmacology. J Pract Nurs. 1994;44(1):47–57.
6. Reppert SM, Weaver DR. Melatonin madness. Cell. 1995;83(7):1059–62.
7. Brage DG. Adolescent depression: a review of the literature. Arch Psychiatr Nurs. 1995;9(1):45–55.
8. Bjorksten KS, Basun H, Wetterberg L. Disorganized sleep-wake schedule associated with neuroendocrine abnormalities in dementia: a clinical study. Int J Geriatr Psychiatry. 1995;10(2):107–13.
9. Morrison DN, McGee R, Stanton WR. Sleep problems in adolescence. J Am Acad Child Adolesc Psychiatry. 1992;31(1):94–9.
10. Dahl RE, Carskadon MA. Sleep and its disorders in adolescence. Principles and practice of sleep medicine in the child, vol. 2. Philadelphia: W.B. Saunders; 1995.
11. Carskadon MA, Harvey K, Duke P, Anders TF, Litt IF, Dement WC. Pubertal changes in daytime sleepiness. Sleep. 1980;2:453–60.
12. Carskadon MA, Vieira C, Acebo C. Association between puberty and a circadian phase delay. Sleep. 1993;16(3):258–62.
13. Giannotti F, Cortesi F, Sebastiani T, Ottaviano S. Circadian preference, sleep and daytime behaviour in adolescence. J Sleep Res. 2002;11(3):191–9.
14. Carskadon MA, Acebo C, Richardson GS, Tate BA, Seifer R. An approach to studying circadian rhythms of adolescent humans. J Biol Rhythm. 1997;12(3):278–89.
15. Carskadon MA. When worlds collide: adolescent need for sleep versus societal demands. Phi Delta Kappan. 1999;80(5):348–53.
16. Andrade MM, Benedito-Silva AA, Domenice A, Arnhold IPJ, Menna-Barreto L. Sleep characteristics of adolescents: a longitudinal study. J Adolesc Health. 1993;14:401–6.
17. Giannotti F, Cortesi F. Sleep patterns and daytime function in adolescence: an epidemiological survey of an Italian high school student sample. In: Carskadon M, editor. Adolescent sleep patterns: biological, social, and psychological influences. Cambridge: Cambridge University Press; 2002.
18. LeBourgeois MK, Giannotti F, Cortesi F, Wolfson A, Harsh J. Sleep hygiene and sleep quality in Italian and American adolescents. Ann N Y Acad Sci. 2004;1021:352–4.
19. Sadeh A, Raviv R, Gruber R. Sleep patterns and sleep disruptions in school-age children. Dev Psychol. 2000;36(3):291–301.
20. Tzischinsky O, Shochat T. Eveningness, sleep patterns, daytime functioning, and quality of life in Israeli adolescents. Chronobiol Int. 2011;28(4):338–43.
21. Carskadon MA. The second decade. In: Guilleminault C, editor. Sleeping and waking disorders: indications and techniques. Menlo Park: Addison Wesley; 1982.
22. Wolfson AR, Tzischinsky O, Brown C, Darley C, Acebo C, Carskadon MA. Sleep, behavior, and stress at the transition to senior high school. J Sleep Res. 1995;24:115.
23. Dahl RE. The consequences of insufficient sleep for adolescents: links between sleep and emotional regulation. Phi Delta Kappan. 1999;80(5):354–9.
24. Winsler A, Deutsch A, Vorona RD, Payne PA, Szklo-Coxe M. Sleepless in Fairfax: the difference one more hour of sleep can make for teen hopelessness, suicidal ideation, and substance use. J Youth Adolesc. 2015;44(2):362–78.

25. Shochat T, Barker DH, Sharkey KM, Van Reen E, Roane BM, Carskadon MA. An approach to understanding sleep and depressed mood in adolescents: person-centered sleep classification. J Sleep Res. 2017;26(6):709–17.
26. Noland H, Price JH, Dake J, Telljohann SK. Adolescents' sleep behaviors and perceptions of sleep. J Sch Health. 2009;79(5):224–30.
27. Allen RP, Mirabile J. Self-reported sleep-wake patterns for students during the school year from two different senior high schools. J Sleep Res. 1989;18:132.
28. Allen RP. School week sleep lag: sleep problems with earlier starting of senior high schools. J Sleep Res. 1991;20:198.
29. Allen RP. Social factors associated with the amount of school week sleep lag for seniors in an early starting suburban high school. J Sleep Res. 1992;21:114.
30. Kowalski N, Allen R. School sleep lag is less but persists with a very late starting high school. J Sleep Res. 1995;24:124.
31. Wolfson A, Carskadon MA. Sleep schedules and daytime functioning in adolescents. Child Dev. 1998;69(4):875–87.
32. Carskadon MA, Wolfson AR, Acebo C, Tzischinsky O, Seifer R. Adolescent sleep patterns, circadian timing, and sleepiness at a transition to early school days. Sleep. 1998;21(8): 871–81.
33. Callan RJ. Early morning challenge: the potential effects of chronobiology on taking the scholastic aptitude test. Clearing House. 1995;68(3):174–6.
34. Wahlstrom KL. School start time study final report, volume 2: analysis of student survey data. Minneapolis: University of Minnesota, Center for Applied Research and Educational Improvement; 1998. http://hdl.handle.net/11299/4249. Accessed 10 Aug 2019.
35. Wahlstrom K. Changing times: findings from the first longitudinal study of later high school start times. NASSP Bull. 2002;86(633):3–21.
36. Drake C, Nickel C, Burduvali E, Roth T, Jefferson C, Pietro B. The pediatric daytime sleepiness scale (PDSS): sleep habits and school outcomes in middle-school children. Sleep. 2003;26(4):455–8.
37. Fredriksen K, Rhodes J, Reddy R, Way N. Sleepless in Chicago: tracking the effects of adolescent sleep loss during the middle school years. Child Dev. 2004;75(1):84–95.
38. Pollak CP, Bright D. Caffeine consumption and weekly sleep patterns in US seventh-, eighth-, and ninth-graders. J Pediatr. 2003;111(1):42–6.
39. Ludden AB, Wolfson AR. Understanding adolescent caffeine use: connecting use patterns with expectancies, reasons, and sleep. Health Educ Behav. 2010;37(3):330–42.
40. McKnight-Eily LR, Eaton DK, Lowry R, Croft JB, Presley-Cantrell L, Perry GS. Relationships between hours of sleep and health-risk behaviors in US adolescent students. Prev Med. 2011;53:271–3.
41. Danner F, Phillips B. Adolescent sleep, school start times, and teen motor vehicle crashes. J Clin Sleep Med. 2008;4(6):533–5.
42. Vorona RD, Szklo-Coxe M, Wu A, Dubik M, Zhao Y, Ware JC. Dissimilar teen crash rates in two neighboring southeastern Virginia cities with different high school start times. J Clin Sleep Med. 2011;7(2):145–51.
43. Borlase BJ, Gander PH, Gibson RH. Effects of school start times and technology use on teenagers' sleep: 1999–2008. Sleep Biol Rhythms. 2013;11(1):46–54.
44. Do YK, Shin E, Bautista MA, Foo K. The associations between self-reported sleep duration and adolescent health outcomes: what is the role of time spent on internet use? Sleep Med. 2013;14(2):195–200.
45. Wahlstrom KL. Teen sleep habits survey. University of Minnesota. http://innovation.umn.edu/teen-sleep/. Accessed 10 Aug 2019.
46. Wahlstrom KL, Dretzke B, Gordon M, Peterson K, Edwards K, Gdula J. Examining the impact of later high school start times on the health and academic performance of high school students: a multi-site study. Minneapolis: University of Minnesota: Center for Applied Research and Educational Improvement; 2014. http://hdl.handle.net/11299/162769. Accessed 10 Aug 2019.

47. American Medical Association. Insufficient sleep in adolescents H-60.930. 2016. https://policysearch.ama-assn.org/policyfinder/detail/school%20start%20time?uri=%2FAMADoc%2FHOD.xml-0-5024.xml, H-60,930. Accessed 10 Aug 2019.
48. American Psychological Association. Later school start times promote adolescent well-being. https://www.apa.org/pi/families/resources/school-start-times.pdf. Accessed 10 Aug 2019.
49. Wahlstrom KL. The prickly politics of school starting times. Phi Delta Kappan. 1999;80(5):344–7.
50. Dunster GP, de la Iglesia L, Ben-Hamo M, Nave C, Fleischer JG, Panda S, et al. Sleepmore in Seattle: later school start times are associated with more sleep and better performance in high school students. Sci Adv. 2018;9(12):eaau6200.
51. Wheaton AG, Chapman DP, Croft JB. School start times, sleep, behavioral, health, and academic outcomes: a review of the literature. J Sch Health. 2016;86(5):363–81.
52. Marx R, Tanner-Smith EE, Davison CM, Ufholz LA, Freeman J, Shankar R, et al. Later school start times for supporting the education, health, and well-being of high school students. Cochrane Database Syst Rev. 2017;7:CD009467.
53. Morgenthaler TI, Hashmi S, Croft JB, Dort L, Heald JL, Mullington J. High school start times and the impact on high school students: what we know, and what we hope to learn. J Clin Sleep Med. 2016;12(12):1681–9.
54. Carrell SE, Maghakian T, West JE. A's from Zzzz's? The causal effect of school start time on the academic achievement of adolescents. Am Econ J Econ Policy. 2011;3:62–81.
55. Paksarian D, Rudolph KE, He JP, Merikangas KR. School start time and adolescent sleep patterns: results from the U.S. National Comorbidity Survey--Adolescent supplement. Am J Public Health. 2015;105(7):1351–7.
56. Wheaton AG, Olsen EO, Miller GF, Croft JB. Sleep duration and injury-related risk behaviors among high school students--United States, 2007-2013. MMWR Morb Mortal Wkly Rep. 2016;65(13):337–41.
57. Lewin DS, Wang G, Chen YI, et al. Variable school start times and middle school student's sleep health and academic performance. J Adolesc Health. 2017;61(2):205–11.
58. Eaton DK, McKnight-Eily LR, Lowry R, Perry GS, Presley-Cantrell L, Croft JB. Prevalence of insufficient, borderline, and optimal hours of sleep among high school students - United States, 2007. J Adolesc Health. 2010;46(4):399–401.
59. Watson NF, Martin JL, Wise MS, Carden KA, Kirsch DB, Kristo DA, et al. Delaying middle school and high school start times promotes student health and performance: an American Academy of Sleep Medicine position statement. J Clin Sleep Med. 2017;13(4):623–5.
60. Owens J, Drobnich D, Baylor A, Lewin D. School start time change: an in-depth examination of school districts in the United States. Mind Brain Educ. 2014;8(4):182–213.
61. Owens JA, Dearth-Wesley T, Lewin D, et al. Self regulation and sleep duration, sleepiness, and chronotype in adolescents. Pediatrics. 2016;138(6):e20161406.
62. Escribano C, Diaz-Morales JF. Daily fluctuations in attention at school considering starting time and chronotype: an exploratory study. Chronobiol Int. 2014;31(6):761–9.
63. Martin JS, Gaudreault MM, Perron M, Laberge L. Chronotype, light exposure, sleep, and daytime functioning in high school students attending morning or afternoon school shifts: an actigraphic study. J Biol Rhythm. 2016;31(2):205–17.
64. Feinberg I, Davis NM, de Bie E, Grimm KJ, Campbell IG. The maturational trajectories of NREM and REM sleep durations differ across adolescence on both school-night and extended sleep. Am J Physiol Regul Integr Comp Physiol. 2012;302(5):R533–40.
65. Suglia SF, Kara S, Robinson WR. Sleep duration and obesity among adolescents transitioning to adulthood: do results differ by sex? J Pediatr. 2014;165(4):750–4.
66. Beebe DW, Simon S, Summer S, Hemmer S, Strotman D, Dolan LM. Dietary intake following experimentally restricted sleep in adolescents. Sleep. 2013;36(6):827–34.
67. Wahlstrom K, Owens J. School start time effects on adolescent learning, mood and behavior. Curr Opin Psychiatry. 2017;30(6):485–90.

68. Johansson AEE, Petrisko MA, Chasens ER. Adolescent sleep and the impact of technology use before sleep on daytime function. J Pediatr Nurs. 2016;31(5):498–504.
69. Dupuis DN. The association between elementary school start time and students' academic achievement in Wayzata Public Schools. Minneapolis: University of Minnesota: Center for Applied Research and Educational Improvement; 2015. http://hdl.handle.net/11299/181187. Accessed 10 Aug 2019.
70. Lund HG, Reider BD, Whiting AB, Prichard JR. Sleep patterns and predictors of disturbed sleep in a large population of college students. J Adolesc Health. 2010;46(2):124–32.
71. Hartmann ME, Prichard JR. Calculating the contribution of sleep problems to undergraduates' academic success. Sleep Health. 2018;4(5):463–71.
72. US Congresswoman Zoe Lofgren (CA). "ZZZ's to A's Act" –H.R 5678 113th Congress. This bill was initially introduced on March 24, 1999, in a previous session of Congress, but was not enacted. It was re-introduced in 2017. It authorizes the US Secretary of Education to make grants to local educational agencies that agree to begin school for secondary students after nine o'clock in the morning.
73. State Senator Anthony Portantino (CA). *SB 328- School Start Time*: This bill requires the school day for high schools in California to begin no earlier than 8:30 a.m. as of 2022. The bill was passed by both houses of the California Assembly and signed into law in October, 2019. http://leginfo.legislature.ca.gov/faces/billTextClient.xhtml?bill_id=201920200SB328.

Chapter 8
Advanced Sleep-Wake Rhythm Disorder

Elliott Kyung Lee

Advance-Related Sleep Complaints and Advanced Sleep-Wake Phase Disorder (ASWPD)

The term circadian rhythm sleep-wake disorder (CRSWD) is used to encompass a wide variety of conditions in which there is significant misalignment between the innately preferred sleep/wake schedule and the 24-hour light/dark cycle [1]. This chapter focuses on advance-related sleep complaints, when individuals have habitual sleep onset and/or offset times that are markedly earlier than desired. These patients may present with complaints of early evening sleepiness in conjunction with early-morning awakenings. Alternatively, individuals may be obligated to maintain a relatively delayed/conventional bedtime (or fail to recognize/report inadvertent evening sleep bouts that occur prior to their defined "bedtimes"), but persist with undesirably early rise times, leading to chronic insufficient sleep and daytime sleepiness [2]. Although sleep complaints can be an issue for these persons, some data suggest an advanced sleep phase or morningness traits may be more socially acceptable and possibly confer increased resilience and optimism, which may contribute to these persons not seeking clinical attention.

Advance-Related Sleep Complaints

Table 8.1 identifies criteria for advanced sleep-wake phase disorder (ASWPD), whereby patients identify simultaneous nighttime and morning complaints. The International Classification of Sleep Disorders Third Edition (ICSD-3) states that

E. K. Lee (✉)
Department of Psychiatry, Royal Ottawa Mental Health Center and Institute for Mental Health Research, University of Ottawa, Ottawa, ON, Canada
e-mail: elliott.lee@theroyal.ca

R. R. Auger (ed.), *Circadian Rhythm Sleep-Wake Disorders*,
https://doi.org/10.1007/978-3-030-43803-6_8

Table 8.1 Advanced sleep-wake phase disorder diagnostic criteria

Diagnostic criteria	Description (criteria A–E must be met)
A	There is an advance (early timing) in the phase of the major sleep episode in relation to the desired or required sleep time and wake-up time, as evidenced by a chronic or recurrent complaint of difficulty staying awake until the required or desired conventional bedtime, together with an inability to remain asleep until the required or desired time for awakening
B	Symptoms are present for at least 3 months
C	When patients are allowed to sleep in accordance with their internal biological clock, sleep quality and duration are improved with a consistent but advanced timing of the major sleep episode
D	Sleep logs and, whenever possible, actigraphy monitoring for at least 7 days (preferably 14 days) demonstrate a stable advance in the timing of the habitual sleep period. Both workdays/school days and free days must be included within this monitoring
E	The sleep disturbance is not better explained by another current sleep disorder, medical or neurological disorder, mental disorder, medication use, or substance use disorder

Reprinted with permission from: American Academy of Sleep Medicine [91]
Alternate names: Advanced sleep phase type, advance sleep phase disorder, advance sleep phase syndrome
Notes
1. Standardized chronotype questionnaires are useful tools to assess the chronotype of eveningness and morningness. Individuals with advanced sleep phase score as morning types
2. Demonstration of an advance (typically greater than 2 hours) in the timing of other circadian rhythms such as dim light melatonin onset (DLMO) or urinary 6-sulfatoxymelatonin is desirable to confirm the advanced circadian phase

such patients will "typically" exhibit sleep onset and offset times between 1800 and 2100 hours and 0200 and 0500 hours, respectively [3]. Accordingly, afflicted patients present with difficulties remaining awake in the late afternoon/early evening, in addition to endorsement of early-morning awakenings. This innate circadian preference makes it difficult or impossible to adhere to a socially desirable sleep/wake schedule [4]. While the ICSD-3 stipulates a 3-month duration criterion, the Diagnostic and Statistical Manual of Mental Disorders (DSM-5) requires only 1 month [2, 5]. There are select physiologic data that demonstrate earlier timing of circadian biomarkers (melatonin, core body temperature) of ASWPD patients (2–4 hours earlier than unaffected subjects) [6].

Clinicians are unlikely to encounter patients meeting strict ICSD-3 criteria for ASWPD, however, in part because the affiliated sleep/wake schedule infrequently presents with marked social or occupational conflicts. Indeed, such behavior is often rewarded in occupational settings. In addition, one can voluntarily delay sleep onset times and/or fail to report inadvertent sleep that occurs out of the bedroom, making it difficult to identify individuals with an early evening sleepiness complaint required for the diagnosis [7]. As such, a broader consideration of advance-related sleep complaints is required, whereby sole sleep maintenance difficulties are inferred to be due to a phase advance in the circadian cycle and, by relation, responsive to circadian interventions [10].

Epidemiology

Prevalence statistics on ASWPD are varied. One large Norwegian study of 7700 individuals using strict ICSD criteria did not identify a single subject meeting diagnostic criteria [11, 12]. Other studies have suggested a population prevalence of 0.5%–1% [11, 13]. A separate study from New Zealand by Paine and colleagues described a prevalence of 0.25%–7.13%, depending upon the definition used, with men and older individuals more likely to be affected [14]. Another study by Ando and colleagues suggested that up to 7.4% of the general population may have advance-related sleep *complaints* [15], based on telephone surveys administered to random participants in San Diego. A 2019 study by Curtis and colleagues evaluated the prevalence of advanced sleep phase (ASP), familial advanced sleep phase (FASP), and ASWPD in 2422 patients seen in a Utah sleep clinic over a span of almost 10 years [8]. Assessments included the Morningness-Eveningness Questionnaire (MEQ), structured clinical interviews and assessments and, when possible, polysomnography, 10-day ambulatory actigraphy, sleep logs, and salivary dim light melatonin onset (DLMO). Their results showed an ASP prevalence of 0.33%, an FASP prevalence of 0.21%, and a ASWPD prevalence of 0.04% in patients referred to a North American sleep clinic for an assessment [8].

Etiology and Risk Factors

Genetics

Advance-related sleep complaints have a strong heritability component. The first familial study was done by Jones et al. in 1999 [6]. In this study, 29 out of 75 evaluated family members of Northern European descent were shown to have significant advanced, or "morning lark" traits, with autosomal dominant transmission and high penetrance. While the youngest of these subjects was 8 years old, most subjects knew by age 30 that they had advanced traits. A 3–4 hour advance in melatonin and body temperature rhythms was documented in comparison to controls. One subject also demonstrated a shorter circadian period (23.3 hours) when evaluated in temporal isolation. Further analysis identified an autosomal dominant inherited missense mutation (serine to glycine substitution at amino acid 662 – S662G) at the Period 2 (h*Per*2) gene located on the short arm of chromosome 2 [16], resulting in decreased phosphorylation of the h*Per*2 protein by casein kinase epsilon (CK1ε). Phosphorylation normally promotes degradation of the Per protein, preventing subsequent dimerization with the Cryptochrome (Cry) protein and leading to moderation of nuclear accumulation. As a result, h*Per*2 degradation is impaired, leading to increased accumulation and positive regulation of BMAL1. The BMAL1/Clock heterodimer normally drives the production of protein products of Per, Cry, and clock-controlled genes (CCGs) (see Fig. 8.1). These processes lead to a secondary increased transcription of BMAL1 and subsequent phase advancement. The mutation is not ubiquitous, however, as another

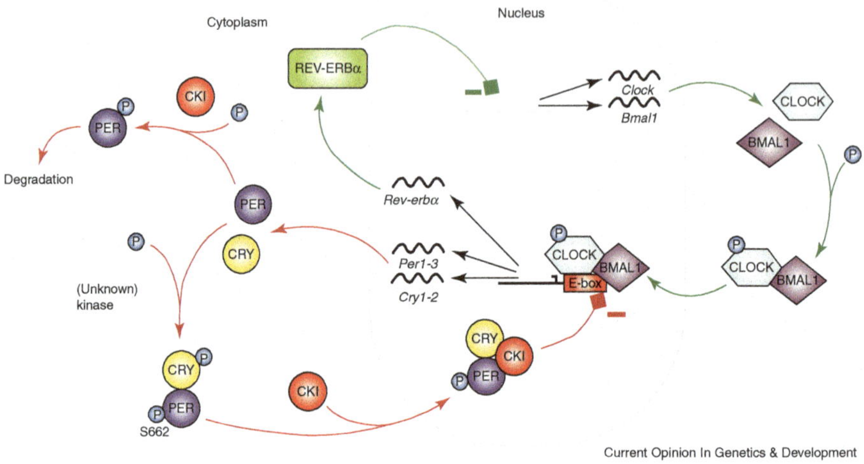

Fig. 8.1 Schematic representation of the basic components of the molecular circadian clock from Tafti et al. (2007) [16]. This molecular clock is based on several interacting positive (shown in green) and negative (shown in red) transcriptional loops, resulting in oscillating RNA and protein levels of key clock components. Transcription factors Clock and BMAL1 heterodimerize and are subsequently phosphorylated and translocated across the nucleus to activate transcription of 3 period genes (*Per*1-3) and two cryptochrome genes (*Cry*1-2). The protein products of the *Per* and *Cry* genes subsequently dimerize outside the nucleus in several combinations, and may undergo phosphorylation by casein kinase I (CKI) and translocate across the nucleus to inhibit the transcriptional activation by the Clock/BMAL1 heterodimer (negative loop, exerting autoregulation of their own transcription). CKI also phosphorylates *Per* proteins tagging them for degradation. Mutations in *Per*2 and *Per*3 result in impairment of phosphorylation by CKI, and have been identified as etiologies for phase advancement. On the other hand, the Clock-BMAL1 heterodimer also activates transcription of REV-ERBα. These proteins then translocate across the nucleus to activate transcription of Clock and BMAL1 proteins (positive feedback). Other genes implicated in phase advancement include Dec2, *Cry*2, and CKIδ (not shown). Per *Per*1-3, Cry *Cry*1-2, CKI CK1ε, and CKIδ. (Reprinted from Tafti et al. [16], with permission from Elsevier)

phenotypically similar Japanese family was identified, that did not exhibit the same mutation [7]. Figure 8.1 provides an example of clock gene mechanics within a cell.

Other studies support genetic heterogeneity of ASWPD. One study highlighted involvement of the Period 3 (Per3) gene, with two rare missense mutations (Per3-P415A/H417R) found in association with a familial advanced sleep phase and seasonal affective disorder, suggesting a genetic pathway for a connection between circadian rhythm and mood regulation [17]. Other implicated mutations have been found in casein kinase I delta (CKIδ), Basic Helix Loop Helix E41 (BHLHE41, i.e., Dec2), and Cryptochrome 2 (Cry2) proteins [13, 18–20]. Advanced sleep-wake phase disorder has also been identified in Smith Magenis syndrome, a congenital condition associated with deletion of chromosome 17 band p11.2, which includes the RAI1 gene [21–23]. In the absence of discretely identified genetic causes, some have speculated that patients with advance-related sleep complaints may have higher sensitivity and/or exposure to morning light and accompanying advancing effects, or decreased sensitivity and/or exposure to evening light and accompanying delaying effects [24].

Risk Factors

Age

Children born preterm have been found to have an advanced sleep phase during subsequent adolescence [25]. It has been suggested that this is due to increased neonatal stress including hypoxia, nonideal nutrition, and chronic exposure to light (e.g., in an intensive care unit), which may result in compromised development of the suprachiasmatic nucleus [25, 26]. Conversely, older individuals tend to shift toward a morning preference and may be less sensitive to the circadian effects of light compared to younger adults [27–29], although not all studies agree [30]. Increasing age is accompanied by an advance in circadian phase, including peak melatonin concentration and wake time [9, 19, 31, 32]. There has been increasing interest in the potential effect of ethnicity on circadian phase, with some investigators suggesting persons of African American descent are predisposed to an increased sensitivity to the phase-advancing effects of light, and may possess a shorter innate circadian period, leading to a higher risk of having an advanced sleep phase [2, 19, 33].

Diagnosis

A diagnosis of ASWPD or advance-related sleep complaints requires a thorough clinical history, ideally with collateral information [3]. Lack and colleagues described 25 patients with advance-related sleep complaints, categorized with the use of sleep diaries, a sleep questionnaire, and a Beck Depression Inventory (the latter to rule out a depressive disorder). This phenotype was subsequently validated with core body temperature measurements and actigraphy [34]. Further useful assessment tools include validated chronotype questionnaires, such as the Morningness-Eveningness Questionnaire (MEQ) or the Munich Chronotype Questionnaire (MCTQ). The MEQ has been validated against core body temperature minimum (CBT_{min}) [35]. Both the MEQ, midpoint of sleep on work-free days, and sleep corrected score (MSF_{sc}) of the MCTQ have been shown to be correlated with the dim light melatonin onset (DLMO) [36, 37]. Palmer and colleagues recruited 47 patients prescreened for advance-related sleep complaints, 91% of whom confirmed morningness traits on the MEQ, with correlation with urinary 6-sulfatoxymelatonin (aMT6s) levels [38]. In Jones' study of familial ASWPD, MEQ scores of affected probands were dramatically higher (average score 77, where MEQ scores >59 suggest moderate morning type and scores >69 indicate definite morning type) than unaffected relatives (average score = 48.2, where scores between 42 and 58 indicate intermediate type) [6].

Actigraphy is an additional clinical tool to longitudinally assess the stability of sleep-wake complaints. Data should include at least 7 days (including both "free" and work/school days) and preferably 14 days for adequate interpretation [3, 5, 24]. Other confounding conditions need to be excluded, most notably mood disorders and inadequate sleep hygiene. While a major depressive disorder may present with early-morning awakenings, other associated symptoms such as low mood and anhedonia are not associated with an advanced sleep phase [24]. Physiologic phase markers such as salivary DLMO may also be useful as a marker of circadian rhythm, if feasible to obtain [5]. While normative data are not available, several studies have suggested its use to diagnose CRSWDs (reviewed by Keijzer et al. [39]). These patients may be more vulnerable to abusing substances such as alcohol or other hypnotics in an attempt to stay asleep longer at night and/or may use stimulants in the early evenings to reduce sleepiness [2]. Other differential diagnoses to consider include free running or non-24-hour sleep/wake rhythm disorder. A careful history can usually clarify. Most patients afflicted with this condition are blind, and sleep-related complaints vary in time and nature, depending upon the alignment of their circadian rhythm with the light/dark cycle [3].

Treatment Options

Using the GRADE approach (Grading of Recommendations Assessment, Development and Evaluation), the American Academy of Sleep Medicine (AASM) published updated practice parameters for the treatment of circadian rhythm sleep-wake disorders in 2015 [5]. Recommendations can be divided into four main categories for practitioners' consideration: behavioral options, hypnotics and stimulant medications, strategically timed melatonin, and strategic use of light therapy.

Behavioral Options

Sleep scheduling There is insufficient evidence at this time to recommend sleep scheduling as a primary means of treating advance-related sleep complaints. A singular case report by Moldofsky et al. from the 1980s (62-year-old male patient) described a protocol of phase advancing the sleep initiation time by 3 hours every 2 days over a period of 2 weeks [40]. Referred to as chronotherapy (i.e., changing the sleep and wake time gradually in a manner that favors the individual's circadian preference, with subsequent strict adherence to the achieved/desired sleep/wake schedule), the intervention successfully changed the sleep onset time from 1830 to

2300 hours and sleep offset time from 0230 to 0600 hours. Previous daytime and evening somnolence resolved with chronotherapy as well, with maintained benefits over 5 months. Initial polysomnography 1 week following chronotherapy completion revealed a longer total sleep time, but more wakefulness after sleep onset and less deep sleep and rapid-eye-movement (REM) sleep compared to baseline polysomnography. These findings were thought to be related to adaptation effects to the initial chronotherapy. At 5 months, however, the prolonged wakefulness had resolved and the patient endorsed increased alertness and energy during the midday hours [40]. While not tested as an intervention directly for patients with advance-related sleep complaints, avoiding evening naps is also a routine intervention that can be suggested. Evening naps have been shown to advance the sleep phase. Buxton and colleagues performed a 1-week study on the effect of daytime and evening naps on circadian phase in 25 normal male subjects ages 20–30 years old. Their results showed that daytime naps can phase shift circadian rhythms in normal subjects, with evening naps (1900–0100 hours) showing the largest phase advancement of circadian rhythms (44 min ± 17 min) as measured by DLMO and nocturnal thyroid-stimulating hormone (TSH) secretion [41]. Another study by Yoon and colleagues showed that evening naps resulted in earlier sleep offset times and advances in sleep phase as measured by urinary 6-sulfatoxymelatonin [42]. Consequently, although it has not been studied as an ameliorative measure, the avoidance of evening naps is a reasonable recommendation to consider in patients with advance-related sleep complaints.

Hypnotic and Stimulant Medications

Neither hypnotic nor stimulant medications have been studied for the treatment of advance-related sleep complaints. While hypnotics may be reasonable to consider for sleep maintenance difficulties or early-morning awakenings, the known side effects and risks of these medications warrant careful consideration. Drugs such as benzodiazepines, as well as the "z" drugs including eszopiclone and zolpidem, can increase risk of falls and daytime somnolence, especially among the elderly [43–46]. Cognitive side effects, tolerance, and dependence are also concerns, but the incidence of these complications is not entirely clear, as limited numbers of high-quality studies and frequent variability in methodology and design limit conclusions that can be drawn [47–49]. These risks, however, should be balanced against the risks of sleep disturbances themselves being associated with higher fall risk, as suggested by Avidan and colleagues [50]. In their study of approximately 34,000 nursing home residents with up to a 210-day follow-up, insomnia but not hypnotic use was associated with a higher risk of falls, suggesting that hypnotic medication use in the community may have been a proxy for underlying sleep disturbances [3, 50]. Data on use of stimulants for early evening sleepiness for these patients are also lacking, but could be considered if such symptoms warrant clinical attention.

Melatonin

The phase response curve for melatonin in humans describes phase delays with melatonin administration in the biological morning, but its use has not been studied within relevant patient populations [51]. Moreover, because of its potential soporific effects, caution is warranted for clinical practice [5, 11], with use of the lowest doses and gradual uptitration as needed. Melatonin agonists such as ramelteon or agomelatine could also have therapeutic roles, but have not been studied for these purposes.

Light Therapy

Light administered prior to the CBT_{min} (occurs 3–4 hours prior to habitual sleep offset) will delay the circadian rhythm, while light therapy administered subsequent to this inflection point will advance the circadian rhythm in individuals normally entrained to the light/dark cycle [3, 52, 53]. As a result, light therapy for advance-related sleep complaints is provided during evening hours, as close to the period of sleep onset as possible, to maximize phase delays [11]. There are no evidence-based protocols to inform ongoing maintenance treatment, but maintaining or resuming an effective intervention is a reasonable clinical practice [5].

Campbell and colleagues demonstrated efficacy of light therapy for delaying the sleep phase among ASWPD patients [54]. Their protocol on 16 older patients (ages 62–81) described 2 weeks of light therapy (4000 lux, administered for 2 hours between 2000 and 2300 hours) and demonstrated a 3.13 hours phase delay (as measured by CBT_{min}) in comparison to an 8-minute delay among controls, with accompanying improvements in sleep quality as measured by polysomnography. However, when this group repeated a similar study in 15 older persons subjected to a nearly identical protocol in their homes, there was no accompanying improvement in sleep quality. While subjects demonstrated an initial average phase delay of 94 minutes (CBT_{min}) compared to controls, body temperature rhythms gradually advanced to their normal (preintervention) rhythms despite twice weekly evening (2100–2300 hours) light therapy over a 3-month follow-up period [55]. The authors speculate that possibly the proximity of the timing of the light exposure to CBT_{min} was not optimal and that inferior compliance to light therapy in this study compared to their previous protocol may have contributed to the negative results. Another pilot study by Lack and colleagues on nine patients (mean age 53.4 years) with complaints of early-morning insomnia demonstrated that two evenings of light therapy (2500 lux) from 2000 to 2400 hours resulted in a subsequent delay of 2–4 hours in CBT_{min} and 1–2 hours in urinary melatonin phase markers. While sleep onset was not significantly delayed as measured by actigraphy, total sleep time was increased by over an hour in these subjects due to a 1 hour and 12-minute delay in mean wake-up time [56]. This group did a separate controlled study on 24 subjects (average age 61.2 years) with early-morning awakenings and terminal insomnia, which were presumed to be advance-related sleep complaints. Using an identical protocol of evening light therapy, data yielded average phase delays of 2 hours in these subjects based on rectal temperature and urinary melatonin measurements [34]. Evidence of phase delay persisted during a 4-week follow-up period. Another study by Palmer and colleagues showed a lack of efficacy with light therapy of 265 lux administered for 2–3 hours in the evening (1900–2200 hours) for 4 weeks in 47 older adults (age 60–86) [38]. Other negative studies on the efficacy of light therapy have been reported [38, 57], but there is significant variability in the definitions of an advanced sleep phase, as well as brightness, timing, frequency, and length of exposure to light therapy as well as distance to the light source. These issues make it difficult to provide specific recommendations. Figueiro et al. have explored delivering light therapy in pulses during the night with a light mask (through closed eyelids while patients are asleep) to provide light stimuli during the steepest portion of the phase response curve of light, but with no significant delay in sleep onset occurring in patients with an advanced sleep phase [58].

Reported side effects of light therapy have generally been modest and in a placebo-controlled trial included eye strain and headaches [59]. Other reported side effects include nausea, fatigue, and irritability [60, 61]. Caution should be considered for use in patients who are on photosensitizing medications such as tricyclic antidepressants, antibiotics such as fluoroquinolones or sulfonamide drugs, or acne treatment with isotretinoin [62]. Other conditions in which caution is warranted with use of light therapy include the presence of skin conditions with photosensitivity including lupus, porphyrias, or solar urticaria, as well as migraines, diabetic retinopathy, macular degeneration, or a history of bipolar disorder [62–65]. Light therapy can induce migraines in approximately one-third of those susceptible [66]. A 2017 systematic review by Brower and colleagues found light therapy to be safe for the eyes in the absence of underlying ocular problems [62]. Patients with relevant conditions should have appropriate monitoring of their respective ophthalmologic, dermatologic, and/or psychiatric condition [5]. Finally, although potentially intuitively helpful, strategic avoidance of light (i.e., during a period time of morning when light would be expected to affect phase advances) has not been studied as a method of achieving phase delays among these patients [5].

Morningness and Resilience

Factors other than occupational "rewards" conferred to "early risers" may relate to the infrequency of advance-related sleep complaints in the clinical setting. Lewy first proposed a "phase shift hypothesis" in 1988, suggesting that the therapeutic effects of light therapy for mood disorders, particularly seasonal affective disorder, may be due to the phase-advancing effects of light "correcting" the phase delay among afflicted patients [67]. This hypothesis was later revised to the "phase angle difference hypothesis," whereby the internal phase delay compared to the midpoint of sleep is the determining factor for the therapeutic response [68]. Several data have supported this hypothesis [69, 70], but other data are conflicting [71, 72]. Burnout is a syndrome of emotional exhaustion and cynicism that occurs in people who have an occupation that involves working with other people [73]. Given that eveningness traits have been associated with a higher risk of mood disorders and burnout [74–78], some have speculated that morningness may be associated with counteracting protective factors. One study by Muller and colleagues on 93 nonseasonal depressed inpatients found that morning types were underrepresented in this sample compared to healthy samples [79]. Indeed, morningness has been associated with findings of higher resilience and optimism [80–82]. These findings have been postulated to be related to these persons having relatively more exposure to sunlight, less social jet lag, and as a result a greater likelihood of meeting sleep duration needs [15, 16]. Social jet lag refers to a chronic misalignment between the preferred sleep-wake schedule and the sleep/wake timing imposed by a person's social or occupational schedule [83, 84]. Social jet lag is seen more frequently in subjects with later chronotypes [85, 86]. In addition to an insufficient sleep quantity that is common for these persons because of this circadian and social/occupational schedule misalignment, sleep quality also suffers as it does not occur within the temporal window afforded by the circadian sleep cycle. Related to the latter, morning persons have been shown to have a faster dissipation rate of homeostatic sleep pressure compared to intermediate and evening type persons, leading to a shorter sleep satiation and subsequent lower sleep duration need [87]. Not all studies are consistent, however, as Lemoine and colleagues found in a large sample of psychiatric inpatients that patients with a depressive or psychotic disorder were more likely to be morning types [88]. Furthermore, Lavebratt et al. have found an association with genetic variations of the h*Per2* gene and depression [89], suggesting further research is needed.

Summary

Patients with advance-related sleep complaints may be difficult to recognize in the clinical setting. While they may present with evening sleepiness, sleep maintenance difficulties, and/or early-morning awakenings, many may not view these as treatable problems and choose instead to adjust their lifestyles. Preterm birth may increase the risk of having an advanced sleep phase. Additionally, older age as well as African American heritage may be associated with a higher phenotypic frequency. Some data even suggest patients with these traits may have more resilience and optimism, which may be protective factors against depression and burnout. A thorough clinical history and evaluation is warranted to properly identify patients with advance-related sleep complaints. Additional clinical tools such as sleep logs or questionnaires such as the MEQ, MCTQ, or actigraphy can be helpful. Behavioral recommendations including maintaining proper sleep hygiene, as well as the avoidance of evening naps and early-morning light, are simple recommendations (albeit not evidence-based) that are easily implemented. There is insufficient data to make a recommendation regarding the use of post-awakening melatonin for these patients [90], but melatonin may be reasonable to consider with appropriate precautions about potential soporific side effects. While evening light therapy may offer benefit for some patients, much work needs to be done regarding determination of the optimal timing, intensity, duration, and wavelength of such treatment. Other treatment options for symptomatic relief include hypnotic and stimulant medications, but these have not been studied for this patient population. Their use can be considered on a case-by-case basis, weighing anticipated benefits against predicted risks.

References

1. Martinez D, Lenz MC. Circadian rhythm sleep disorders. Indian J Med Res. 2010;131:141–9.
2. American Psychiatric Association. Diagnostic and statistical manual of mental disorders, fifth edition (DSM-5). Arlington: American Psychiatric Publishing; 2013.
3. Auger RR. Advance-related sleep complaints and advanced sleep phase disorder. Sleep Med Clin. 2009;4(2):219–27.
4. Sack RL, et al. Circadian rhythm sleep disorders: part II, advanced sleep phase disorder, delayed sleep phase disorder, free-running disorder, and irregular sleep-wake rhythm. An American Academy of Sleep Medicine review. Sleep. 2007;30(11):1484–501.
5. Auger RR, et al. Clinical practice guideline for the treatment of intrinsic circadian rhythm sleep-wake disorders: advanced sleep-wake phase disorder (ASWPD), delayed sleep-wake phase disorder (DSWPD), non-24-hour sleep-wake rhythm disorder (N24SWD), and irregular sleep-wake rhythm disorder (ISWRD). An update for 2015: an American Academy of Sleep Medicine clinical practice guideline. J Clin Sleep Med. 2015;11(10):1199–236.
6. Jones CR, et al. Familial advanced sleep-phase syndrome: a short-period circadian rhythm variant in humans. Nat Med. 1999;5(9):1062–5.
7. Satoh K, et al. Two pedigrees of familial advanced sleep phase syndrome in Japan. Sleep. 2003;26(4):416–7.
8. Curtis BJ, Ashbrook LH, Young T, Finn LA, Fu YH, Ptáček LJ, et al. Extreme morning chronotypes are often familial and not exceedingly rare: the estimated prevalence of advanced sleep phase, familial advanced sleep phase, and advanced sleep-wake phase disorder in a sleep clinic population. Sleep. 2019;42(10):zsz148.
9. Roenneberg T, et al. A marker for the end of adolescence. Curr Biol. 2004;14(24):R1038–9.
10. Roenneberg T, Kumar CJ, Merrow M. The human circadian clock entrains to sun time. Curr Biol. 2007;17(2):R44–5.
11. Bjorvatn B, Pallesen S. A practical approach to circadian rhythm sleep disorders. Sleep Med Rev. 2009;13(1):47–60.
12. Schrader H, Bovim G, Sand T. The prevalence of delayed and advanced sleep phase syndromes. J Sleep Res. 1993;2(1):51–5.
13. Hirano A, Shi G, Jones CR, Lipzen A, Pennacchio LA, Xu Y, et al. A Cryptochrome 2 mutation yields advanced sleep phase in humans. elife. 2016;5:e16695.
14. Paine SJ, et al. Identifying advanced and delayed sleep phase disorders in the general population: a national survey of New Zealand adults. Chronobiol Int. 2014;31(5):627–36.
15. Ando K, Kripke DF, Ancoli-Israel S. Delayed and advanced sleep phase symptoms. Isr J Psychiatry Relat Sci. 2002;39(1):11–8.
16. Tafti M, Dauvilliers Y, Overeem S. Narcolepsy and familial advanced sleep-phase syndrome: molecular genetics of sleep disorders. Curr Opin Genet Dev. 2007;17(3):222–7.

17. Zhang L, et al. A PERIOD3 variant causes a circadian phenotype and is associated with a seasonal mood trait. Proc Natl Acad Sci U S A. 2016;113(11):E1536–44.
18. Xu Y, et al. Modeling of a human circadian mutation yields insights into clock regulation by PER2. Cell. 2007;128(1):59–70.
19. von Schantz M. Natural variation in human clocks. Adv Genet. 2017;99:73–96.
20. He Y, et al. The transcriptional repressor DEC2 regulates sleep length in mammals. Science. 2009;325(5942):866–70.
21. Kocher L, et al. Phase advance of circadian rhythms in Smith-Magenis syndrome: a case study in an adult man. Neurosci Lett. 2015;585:144–8.
22. De Leersnyder H, et al. Inversion of the circadian rhythm of melatonin in the Smith-Magenis syndrome. J Pediatr. 2001;139(1):111–6.
23. De Leersnyder H, et al. Inversion of the circadian melatonin rhythm in Smith-Magenis syndrome. Rev Neurol (Paris). 2003;159(11 Suppl):6S21–6.
24. Abbott SM, Reid KJ, Zee PC. Circadian rhythm sleep-wake disorders. Psychiatr Clin North Am. 2015;38(4):805–23.
25. Hibbs AM, et al. Advanced sleep phase in adolescents born preterm. Behav Sleep Med. 2014;12(5):412–24.
26. Kennaway DJ. Programming of the fetal suprachiasmatic nucleus and subsequent adult rhythmicity. Trends Endocrinol Metab. 2002;13(9):398–402.
27. Dijk DJ, Archer SN. PERIOD3, circadian phenotypes, and sleep homeostasis. Sleep Med Rev. 2010;14(3):151–60.
28. Klerman EB, et al. Absence of an increase in the duration of the circadian melatonin secretory episode in totally blind human subjects. J Clin Endocrinol Metab. 2001;86(7):3166–70.
29. Carrier J, et al. Sleep and morningness-eveningness in the 'middle' years of life (20–59 y). J Sleep Res. 1997;6(4):230–7.
30. Kim SJ, et al. Phase-shifting response to light in older adults. J Physiol. 2014;592(1):189–202.
31. Duffy JF, Czeisler CA. Age-related change in the relationship between circadian period, circadian phase, and diurnal preference in humans. Neurosci Lett. 2002;318(3):117–20.
32. Duffy JF, et al. Peak of circadian melatonin rhythm occurs later within the sleep of older subjects. Am J Physiol Endocrinol Metab. 2002;282(2):E297–303.
33. Eastman CI, et al. Circadian rhythm phase shifts and endogenous free-running circadian period differ between African-Americans and European-Americans. Sci Rep. 2015;5:8381.
34. Lack L, et al. The treatment of early-morning awakening insomnia with 2 evenings of bright light. Sleep. 2005;28(5):616–23.
35. Duffy JF, Rimmer DW, Czeisler CA. Association of intrinsic circadian period with morningness-eveningness, usual wake time, and circadian phase. Behav Neurosci. 2001;115(4):895–9.
36. Kitamura S, et al. Validity of the Japanese version of the Munich ChronoType questionnaire. Chronobiol Int. 2014;31(7):845–50.
37. Kantermann T, Sung H, Burgess HJ. Comparing the Morningness-Eveningness questionnaire and Munich ChronoType questionnaire to the dim light melatonin onset. J Biol Rhythm. 2015;30(5):449–53.
38. Palmer CR, et al. Efficacy of enhanced evening light for advanced sleep phase syndrome. Behav Sleep Med. 2003;1(4):213–26.
39. Keijzer H, et al. Why the dim light melatonin onset (DLMO) should be measured before treatment of patients with circadian rhythm sleep disorders. Sleep Med Rev. 2014;18(4):333–9.
40. Moldofsky H, Musisi S, Phillipson EA. Treatment of a case of advanced sleep phase syndrome by phase advance chronotherapy. Sleep. 1986;9(1):61–5.
41. Buxton OM, et al. Daytime naps in darkness phase shift the human circadian rhythms of melatonin and thyrotropin secretion. Am J Physiol Regul Integr Comp Physiol. 2000;278(2):R373–82.
42. Yoon IY, et al. Age-related changes of circadian rhythms and sleep-wake cycles. J Am Geriatr Soc. 2003;51(8):1085–91.
43. Allain H, et al. Postural instability and consequent falls and hip fractures associated with use of hypnotics in the elderly: a comparative review. Drugs Aging. 2005;22(9):749–65.

44. Gunja N. In the Zzz zone: the effects of Z-drugs on human performance and driving. J Med Toxicol. 2013;9(2):163–71.
45. Leufkens TR, Vermeeren A. Highway driving in the elderly the morning after bedtime use of hypnotics: a comparison between temazepam 20 mg, zopiclone 7.5 mg, and placebo. J Clin Psychopharmacol. 2009;29(5):432–8.
46. Verster JC, et al. Zopiclone as positive control in studies examining the residual effects of hypnotic drugs on driving ability. Curr Drug Saf. 2011;6(4):209–18.
47. Riemann D, et al. European guideline for the diagnosis and treatment of insomnia. J Sleep Res. 2017;26(6):675–700.
48. Sateia MJ, et al. Clinical practice guideline for the pharmacologic treatment of chronic insomnia in adults: an American Academy of Sleep Medicine clinical practice guideline. J Clin Sleep Med. 2017;13(2):307–49.
49. Schonmann Y, et al. Chronic hypnotic use at 10 years-does the brand matter? Eur J Clin Pharmacol. 2018;74(12):1623–31.
50. Avidan AY, et al. Insomnia and hypnotic use, recorded in the minimum data set, as predictors of falls and hip fractures in Michigan nursing homes. J Am Geriatr Soc. 2005;53(6):955–62.
51. Lewy AJ. Clinical applications of melatonin in circadian disorders. Dialogues Clin Neurosci. 2003;5(4):399–413.
52. Czeisler CA, et al. Bright light resets the human circadian pacemaker independent of the timing of the sleep-wake cycle. Science. 1986;233(4764):667–71.
53. Czeisler CA, et al. Bright light induction of strong (type 0) resetting of the human circadian pacemaker. Science. 1989;244(4910):1328–33.
54. Campbell SS, Dawson D, Anderson MW. Alleviation of sleep maintenance insomnia with timed exposure to bright light. J Am Geriatr Soc. 1993;41(8):829–36.
55. Suhner AG, Murphy PJ, Campbell SS. Failure of timed bright light exposure to alleviate age-related sleep maintenance insomnia. J Am Geriatr Soc. 2002;50(4):617–23.
56. Lack L, Wright H. The effect of evening bright light in delaying the circadian rhythms and lengthening the sleep of early morning awakening insomniacs. Sleep. 1993;16(5):436–43.
57. Pallesen S, et al. Bright light treatment has limited effect in subjects over 55 years with mild early morning awakening. Percept Mot Skills. 2005;101(3):759–70.
58. Figueiro MG, et al. Impact of an individually tailored light mask on sleep parameters in older adults with advanced phase sleep disorder. Behav Sleep Med. 2020;18:226–40.
59. Botanov Y, Ilardi SS. The acute side effects of bright light therapy: a placebo-controlled investigation. PLoS One. 2013;8(9):e75893.
60. Dauphinais DR, et al. Controlled trial of safety and efficacy of bright light therapy vs. negative air ions in patients with bipolar depression. Psychiatry Res. 2012;196(1):57–61.
61. Kogan AO, Guilford PM. Side effects of short-term 10,000-lux light therapy. Am J Psychiatry. 1998;155(2):293–4.
62. Brouwer A, et al. Light therapy: is it safe for the eyes? Acta Psychiatr Scand. 2017;136(6):534–48.
63. Vanagaite J, et al. Light-induced discomfort and pain in migraine. Cephalalgia. 1997;17(7):733–41.
64. Tekatas A, Mungen B. Migraine headache triggered specifically by sunlight: report of 16 cases. Eur Neurol. 2013;70(5–6):263–6.
65. Sit D, et al. Light therapy for bipolar disorder: a case series in women. Bipolar Disord. 2007;9(8):918–27.
66. Ulrich V, et al. Possible risk factors and precipitants for migraine with aura in discordant twin-pairs: a population-based study. Cephalalgia. 2000;20(9):821–5.
67. Lewy AJ, et al. Winter depression and the phase-shift hypothesis for bright light's therapeutic effects: history, theory, and experimental evidence. J Biol Rhythm. 1988;3(2):121–34.
68. Lewy AJ, et al. The circadian basis of winter depression. Proc Natl Acad Sci U S A. 2006;103(19):7414–9.
69. Terman JS, et al. Circadian time of morning light administration and therapeutic response in winter depression. Arch Gen Psychiatry. 2001;58(1):69–75.

70. Terman M, Terman JS. Light therapy for seasonal and nonseasonal depression: efficacy, protocol, safety, and side effects. CNS Spectr. 2005;10(8):647–63; quiz 672.
71. LeGates TA, Fernandez DC, Hattar S. Light as a central modulator of circadian rhythms, sleep and affect. Nat Rev Neurosci. 2014;15(7):443–54.
72. Knapen SE, Gordijn MC, Meesters Y. The relation between chronotype and treatment outcome with light therapy on a fixed time schedule. J Affect Disord. 2016;202:87–90.
73. Maslach C, Jackson SE. The measurement of experienced burnout. J Organ Behav. 1981;2:99–113.
74. Merikanto I, et al. Circadian preference links to depression in general adult population. J Affect Disord. 2015;188:143–8.
75. Melo MC, et al. Sleep and circadian alterations in people at risk for bipolar disorder: a systematic review. J Psychiatr Res. 2016;83:211–9.
76. Merikanto I, et al. Eveningness relates to burnout and seasonal sleep and mood problems among young adults. Nord J Psychiatry. 2016;70(1):72–80.
77. Melo MCA, et al. Chronotype and circadian rhythm in bipolar disorder: a systematic review. Sleep Med Rev. 2017;34:46–58.
78. Togo F, Yoshizaki T, Komatsu T. Association between depressive symptoms and morningness-eveningness, sleep duration and rotating shift work in Japanese nurses. Chronobiol Int. 2017;34(3):349–59.
79. Muller MJ, et al. Chronotypes in patients with nonseasonal depressive disorder: distribution, stability and association with clinical variables. Chronobiol Int. 2015;32(10):1343–51.
80. Tafoya SA, et al. Resilience, sleep quality and morningness as mediators of vulnerability to depression in medical students with sleep pattern alterations. Chronobiol Int. 2018;36:1–11.
81. Lee SJ, et al. Association between morningness and resilience in Korean college students. Chronobiol Int. 2016;33(10):1391–9.
82. Antunez JM, Navarro JF, Adan A. Circadian typology is related to resilience and optimism in healthy adults. Chronobiol Int. 2015;32(4):524–30.
83. Wittmann M, et al. Social jetlag: misalignment of biological and social time. Chronobiol Int. 2006;23(1–2):497–509.
84. McMahon DM, et al. Persistence of social jetlag and sleep disruption in healthy young adults. Chronobiol Int. 2018;35(3):312–28.
85. Roenneberg T, et al. Social jetlag and obesity. Curr Biol. 2012;22(10):939–43.
86. Levandovski R, et al. Depression scores associate with chronotype and social jetlag in a rural population. Chronobiol Int. 2011;28(9):771–8.
87. Mongrain V, Carrier J, Dumont M. Circadian and homeostatic sleep regulation in morningness-eveningness. J Sleep Res. 2006;15(2):162–6.
88. Lemoine P, Zawieja P, Ohayon MM. Associations between morningness/eveningness and psychopathology: an epidemiological survey in three in-patient psychiatric clinics. J Psychiatr Res. 2013;47(8):1095–8.
89. Lavebratt C, et al. PER2 variantion is associated with depression vulnerability. Am J Med Genet B Neuropsychiatr Genet. 2010;153B(2):570–81.
90. Burgess HJ, Emens JS. Drugs used in circadian sleep-wake rhythm disturbances. Sleep Med Clin. 2018;13(2):231–41.
91. American Academy of Sleep Medicine. International classification of sleep disorders. 3rd ed. Darien: American Academy of Sleep Medicine; 2014.

Chapter 9
Non-24-Hour Sleep-Wake Rhythm Disorder

Jonathan Emens

Introduction

Basic Concepts

Non-24-hour sleep/wake rhythm disorder (non-24) is characterized by a relapsing and remitting pattern of insomnia and/or daytime somnolence as well as sleep and wake times that can drift progressively earlier or later each day [1]. The International Classification of Sleep Disorders, Third Edition (ICSD-3), requires that the latter criteria be documented with at least 14 days of sleep diary or actigraphy data. These criteria and the unique pathophysiology of non-24 (reviewed below) set it apart from other circadian rhythm sleep disorders (CRSDs) and yet non-24 epitomizes the disruptions to sleep and wakefulness that occur in CRSDs. Before discussing non-24 in detail, some of the basic circadian physiology that was reviewed in previous chapters bears repeating. In particular, the concept of *entrainment*, or synchronization of the hypothalamic circadian pacemaker (~24-hour biological clock), is central to this disorder.

The circadian pacemaker, located in the suprachiasmatic nuclei (SCN) of the hypothalamus, acts to internally synchronize the disparate molecular clockwork found throughout the human body. The pacemaker is itself synchronized, or *entrained*, to the external 24-hour day by time cues, or zeitgebers ("time givers"), that act to reset the clock. Primary among these zeitgebers is light which resets the clock via intrinsically photosensitive retinal ganglion cells (as well as rods and cones) that have a monosynaptic projection to the SCN (the retinohypothalamic tract) [2]. Light in the biological morning resets the timing of the pacemaker

J. Emens (✉)
Department of Psychiatry, Oregon Health and Science University, Division of Mental Health and Clinical Neurosciences, VA Portland Health Care System, Portland, OR, USA
e-mail: emensj@ohsu.edu

R. R. Auger (ed.), *Circadian Rhythm Sleep-Wake Disorders*,
https://doi.org/10.1007/978-3-030-43803-6_9

(referred to as *circadian phase*) to an earlier time (termed a *phase advance*), while light exposure in the biological evening and early night resets the pacemaker to a later time (termed a *phase delay*) [3–5]. The pacemaker requires regular resetting because it does not keep perfect time: it may either "run fast" with a periodicity of less than 24 hours or "run slow" with a periodicity of greater than 24 hours. On average, the human circadian pacemaker has a periodicity (*circadian period*) of about 24 hours and 9 minutes with a standard deviation of 12 minutes; circadian period is shorter on average in women and women are more likely to have a period length that is less than 24 hours (a finding that has important treatment implications in non-24 as discussed below) [6, 7].

Circadian Period and Non-entrainment

Another way of conceptualizing circadian period is that the pacemaker takes 24 hours and 9 minutes, on average, to complete one full cycle (biological day). In the absence of any resetting a pacemaker with an average period would complete a full circadian cycle every 24 hours and 9 minutes and would be "set" about 9 minutes later (phase delay) each day relative to the outside 24-hour world. Figure 9.1 is a schematic showing the timing of a pacemaker that is no longer being reset in just such a manner. On day 1 the pacemaker is synchronized to the outside world (both clocks read 9:00 pm). However, the pacemaker, in the absence of any resetting, is set 9 minutes later each day. On day 2 the timing of the pacemaker shifts later and "reads" 9:09 pm at 9:00 pm. Such shifting to a later time persists until, eventually, after 61 days and a total of 9 hours and 9 minutes of phase delay, the pacemaker is set to 6:09 am at 9:00 pm. A pacemaker with a circadian period that was less than 24 hours would, in the absence of any resetting, shift to an earlier time (phase advance).

The circadian pacemaker illustrated in Fig. 9.1 is no longer synchronized, or *entrained*, to the 24-hour day. As a result, the timing of the pacemaker and the timing of the multitude of biologically important rhythms under its control no longer occur at the appropriate time. It is this loss of entrainment that forms the pathophysiological basis of non-24: sleep/wake propensity is under significant circadian control and as the timing of the pacemaker drifts progressively earlier or later each day the circadian drives for sleep and wakefulness similarly drift earlier or later. The result is either a relapsing and remitting pattern of insomnia and daytime sleepiness as an individual attempts to maintain consistent sleep and wake times in "opposition" to their shifting biological clock *or* a pattern of sleep and wakefulness that similarly drifts progressively later or earlier each day, tracking the timing of the biological clock. In most instances of non-24, the timing of the pacemaker continues to drift relatively unabated and the clock moves in and out of alignment with the external world with variable frequency from patient to patient. In an individual with a circadian period of 24 hours and 30 minutes, for example, circadian phase would drift about 30 minutes later each day, taking 48 days to shift a full 24 hours (24 hours divided by 0.5 hours per day). While a disparity between external and internal time

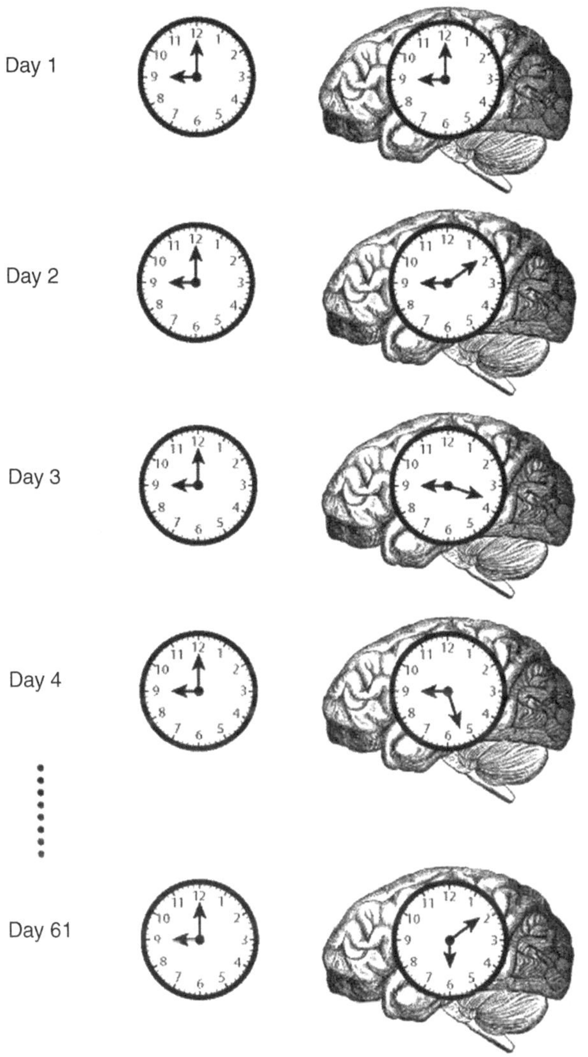

Fig. 9.1 Schematic of non-entrainment. The clocks on the left-hand side represent the local clock hour, while the clocks superimposed on the image of the brain represent the timing of the circadian pacemaker (circadian phase). The figure supposes an individual whose biological clock is no longer reset by time cues in the environment and who has an average period of 24 hours and 9 minutes. On day one, the two clocks happen to be synchronized (both read 9:00 pm) but with each day the timing of the circadian pacemaker shifts later (phase delays) by 9 minutes. By day 61 the circadian pacemaker has shifted a total of 9 hours and 9 minutes later and is "set" to 6:09 am

on the order of minutes (e.g., days 1–3 in Fig. 9.1) may be trivial, much larger disparities are more problematic: it is clear that it might prove difficult to fall asleep and stay asleep at 9:00 pm when the pacemaker is set to 6:09 am and the circadian drive for wakefulness is increasing (day 61, Fig. 9.1).

Circadian Phase

It is not currently possible to directly measure the timing of the hypothalamic circadian pacemaker in patients with non-24 in order to determine circadian phase. Instead, it is only possible to measure the "hands of the clock" (i.e., a downstream marker of the central pacemaker's output). Measurement of a marker of circadian timing can prove difficult since the endogenous rhythms of many variables are masked by evoked changes (e.g., the endogenous rhythm in body temperature is masked by changes in posture, activity, sleep, and emotional state). Currently, the most common measure of circadian phase is the onset of melatonin secretion under dim light conditions in order to avoid the suppressant effect of light (the dim light melatonin onset or DLMO) [8–10]. Under everyday conditions of electrical light, the DLMO typically occurs 2–3 hours before habitual lights out/bedtime [11–14]. Melatonin levels remain elevated during the night, decline in the morning, and remain low throughout the light/wake period. In this sense the onset of melatonin can be thought of as marking the beginning of the "biological night," the period where melatonin levels are elevated denotes the "biological night" itself, and the period of time melatonin levels are low represents the "biological day." Throughout this chapter the timing of events will be described either in terms of clock hour, a specific time relative to a marker of circadian phase such as the DLMO or more generally relative to circadian timing (e.g., as occurring during the biological day or night as described above).

Etiology

Blind Individuals

As noted in the ICSD-3, the signs and symptoms of non-24 are directly attributable to the lack of entrainment already described. Because light is the primary synchronizer of the pacemaker, a loss of photic input to the biological clock is generally considered the main etiology of this disorder [15]. As a result, non-24 commonly presents in totally blind individuals (e.g., those lacking any conscious light perception). There is no particular cause of blindness that has been reliably shown to be more associated with non-24, but any pathology resulting in sufficient, bilateral damage to the photoreceptive retinal ganglion cells or interruption of the retinohypothalamic tract would be expected to result in the disorder [16].

It should be noted here that some totally blind individuals maintain circadian photoreception, and therefore entrainment, even in the absence of conscious vision, negative electroretinographic responses, and negative visual evoked potentials [17, 18]. Therefore, one potential iatrogenic cause of non-24 would be bilateral enucleation in totally blind individuals for reasons such as intractable ocular pain or infection. The removal of what are presumed to be nonfunctioning organs could very well precipitate the onset of non-24.

It is also possible that a combination of decreased circadian photoreception and decreased *exposure* to light might contribute to the development of non-24. For example, we have documented limited circadian photoreception in a blind individual who nonetheless suffered from non-24 and this may have been at least partially due to his chronically low level of light exposure.

Finally, while some totally blind individuals maintain entrainment via preservation of circadian photoreception, it is also possible that others might do so as a result of the resetting effects of non-photic zeitgebers: we [15], and others [16], have found individuals who were able to maintain entrainment despite lacking eyes. Therefore a lack of exposure or response to known non-photic time cues (e.g., physical activity) might also play some role in the development of non-24 among the blind.

Sighted Individuals

More uncertain is the etiology of non-24 among sighted individuals. One possible cause is a loss of photic input to the pacemaker as occurs in the blind but with preservation of conscious light perception. This would require a mechanism that selectively damaged or rendered nonfunctional the intrinsically photosensitive retinal ganglion cells (IPRGCs) while preserving input to the visual cortex. It is difficult to imagine a pathological process that would eliminate the retinal ganglion cells while preserving conscious vision: firstly, the efferent pathway from the rods and cones to the visual cortex passes through the retinal ganglion cells, and, secondly, the input requirements for complex image formation are much greater than those for simple circadian light detection. However, an absence of the novel photopigment melanopsin, which forms the basis of the IPRGCs circadian photoreception, could provide the necessary pathological mechanism. Just this type of mutant animal model has been developed and in such animals conscious vision is preserved while circadian photoreception is eliminated [19]. However, such a selective loss of photic input to the biological clock that maintains input to the visual cortex has yet to be demonstrated in humans. Indeed, we have found that photic input to the circadian pacemaker, assessed via downstream melatonin suppression in response to bright light, was well preserved in a sighted individual with non-24 [20]. Even with preserved photic input to the circadian pacemaker, it still remains possible that sighted individuals with non-24 may differentially respond to light in a way that increases their risk of losing entrainment (e.g., a decreased response to the phase-advancing effects of morning light exposure and/or an increased response to the phase-delaying effects of evening light exposure). To date, no trials have been done comparing the circadian resetting response to light in sighted individuals with non-24 to that in matched controls.

Another possibility is that sighted individuals with non-24 may have a circadian period that is too different from 24 hours to entrain to the 24-hour day. This would fit with the very long sleep/wake periods that such individuals sometimes demonstrate (e.g., 28 hours or longer) [21, 22]. It was indeed shown that sighted individuals with non-24 had circadian periods longer than those found in entrained, healthy, unmatched controls in a small study [23]. However, this was only the case when the patients with non-24 were compared to individuals with neither a morning nor evening diurnal preference; when compared to *entrained* healthy controls who had an evening diurnal preference the individuals with non-24 did *not* have longer circadian periods. Furthermore, the circadian periods found in the patients with non-24, with perhaps one exception, were all within the range of periods found among healthy historical control subjects [6]. We have also found, in one sighted individual with non-24, that the intrinsic circadian period assessed in the laboratory setting both falls within the range of normal controls and is significantly shorter than the observed sleep/wake period [20]. These data strongly suggest that sighted individuals with non-24 do not have circadian periods that are too long or too short to entrain to the 24-hour day.

The finding that the intrinsic circadian periods measured in the laboratory do not match the much longer and more variable observed sleep/wake periods seen in ambulatory sleep/wake diaries and wrist actigraphy data hint at one final possible etiology of non-24 in the sighted. It is possible that the self-selected light/dark cycles of individuals with non-24 result in consistent, daily circadian phase shifts that shift the timing of the pacemaker progressively later, or earlier, each day.

Such a situation would be analogous to healthy normal individuals living in the laboratory in isolation from external time cues [24, 25]. Studies have shown that such individuals tend to initiate sleep much later in the biological night [25]. As a result, they "shield" with darkness that part of the biological morning where light would cause phase advances and newly expose to light that part of the biological night where light causes phase delays [3–5, 7]. The results are consistent phase delays and sleep/wake schedules with periods of up to 65 hours [24].

A more commonplace example of self-selected light/dark schedules shifting sleep/wake and circadian timing later occurs on weekends and work-free days. Sleep/wake timing can drift up to 7 hours later on weekends and work-free days [26] and circadian phase will similarly shift later [27]. It has been shown that evening, artificial dim light exposure is responsible for these weekend shifts in circadian and sleep/wake timing to a later time: in the absence of artificial light, such shifts are eliminated [27]. Thus, self-selected patterns of light/dark exposure of the type described above would be sufficient to explain non-24 among sighted individuals.

One final comment regarding etiology among the sighted concerns the use of "chronotherapy" for the treatment of delayed sleep/wake phase disorder (DSWPD) [28]. Chronotherapy involves intentionally shifting sleep/wake timing to a later time each day until the patient's sleep timing "wraps around the clock" and is occurring at a relatively earlier clock hour. There is concern that chronotherapy may be an iatrogenic cause of non-24 among sighted individuals: by intentionally shifting sleep/wake timing later, a pattern of light/dark exposure could be created that results in consistent phase delays by the mechanisms discussed above.

Epidemiology

Most case series are relatively small, but entrainment has been shown to be lost in approximately 55–70% of totally blind individuals (i.e., those lacking conscious light perception) [29, 30]. That said, no large-scale assessments of circadian entrainment status have been done among the totally blind. It is possible that the research to date has resulted in a selection bias of individuals who are symptomatic and may therefore have overestimated the percentage of totally blind patients with non-24.

Presentation and Diagnosis

As noted above, the ICSD-3 requires a pattern of insomnia and/or daytime somnolence that alternates with asymptomatic periods, symptoms of at least 3 months in duration, and at least 14 days of sleep diary or actigraphy data showing a shift in sleep/wake timing, typically later, from day to day [1].

In practice, blind patients with non-24 often present with a complaint of relapsing and remitting nighttime insomnia and daytime somnolence [31, 32]. The patient may not be aware of the periodicity or may have a clear sense of the pattern of relapses and remissions. Less commonly, blind subjects will present with a frank non-24-hour sleep/wake pattern if social and work obligations permit such a schedule. It has been shown that blind individuals with non-24 can spontaneously entrain, albeit at an abnormally delayed or advanced phase, for up to many months at a time [15]. During this period the patient might be asymptomatic or present with signs and symptoms consistent with DSWPD or advanced sleep/wake phase disorder (ASWPD). The result is a great deal of inter- and intra-patient variability in the subjective and physiological presentations of blind individuals with non-24 [15]. In contrast, sighted individuals present with a clear non-24-hour pattern of sleep/wake timing with varying degrees of success when attempts are made to maintain a consistent, 24-hour sleep/wake schedule [1, 21].

The ICSD-3 specifies that the etiology of non-24 is the "non-entrained endogenous circadian" pacemaker [1]. Unfortunately, the routine measurement of circadian phase remains confined to the research setting and there is no clinical test of circadian phase (e.g., a clinical DLMO) that would allow entrainment status to be determined. The demonstration that home assessments of salivary DLMOs are equivalent to those obtained in the laboratory [33] raises the possibility that clinical assessments of circadian phase could exist in the future, but currently the diagnosis of non-24 is based solely on the clinical history [1].

Morbidity

Non-24 in the blind results in both subjective and objective changes in sleep and wakefulness [31, 32, 34, 35]. Polysomnographic studies of blind individuals with non-24 show decreases in sleep efficiency and increases in wakefulness after sleep onset when individuals attempt to sleep in opposition to their biological clock [34]. Similarly, blind individuals with non-24 have been found, with polysomnography, to have decreased total sleep time, sleep efficiency, and rapid-eye-movement (REM) sleep compared to sighted control subjects [35].

More recently it has been shown in both animals and humans that circadian misalignment (i.e., misalignment between the timing of the circadian pacemaker and biologically relevant behaviors such as sleeping and feeding) has a variety of adverse effects [19, 36–40]. These findings raise the prospect that individuals with non-24 might be at increased risk of cardiometabolic, psychiatric, reproductive, and oncology-related sequelae as a result of chronic circadian misalignment. In this way, patients with untreated non-24 may face risks similar to those seen in chronic shift workers [38].

Treatment

Successful treatment of non-24 requires a further understanding of the dynamics of circadian resetting that were introduced above. As noted, light is the primary zeitgeber for the circadian system and, like all zeitgebers, it will reset the timing of the pacemaker in different directions and by different amounts depending on the biological time (circadian phase) that it occurs. Experimental presentation of zeitgebers at different circadian phases allows for the construction of a phase response curve or PRC; just as dose-response curves indicate the therapeutic response for a given *dose* of drug, a PRC indicates the magnitude and direction of a phase shift for a given *time* of administration. By convention phase shifts to an earlier time are positive, while phase shifts to a later time are negative. A variety of PRCs to light have been constructed and they generally show that light exposure in the biological evening causes phase delays, while light exposure in the biological morning causes phase advances [3–5].

PRCs have also been constructed for exogenous administration of melatonin and these generally have a profile that is somewhat the opposite of the PRCs to light: melatonin administration in the biological afternoon and evening causes phase advances (with a maximum effect about 5–7 hours before habitual bedtime/lights out), while administration in the morning causes phase delays (with a maximum effect around habitual wake time/lights on) [41, 42].

Treatment: Blind

Multiple controlled [30, 34, 43, 44] and uncontrolled studies [45–49] have conclusively shown that both melatonin [34, 43–50] and the melatonin agonist tasimelteon [30] can successfully entrain the circadian pacemaker in blind individuals with non-24 using established markers of circadian phase such as the DLMO. Subsequent work has shown that melatonin dose and timing of administration can have an impact on treatment outcomes as discussed below [45–49].

Dose of Administration: Demonstration of Entrainment

Sack and colleagues were the first to successfully demonstrate successful treatment of non-24 [34]. In a small placebo-controlled study of totally blind individuals they entrained six out of seven individuals after the administration of 10 mg of melatonin 1 hour prior to bedtime. Later, it was found that higher doses of melatonin or longer durations of treatment were unsuccessful in entraining the seventh individual and that it was only when the dose was *lowered* to 0.5 mg that entrainment was achieved [47]. Indeed, melatonin doses as low as 0.02 mg (20 μg) have been found capable of entraining the system with a clear dose-response relationship for doses below 0.5 mg [49]. What explains this therapeutic window for melatonin? Low doses of melatonin may be effective for individuals who have circadian periods close to 24 hours and require minimal daily resetting or because non-photic time cues (e.g., physical activity, feeding, or sleep) provide additional resetting [15]. It has been hypothesized that higher doses of melatonin (i.e., above 10 mg) provide a lesser resetting effect since blood levels of melatonin remain elevated for longer periods of time and stimulate both the phase advance and phase delay regions of the melatonin PRC. Such higher doses would therefore cause both phase advances and phase delays with a resulting smaller net resetting effect (i.e., exogenous melatonin levels "spillover" from the advance to the delay regions of the PRC) [48].

With these caveats, melatonin is generally effective in a wide variety of blind patients with non-24. A meta-analysis of the controlled studies of melatonin for non-24 showed that 67% of patients entrained to doses of 0.5 or 10 mg administered either at the fixed time of 21:00 or 1 hour prior to bedtime for a period of approximately 4–12 weeks. The resulting odds ratio for entrainment was 21.18 (95% CI of 3.22–39.17) [50]. Melatonin doses of 0.5 mg and 3 mg have been shown to have roughly equivalent resetting effects in healthy control subjects [42] and starting doses within this range are therefore recommended for the treatment of non-24. If unwanted soporific effects occur, the dose can be lowered.

In comparison, the melatonin agonist tasimelteon, when given at a 20 mg dose 1 hour prior to bedtime for a period of 4 weeks, resulted in entrainment of 20% of blind patients with non-24, while 50% of patients entrained during the 12–18-week open-label phase and an exploratory analysis showed that 59% entrained after 7 months [30]. Some of the relatively lower success rates with tasimelteon can be attributed to the duration of the treatment trials. It may also be possible that "spillover" effects exist for tasimelteon as well and that a similar therapeutic window exists for the drug.

Time of Administration: Demonstration of Entrainment at the Correct Time

While the studies administering melatonin or melatonin agonists an hour prior to bedtime [30, 34] or at a fixed clock hour in the evening (e.g., 21:00) [43, 44] achieved entrainment, they did not necessarily achieve entrainment at a normal phase (i.e., the DLMO occurring ~2–3 hours prior to bedtime as described above). In most patients with non-24, the circadian period is greater than 24 hours and corrective phase *advances* are required. In such individuals it has been shown that, after successful treatment with melatonin, the timing of the entrained DLMO is between about 0 and 5 hours *after* the clock hour of melatonin administration [48, 49]. An analogy that could be provided to patients is that of a speeding car that suddenly breaks: the car will travel some distance past the point where the breaks were applied. As a result, if melatonin is administered around bedtime the entrained DLMO will occur abnormally late (i.e., after bedtime) and the likely clinical result is DSWPD. Therefore, to achieve entrainment at the correct time melatonin should be administered about *6 hours prior to the desired bedtime*. If symptoms of DSWPD occur (e.g., sleep onset insomnia and morning hypersomnolence) then the administration time can be moved earlier. Conversely if symptoms of ASWPD occur (e.g., evening hypersomnolence and terminal insomnia), then the time of administration can be moved later.

Special consideration should be given to individuals with non-24 who have circadian periods less than 24 hours. As noted above, this is most commonly the case in females [6, 15]. In such patients circadian phase will drift progressively earlier from day to day and corrective phase *delays* are needed. In such cases we have found that, after successful treatment with melatonin, the timing of the entrained DLMO is *before* the clock hour of melatonin administration [51]. This is similar to the lay analogy presented above except the speeding car is moving backward when the brakes are applied. Therefore, to achieve entrainment at the correct time (again, with the DLMO occurring 2–3 hours before desired bedtime) melatonin should be administered *upon awakening* [51].

Treatment: Sighted

There are no placebo-controlled studies for the treatment of non-24 in the sighted [50], but there are case reports of light [52–55], melatonin [54, 56, 57], and the melatonin agonist ramelteon [58] being used to treat non-24 in an open-label fashion. Treatment of sighted individuals with melatonin or melatonin agonists should take into account the same issues of dose and timing of administration discussed above.

Similarly, the use of light to treat non-24 in the sighted must consider the PRCs to light. Sighted individuals with non-24 in whom sleep/wake timing drifts progressively later require corrective phase advances, which would be achieved when light exposure administered at desired wake time corresponds with the patient's biological morning, while those in whom sleep/wake timing drifts progressively earlier require corrective phase delays, which would be achieved when light exposure administered just before desired bedtime corresponds with the patient's early biological night (see Fig. 9.1) [3–5].

The necessary intensity of light depends on an individuals' prior history of light exposure: maximal resetting effects can occur at levels as low as 550 lux [59] in individuals living under conditions of very dim light [60]. Sighted non-24 patients who were living under conditions of bright indoor electrical lighting or outdoor light might require greater light intensities to obtain maximal resetting effects.

Treatment in sighted individuals must include consideration of the impact of the patient's existing self-selected light/dark schedules and the fact that sleep timing "gates" the timing of light exposure. The clinician should remember that even low intensity [27, 59, 61] and very short duration light [5, 62–64] exposure can have a resetting effect on the circadian pacemaker. In practice, many clinical trials of light therapy in other CRSDs have used exposures of ≥1000 lux for ≥30 minutes per day [50].

References

1. Medicine AAoS. International classification of sleep disorders. American Academy of Sleep Medicine: Darien; 2014.
2. Lucas RJ, Lall GS, Allen AE, Brown TM. How rod, cone, and melanopsin photoreceptors come together to enlighten the mammalian circadian clock. Prog Brain Res. 2012;199:1–18.
3. Czeisler CA, Kronauer RE, Allan JS, et al. Bright light induction of strong (Type O) resetting of the human circadian pacemaker. Science. 1989;244:1328–33.
4. Khalsa SB, Jewett ME, Cajochen C, Czeisler CA. A phase-response curve to single bright light pulses in human subjects. J Physiol. 2003;549:945–52.
5. St Hilaire MA, Gooley JJ, Khalsa SB, Kronauer RE, Czeisler CA, Lockley SW. Human phase response curve to a 1h pulse of bright white light. J Physiol. 2012;590:3035–45.
6. Duffy JF, Cain SW, Chang AM, et al. Sex difference in the near-24-hour intrinsic period of the human circadian timing system. Proc Natl Acad Sci U S A. 2011;108:15602–8.
7. Czeisler CA, Duffy JF, Shanahan TL, et al. Stability, precision and near-24-hour period of the human circadian pacemaker. Science. 1999;284:2177–81.
8. Lewy AJ, Sack RL. The dim light melatonin onset as a marker for circadian phase position. Chronobiol Int. 1989;6:93–102.
9. Lewy AJ, Cutler NL, Sack RL. The endogenous melatonin profile as a marker for circadian phase position. J Biol Rhythm. 1999;14:227–36.
10. Klerman EB, Gershengorn HB, Duffy JF, Kronauer RE. Comparisons of the variability of three markers of the human circadian pacemaker. J Biol Rhythm. 2002;17:181–93.
11. Martin SK, Eastman CI. Sleep logs of young adults with self-selected sleep times predict the dim light melatonin onset. Chronobiol Int. 2002;19:695–707.

12. Burgess HJ, Eastman CI. The dim light melatonin onset following fixed and free sleep schedules. J Sleep Res. 2005;14:229–37.
13. Burgess HJ, Fogg LF. Individual differences in the amount and timing of salivary melatonin secretion. PLoS One. 2008;3:e3055.
14. Emens JS, Yuhas K, Rough J, Kochar N, Peters D, Lewy AJ. Phase angle of entrainment in morning- and evening-types under naturalistic conditions. Chronobiol Int. 2009;26:474–93.
15. Emens JS, Laurie AL, Songer JB, Lewy AJ. Non-24-hour disorder in blind individuals revisited: variability and the influence of environmental time cues. Sleep. 2013;36:1091–100.
16. Lockley SW, Skene DJ, Arendt J, Tabandeh H, Bird AC, Defrance R. Relationship between melatonin rhythms and visual loss in the blind. J Clin Endocrinol Metab. 1997;82:3763–70.
17. Czeisler CA, Shanahan TL, Klerman EB, et al. Suppression of melatonin secretion in some blind patients by exposure to bright light. New Engl J Med. 1995;332:6–11.
18. Klerman EB, ShanahanTL, Brotman DJ, et al. Photic resetting of the human circadian pacemaker in the absence of conscious vision. J Biol Rhythm. 2002;17:548–55.
19. LeGates TA, Fernandez DC, Hattar S. Light as a central modulator of circadian rhythms, sleep and affect. Nat Rev Neurosci. 2014;15:443–54.
20. Emens JS, Brotman DJ, Czeisler CA. Evaluation of the intrinsic period of the circadian pacemaker in a patient with a non-24-hour sleep-wake schedule disorder. Sleep Res. 1994;23:256.
21. Uchiyama M, Lockley SW. Non-24-hour sleep-wake syndreom in sighted and blind patients. Sleep Med Clin. 2009;4:195–211.
22. Hayakawa T, Uchiyama M, Kamei Y, et al. Clinical analyses of sighted patients with non-24-hour sleep-wake syndrome: a study of 57 consecutively diagnosed cases. Sleep. 2005;28:945–52.
23. Kitamura S, Hida A, Enomoto M, et al. Intrinsic circadian period of sighted patients with circadian rhythm sleep disorder, free-running type. Biol Psychiatry. 2013;73:63–9.
24. Wever R. The circadian system of man: results of experiments under temporal isolation. New York, NY: Springer-Verlag; 1979.
25. Czeisler CA, Weitzman ED, Moore-Ede MC, Zimmerman JC, Knauer RS. Human sleep: its duration and organization depend on its circadian phase. Science. 1980;210:1264–9.
26. Wittmann M, Dinich J, Merrow M, Roenneberg T. Social jetlag: misalignment of biological and social time. Chronobiol Int. 2006;23:497–509.
27. Stothard ER, McHill AW, Depner CM, et al. Circadian entrainment to the natural light-dark cycle across seasons and the weekend. Curr Biol. 2017;27:508–13.
28. Czeisler CA, Richardson GS, Coleman RM, et al. Chronotherapy: resetting the circadian clocks of patients with delayed sleep phase insomnia. Sleep. 1981;4:1–21.
29. Sack RL, Lewy AJ, Blood ML, Keith LD, Nakagawa H. Circadian rhythm abnormalities in totally blind people: incidence and clinical significance. J Clin Endocrinol Metab. 1992;75:127–34.
30. Lockley SW, Dressman MA, Licamele L, et al. Tasimelteon for non-24-hour sleep–wake disorder in totally blind people (SET and RESET): two multicentre, randomised, double-masked, placebo-controlled phase 3 trials. Lancet. 2015;386:1754–64.
31. Lockley SW, Skene DJ, Butler LJ, Arendt J. Sleep and activity rhythms are related to circadian phase in the blind. Sleep. 1999;22:616–23.
32. Sack RL, Lewy AJ, Blood ML, Keith LD, Nakagawa H. Circadian rhythm abnormalities in totally blind people: incidence and clinical significance. J Clin Endocrinol Metab. 1992;75:127–34.
33. Burgess HJ, Wyatt JK, Park M, Fogg LF. Home circadian phase assessments with measures of compliance yield accurate dim light melatonin onsets. Sleep. 2015;38:889–97.
34. Sack RL, Brandes RW, Kendall AR, Lewy AJ. Entrainment of free-running circadian rhythms by melatonin in blind people. New Engl J Med. 2000;343:1070–7.
35. Leger D, Guilleminault C, Defrance R, Domont A, Paillard M. Prevalence of sleep/wake disorders in persons with blindness. Clin Sci (Lond). 1999;97:193–9.
36. Scheer FAJL, Hilton MF, Mantzoros CS, Shea SA. Adverse metabolic and cardiovascular consequences of circadian misalignment. Proc Natl Acad Sci U S A. 2009;106:4453–8.

37. Ruger M, Scheer FAJL. Effects of circadian disruption on the cardiometabolic system. Rev Endocr Metab Disord. 2009;10:245.
38. Vogel M, Braungardt T, Meyer W, Schneider W. The effects of shift work on physical and mental health. J Neural Trans. 2012;119:1121–32.
39. Jagannath A, Peirson SN, Foster RG. Sleep and circadian rhythm disruption in neuropsychiatric illness. Curr Opin Neurobiol. 2013;23:888–94.
40. Kelleher FC, Rao A, Maguire A. Circadian molecular clocks and cancer. Cancer Lett. 2014;342:9–18.
41. Lewy AJ, Bauer VK, Ahmed S, et al. The human phase response curve (PRC) to melatonin is about 12 hours out of phase with the PRC to light. Chronobiol Int. 1998;15:71–83.
42. Burgess HJ, Revell VL, Molina TA, Eastman CI. Human phase response curves to three days of daily melatonin: 0.5 mg versus 3.0 mg. J Clin Endocrinol Metab. 2010;95:3325–31.
43. Lockley SW, Skene DJ, James K, Thapan K, Wright J, Arendt J. Melatonin administration can entrain the free-running circadian system of blind subjects. J Endocrinol. 2000;164:R1–6.
44. Hack LM, Lockley SW, Arendt J, Skene DJ. The effects of low-dose 0.5-mg melatonin on the free-running circadian rhythms of blind subjects. J Biol Rhythm. 2003;18:420–9.
45. Lewy AJ, Bauer VK, Hasler BP, Kendall AR, Pires ML, Sack RL. Capturing the circadian rhythms of free-running blind people with 0.5 mg melatonin. Brain Res. 2001;918:96–100.
46. Lewy AJ, Hasler BP, Emens JS, Sack RL. Pretreatment circadian period in free-running blind people may predict the phase angle of entrainment to melatonin. Neurosci Lett. 2001;313:158–60.
47. Lewy AJ, Emens JS, Sack RL, Hasler BP, Bernert RA. Low, but not high, doses of melatonin entrained a free-running blind person with a long circadian period. Chronobiol Int. 2002;19:649–58.
48. Lewy AJ, Emens JS, Bernert RA, Lefler BJ. Eventual entrainment of the human circadian pacemaker by melatonin is independent of the circadian phase of treatment initiation: clinical implications. J Biol Rhythm. 2004;19:68–75.
49. Lewy AJ, Emens JS, Lefler BJ, Yuhas K, Jackman AR. Melatonin entrains free-running blind people according to a physiological dose-response curve. Chronobiol Int. 2005;22:1093–106.
50. Auger RR, Burgess HJ, Emens JS, Deriy LV, Thomas SM, Sharkey KM. Clinical practice guideline for the treatment of intrinsic circadian rhythm sleep-wake disorders: advanced sleep-wake phase disorder (ASWPD), delayed sleep-wake phase disorder (DSWPD), non-24-hour sleep-wake rhythm disorder (N24SWD), and irregular sleep-wake rhythm disorder (ISWRD). An update for 2015. J Clin Sleep Med. 2015;11:1199–236.
51. Emens J, Lewy AJ, Yuhas K, Jackman AR, Johnson KP. Melatonin entrains free-running blind individuals with circadian periods less than 24 hours. Sleep. 2006;29:A62.
52. Eastman CI, Anagnopoulus CA, Cartwright RD. Can bright light entrain a free-runner? Sleep Res. 1988;17:372.
53. Hoban TM, Sack RL, Lewy AJ, Miller LS, Singer CM. Entrainment of a free-running human with bright light? Chronobiol Int. 1989;6:347–53.
54. Hayakawa T, Kamei Y, Urata J, et al. Trials of bright light exposure and melatonin administration in a patient with non-24 hour sleep-wake syndrome. Psychiatry Clin Neurosci. 1998;52:261–2.
55. Watanabe T, Kajimura N, Kato M, Sekimoto M, Hori T, Takahashi K. Case of a non-24 h sleep-wake syndrome patient improved by phototherapy. Psychiatry Clin Neurosci. 2000;54:369–70.
56. McArthur AJ, Lewy AJ, Sack RL. Non-24-hour sleep-wake syndrome in a sighted man: circadian rhythm studies and efficacy of melatonin treatment. Sleep. 1996;19:544–53.
57. Kamei Y, Hayakawa T, Jujiro U, et al. Melatonin treatment for circadian rhythm sleep disorders. Psychiatry Clin Neurosci. 2000;54:381–2.
58. Watanabe A, Hirose M, Arakawa C, Iwata N, Kitajima T. A case of non-24-hour sleep-wake rhythm disorder treated with a low dose of ramelteon and behavioral education. J Clin Sleep Med. 2018;14:1265–7.

59. Zeitzer JM, Dijk DJ, Kronauer RE, Brown EN, Czeisler CA. Sensitivity of the human circadian pacemaker to nocturnal light: melatonin phase resetting and suppression. J Physiol. 2000;526:695–702.
60. Chang AM, Scheer FAJL, Czeisler CA. The human circadian system adapts to prior photic history. J Physiol. 2011;589:1095–102.
61. Czeisler CA, Boivin DB, Duffey JF, Kronauer RE. Dose-response relationships for resetting of human circadian clock by light. Nature. 1996;379:540–1.
62. Gronfier C, Wright KP Jr, Kronauer RE, Jewett ME, Czeisler CA. Efficacy of a single sequence of intermittent bright light pulses for delaying circadian phase in humans. Am J Physiol Endocrinol Metab. 2004;287:E174–81.
63. Najjar RP, Zeitzer JM. Temporal integration of light flashes by the human circadian system. The J Clin Invest. 2016;126:938–47.
64. Rahman SA, St Hilaire MA, Chang AM, et al. Circadian phase resetting by a single short-duration light exposure. JCI Insight. 2017;2:e89494.

Chapter 10
Irregular Sleep-Wake Rhythm Disorder

Danielle Goldfarb and Katherine M. Sharkey

Introduction

Irregular sleep-wake rhythm disorder (ISWRD) is characterized by persistent, erratic sleep-wake patterns. In patients with ISWRD, the typical circadian pattern is severely disrupted, and instead there is sleep fragmentation and a striking lack of sleep consolidation. In the most severe cases, it can be difficult to discern a single "main" sleep period on any given day. ISWRD occurs most commonly in individuals with neurodegenerative and neurodevelopmental disorders, patients with schizophrenia, and among those with traumatic brain injury. Internal and external factors implicated in the development of ISWRD include dysfunction of the suprachiasmatic nucleus (SCN) and its networks, abnormal SCN input, and disrupted environmental time cues (zeitgebers). Often a vicious cycle develops where one factor perpetuates another. For example, individuals with advanced dementia are often institutionalized in settings where they are exposed to less robust environmental cues compared to healthy, age-matched individuals, thereby magnifying the pathology.

Pathophysiology

The pathologic mechanisms of ISWRD have not been fully elucidated. It is hypothesized that disruption of any of the internal drivers (SCN and related networks) and/or external drivers (environmental and behavioral time cues) of circadian rhythms

D. Goldfarb
University of Arizona College of Medicine, Phoenix, AZ, USA

K. M. Sharkey (✉)
Department of Medicine and Psychiatry and Human Behavior, Alpert Medical School of Brown University, Providence, RI, USA
e-mail: katherine_sharkey@brown.edu

R. R. Auger (ed.), *Circadian Rhythm Sleep-Wake Disorders*,
https://doi.org/10.1007/978-3-030-43803-6_10

can lead to its development. The integrity of the SCN and the monosynaptic pathway from the retina to the SCN, known as the retinohypothalamic tract, are crucial for proper circadian function. Abnormal SCN development or degeneration leads to the inability to produce circadian rhythmicity resulting in an inability to consolidate sleep and wakefulness. Much of what we know about the pathophysiology of circadian dysrhythmia derives from studies of neuroanatomy and neurophysiology in patients with dementia. For example, brain autopsies of patients with severe Alzheimer's disease (AD) reveal SCN degeneration, namely, neuronal loss and neurofibrillary tangle formation [1], likely contributing to circadian desynchrony.

Abnormal input to the SCN can be a result of impaired light input pathways and/or abnormal melatonin secretion. Within the retina, a small subset of specialized cells, called intrinsically photosensitive retinal ganglion cells (ipRGCs) [2, 3], which comprise only about 1–2% of the total retinal ganglion cells, play a crucial role. The ipRGCs relay information about ambient light to the SCN via the retinohypothalamic tract and contribute to many functions of the eye including sleep regulation, melatonin secretion and suppression, and the pupillary reflex [4]. Loss and pathology of ipRGCs have been shown in postmortem retinas of patients with AD [5] and in glaucoma [6].

Melatonin secretion by the pineal gland follows a circadian rhythm generated by the SCN and occurs only in dark environmental conditions during the biological night. The melatonin secretion pathway begins with retinal photoreceptors that transduce photic information (i.e., absence of light) via the retinohypothalamic tract to the SCN. In turn, ventral SCN neurons enervate neurons in the paraventricular hypothalamic nucleus (PVH) that traverse through the spinal cord to the superior cervical ganglion (SCG). Finally, noradrenergic SCG projections to the pineal gland act on beta-1-adrenergic receptors on pinealocytes to increase the activity of the rate-limiting enzyme required for melatonin synthesis—arylalkylamine-N-acetyltransferase—thus stimulating melatonin secretion. Melatonin binds to SCN receptors and decreases the circadian alerting signal during darkness, thereby promoting sleep. Conversely, when light is present, melatonin secretion is inhibited by decreasing the activating influences of neurons in the PVH nucleus [7]. Even relatively low ambient light levels can acutely suppress nighttime melatonin production. Suppression of melatonin secretion and phase shifting of the dim light melatonin onset (DLMO) are most sensitive to blue spectrum light ~460 nm [8]. Age-related anatomic and physiologic changes lead to decreased nighttime melatonin production, reduced rhythm amplitude, and changes in the timing of the melatonin rhythm [9]. The postulated causes for melatonin changes with aging are similar to those in ISWRD, including disruption of the SCN and its inputs, along with reduced environmental time cues [10]. Melatonin content in postmortem human pineal glands has been shown to be reduced with age [9].

Characteristic age-related changes in rest-activity circadian rhythms include lower amplitude [11], fragmentation and loss of rhythms [12], and decreased sensitivity to zeitgebers such as light exposure [13]. Aging is also associated with reductions in slow wave sleep and sleep efficiency and increases in nighttime awakenings [14]. Sleep disruption may contribute to behavioral disruption of circadian entrainment by increasing light exposure during typical sleep periods.

The eye is gaining increasing attention as a potential biomarker for neurodegenerative disease, specifically via retinal imaging. Besides loss of ipRGCs in AD, reduced retinal nerve fiber thickness has been identified through optical coherence tomography [15], providing additional support for disrupted light pathways. Furthermore, in their 2011 study, Koronyo-Hamaoui and colleagues demonstrated retinal amyloid beta (Aβ)—the peptide that comprises amyloid plaques found in the brains of AD patients—in postmortem eyes of eight AD patients and five suspected early-stage cases using specialized retinal imaging [16].

Genetic and epigenetic mechanisms that regulate the circadian clock and potentially contribute to dysrhythmias like ISWRD are a topic of active investigation. Generating circadian signals at the cellular level, several clock genes have been identified in mammals, including the positive regulators brain and muscle ARNT-like 1 (BMAL1), *Circadian Locomotor Output Cycles Kaput* (CLOCK), and neuronal PAS domain protein 2 (NPAS2) and the negative regulators cryptochromes 1 and 2 (CRY1/2) and period 1, 2, and 3 (PER1/2/3), all of which are expressed in SCN neurons [17]. Individual mutations in any of these genes cause aberrant circadian rhythmicity. For example, a study of BMAL1 knockout mice demonstrated immediate loss of rhythmicity in the absence of a light-dark cycle [18]. BMAL1 may play a particularly significant role in circadian dysrhythmia related to AD: in mouse models, Aβ induces BMAL1 degradation in neuronal cells [19] and hence may contribute to circadian disruption. Furthermore, Cermakian and colleagues [20] showed asynchronous clock gene expression in other (non-SCN) brain regions in AD patients compared to controls.

Disrupted environmental and behavioral inputs to the clock, i.e., zeitgebers such as light exposure, social cues, activity, and mealtimes, influence the period, phase, and amplitude of circadian rhythms [21]. Without sufficient exposure to timed light, the biological clock becomes desynchronized with the solar day, resulting in deleterious effects on various physiological functions, neurobehavioral performance, and sleep [8]. Older adults and, to an even greater extent, institutionalized elderly, are less likely to be exposed to robust daytime light [22]. Ancoli-Israel and colleagues demonstrated that lower daytime light levels contribute to increasingly abnormal circadian rhythms as measured by actigraphy and are associated with an increase in nighttime awakenings, even after controlling for the level of dementia [23].

Gehrman and colleagues [24] demonstrated that rest-activity patterns decline initially in earlier stages of dementia, with a resurgence of rhythmicity in moderate dementia, followed by subsequent decline in severe dementia. The authors posit that the two sources of synchronization of rhythms, the endogenous output by the SCN and entrainment by the environment, give rise to a three-stage model of rest-activity rhythm changes in dementia. In the early stages, SCN damage results in a decline in rhythmicity. Eventually, environmental cues take on a larger role contributing to a resynchronization of circadian rhythms. When dementia becomes severe, environmental cues lose their potency.

Over the last decade, there has been greater recognition that sleep and circadian rhythm abnormalities may be early manifestations of AD, even preceding cognitive decline. In a large actigraphy study of over 1200 healthy women, Tranah and colleagues [25] demonstrated that decreased amplitude and weaker circadian rest-activity rhythms were associated with an increased risk of mild cognitive impairment or dementia within the subsequent 5 years. In a recent actigraphy study of over 2700 community-dwelling older males, rest-activity rhythm changes, including lower amplitude and rhythm robustness, along with phase-advanced acrophase, were associated with clinically significant cognitive decline [26].

Finally, though still a novel area of investigation, there is increasing interest in the link between circadian system malfunction, early-life insults (i.e., severe gestational stress, maternal immune activation, and fetal hormonal milieu alterations), and the development of neuropsychiatric diseases in adulthood. For an excellent discussion on this topic, see the review by Marco and colleagues [27]. Further understanding of the fetal-maternal environment, brain development, and circadian rhythms may ultimately support the discovery of innovative therapeutic pathways.

Clinical Features, Epidemiology, and Diagnosis

Individuals with ISWRD present with either nighttime insomnia, daytime sleepiness, or both. Sleep patterns are unpredictable, and some patients rarely spend a full hour awake during the day or sleep continuously for more than an hour during the night [28]. Initial caregiver reports may describe frequent brief naps, dozing, or nodding off throughout the daytime, along with difficulty with sleep initiation at a conventional time and increased nighttime wakefulness. Though sleep patterns are erratic, overall total hours of sleep may be normal for age in patients with ISWRD. Nevertheless, the irregular sleep patterns can lead to negative repercussions, including nocturnal wandering and falls [29]. Furthermore, because ISWRD patients' irregular sleep patterns cause significant disruption to caregivers' sleep, the disorder is a common cause of institutionalization [29].

Populations affected by ISWRD are diverse and, in addition to what is described above, include those with neurodevelopmental disorders of childhood and schizophrenia. As well, ISWRD has been identified in individuals with traumatic brain injury. Among caregivers, irregular sleep-wake patterns may not be perceived as a separate issue from the underlying disorder or may simply be described as insomnia. Thus, ISWRD is likely to be underdiagnosed unless clinicians make specific inquiries about it [29].

Making the diagnosis of ISWRD requires sleep-activity monitoring for 7–14 days with logs, typically completed by a caregiver and, if possible, actigraphic monitoring. Findings include significant fragmentation without identification of a major sleep period (see Fig. 10.1). According to the International Classification of Sleep Disorders [29], there must be at least three brief sleep periods during a 24-hour period, and symptoms must be present for at least 3 months to make the diagnosis.

Abnormal rest-activity patterns are a common and progressive feature in neurodegenerative diseases such as AD, and actigraphic findings show that increased fragmentation and decreased amplitude of activity correlate with dementia severity [30]. Nurses caring for hospitalized patients in a geriatrics ward observed a blunted

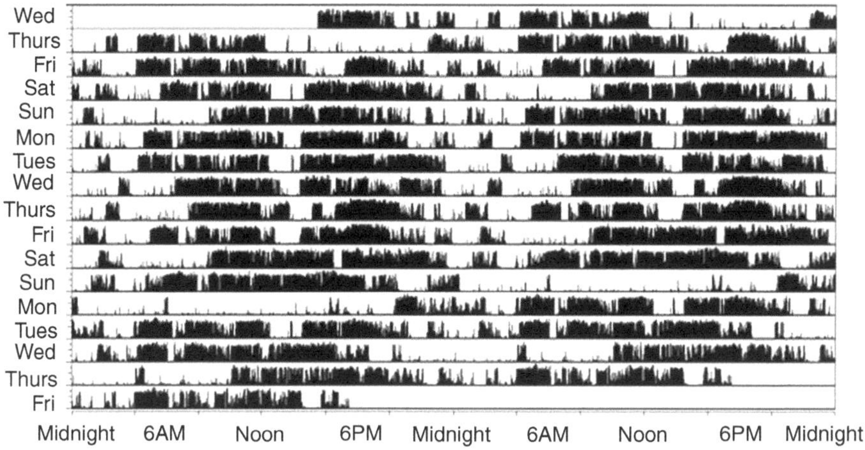

Fig. 10.1 Actigraphy in a patient with ISWRD: Features of irregular sleep-wake rhythm disorder (ISWRD) are illustrated in this double plot of 17 days of wrist actigraphy in a patient with ISWRD associated with neurologic disease. The patient had multiple sleep-wake episodes across 24 hours with multiple wake episodes at night and irregular sleep periods during the day. In this patient, sleep mostly occurred at night, consistent with his efforts to adhere to a conventional sleep-wake schedule

amplitude of rest-activity rhythms that corresponded with abnormal patterns of core body temperature rhythm in patients with vascular dementia or AD compared to age-matched controls [31]. Decreased rest-activity amplitude is also correlated with a higher degree of cognitive impairment in community-dwelling older adults [32].

Sleep difficulties are common in neurodevelopmental disorders of childhood [33], often with co-occurring circadian rhythm disorders. Comorbid ISWRD in youngsters with developmental disorders can be associated with reports from parents and/or caregivers that their child sleeps or naps inappropriately, falls asleep too early at night, awakens too early in the morning, and/or cannot stay awake during the day for activities [29]. Despite the profound impact of irregular sleep-wake patterns on families caring for children with developmental disorders, there is a paucity of research on ISWRD in children and adolescents. Nevertheless, ISWRD has been described in Angelman syndrome [34], Smith-Magenis syndrome [35], neuronal ceroid lipofuscinosis [36], and Williams-Beuren syndrome [37]. A case of ISWRD has also been reported in a congenitally blind, neurodevelopmentally delayed child [38]. Furthermore, abnormalities in melatonin rhythms associated with irregular sleep-wake behavior have been shown in Smith-Magenis syndrome [35] and Williams-Beuren syndrome [37]. Little is known about the clinical course of childhood ISWRD or the consequences of irregular sleep-wake patterns during critical developmental periods. One study of older adults with intellectual disabilities found that, compared to older adults with normal cognitive function, the sleep-wake rhythm in the former group was less stable and more fragmented [39].

ISWRD has been described in schizophrenia, particularly in individuals with "positive" symptoms [40]. Bromundt and colleagues [41] studied 14 middle-aged individuals with schizophrenia treated with antipsychotics using actigraphy and salivary melatonin in addition to neurocognitive testing. Participants with lower circadian amplitude had more fragmented sleep, atypical melatonin secretion patterns, and worse cognitive performance. Vigano and colleagues also observed lower nighttime melatonin levels and blunted melatonin rhythm amplitudes in patients with schizophrenia [42]. It is noted that melatonin levels and rhythmicity in this population are difficult to interpret as the effects of antipsychotic medications on melatonin are poorly understood.

Circadian rhythm disorders also occur after traumatic brain injury (TBI). Ayalon and colleagues [43] studied 42 patients with mild TBI who had complaints of insomnia (34 with difficulties falling asleep or waking up and 8 with sleep maintenance insomnia) and found that 7 had ISWRD and 8 had delayed sleep-wake phase disorder (DSWPD). Compared to DSWPD individuals, those with ISWRD demonstrated a lower amplitude in oral temperature rhythm and 3 of 7 lacked a daily temperature rhythm entirely. The mechanisms of circadian rhythm disturbance in TBI are not well understood; however, there are likely multiple factors at play. Duclos et al. [44] provide a review of possible pathophysiologic mechanisms including a maladaptive immune response, dysregulated clock genes, and confounding effects of acute hospitalization, pain, and anxiety.

Other causes of acquired damage to the SCN contributing to the development of ISWRD have been described. One report detailed posttraumatic irregular sleep-wake rhythm in a 38-year-old woman after she sustained a gunshot wound that damaged her SCN and bilateral optic nerves [45]. Another case report describes an individual who developed ISWRD in association with a prolactin-secreting pituitary microadenoma that impinged upon the SCN [46].

Management

ISWRD treatment aims to restore SCN time cues to decrease sleep fragmentation, consolidate the sleep-wake pattern, and improve circadian rhythm amplitude. Strengthening external circadian signals helps resynchronize the pacemaker, allowing the body to anticipate predictable physiologic and behavioral needs intrinsically tied to day and night. Interventions include light therapy and timed exogenous melatonin administration, as well as directed physical and social activity.

Bright light exposure has been studied to treat irregular sleep-wake patterns in institutionalized older adults, though in most investigations the patient sample was heterogeneous and the participants rarely had formal ISWRD diagnoses. For example, Mishima and colleagues [47] compared bright light (between 3000 and 5000 lux) administered for 2 hours between 9:00 and 11:00 AM in 14 hospitalized patients with dementia and sleep disturbance. After 4 weeks of daily treatment, patients in the bright light group slept less during the day and more at night. Another trial of bright light in 10 elderly patients with severe dementia treated with 5000–8000 lux for 45 minutes daily for 4 weeks between the hours of 8:00–10:00 AM showed behavioral improvements as measured with the Cohen-Mansfield Agitation Inventory (CMAI) and Behavior Pathology in Alzheimer's Disease Rating Scale (BEHAVE-AD) [48].

Augmented light exposure for 45 minutes to 2 hours at levels ranging from 1000 to 8000 lux has been tested in several other studies of elderly nursing home residents with dementia with irregular sleep-wake patterns suggestive of ISWRD, with most studies demonstrating positive effects of bright light, specifically more consolidated sleep at night and less sleep during the day [49–55]. In the 2015 American Academy of Sleep Medicine clinical practice guidelines for the treatment of circadian rhythm sleep wake disorders [56], bright light therapy for ISWRD is recommended among elderly patients with dementia. Side effects from light therapy are generally mild [57] and if it can be provided with relatively low cost and labor for caregivers, the behavioral benefits likely outweigh any associated symptoms.

The AASM clinical practice guidelines specifically recommend against using hypnotics or melatonin to treat ISWRD in elderly patients with dementia [56]. Hypnotic medications increase risk of falls and daytime sleepiness in older adults and there is no data supporting their use in ISWRD; hence, they should generally be avoided. Exogenous melatonin has been tested for ISWRD in older patients with dementia, but has not been shown to regulate sleep-wake patterns better than placebo, and is associated with untoward mood and behavioral outcomes. For example, in a study of 157 AD patients treated with 2.5 mg slow-release melatonin, 10 mg melatonin, or placebo, melatonin failed to improve actigraphically estimated total sleep time (TST) after a 2-month treatment period [57]. This lack of efficacy, combined with another study showing that elders with dementia in an assisted living facility who were treated with 2.5 mg of melatonin in the evening had an increased incidence of negative affect and withdrawal [58], resulted in the recommendation against routine use of melatonin for ISWRD in elderly patients with dementia.

In contrast, data support the use of appropriately timed exogenous melatonin to treat children and adolescents with neurodevelopmental disorders and ISWRD [56]. A double-blind, randomized controlled trial of 16 youngsters ages 3–16 years old with autism spectrum disorder compared 3 months of evening melatonin administration versus placebo in a crossover design [59]. Parents reported shortened sleep onset latency and increased TST with melatonin (dose range 2–10 mg) compared to placebo. Improvement in sleep duration was also observed in a small open trial testing 3 mg of melatonin administered each evening at 6:30 pm to children with severe developmental disabilities and disrupted sleep wake patterns [60]. Finally, in the largest study to date of 125 youngsters with autism spectrum disorder, extended release melatonin, starting at 2 mg and increased to 5 mg, increased parent/caregiver reported sleep duration by nearly 1 hour (compared to 9 minutes with placebo) and shortened latency to sleep onset by 39 minutes, compared to 12.5 minutes with placebo [61]. Most studies using melatonin in youths with ISWRD reported minimal, mild side effects, such as increased daytime sleepiness, with high patient and caregiver treatment acceptability. Nevertheless, clinicians considering melatonin administration to treat ISWRD in children with neurodevelopmental disabilities should be aware of a single nucleotide polymorphism (SNP) of CYP1A2 (Cytochrome P450 Family 1 Subfamily A Member 2) that has been associated with slow melatonin metabolism yielding extremely high daytime levels in youngsters with autism spectrum disorder [62, 63]. Moreover, experts contend that because no long-term safety studies of exogenous melatonin administration in adolescents are available, its use should be reserved for situations with significantly disturbed sleep-wake patterns where the benefits clearly outweigh possible risks [64].

Another strategy for treating irregular sleep-wake patterns associated with severe circadian dysregulation is to halt pineal melatonin secretion by administering a β1-antagonist in the morning and supplementing with exogenous melatonin in the evening. This treatment was tested in 10 children with Smith-Magenis syndrome who displayed a near inversion of typical circadian entrainment at baseline. Use of this regimen increased sleep time at night, improved sleep quality, and reduced problematic behaviors during the day [65].

There are no well-established, evidence-based treatments specifically for ISWRD in patients with schizophrenia, in part because ISWRD is rarely diagnosed formally in patients with psychiatric illness. Circadian-based treatments for irregular sleep-wake patterns have been studied most often in patients with bipolar disorder, and strategies such as bright light therapy and stabilizing sleep patterns are effective both for improving sleep and stabilizing mood [66, 67]. Treatment studies for circadian dysregulation—including ISWRD—in psychiatric illness are needed urgently. Promising therapeutic pathways include bright light therapy, sleep deprivation or restriction, use of the melatonin agonist and serotonin-2C antagonist agomelatine, and further study of the circadian stabilizing properties of established medications, e.g., selective serotonin reuptake inhibitors, lithium, valproic acid, and more novel pharmacologic agents, e.g., ketamine and brexanolone.

Conclusion

ISWRD is a debilitating syndrome in which patients have multiple, irregular sleep-wake episodes across the day. It is observed most commonly in patients with neurodevelopmental and neurodegenerative disorders as well as those with neurologic trauma or schizophrenia. ISWRD is disruptive to successful treatment of comorbid conditions and poses significant challenges to the caregivers of patients with this CRSWD. Disruption of input to and output from the SCN is believed to underlie the circadian desynchrony in ISWRD. Thus, treatment strategies like bright light therapy and exogenous melatonin administration focus on strengthening signaling to and from the internal clock. In addition to better syndrome-specific therapies for ISWRD and for CRSWDs in general, more clinical research is needed to establish optimal "doses" of behavioral and pharmacologic treatments. Best practices and treatment guidelines are needed to guide duration of treatment and the use of combination therapies. Future research should also determine whether prevention and treatment can lead to better outcomes of the comorbid neurologic disorders that often accompany ISWRD.

References

1. Stopa EG, Volicer L, Kuo-Leblanc V, Harper D, Lathi D, Tate B, Satlin A. Pathologic evaluation of the human suprachiasmatic nucleus in severe dementia. J Neuropathol Exp Neurol. 1999;58(1):29–39.
2. Hattar S, Liao HW, Takao M, Berson DM, Yau KW. Melanopsin-containing retinal ganglion cells: architecture, projections, and intrinsic photosensitivity. Science. 2002;295(5557):1065–70.
3. Berson DM, Dunn FA, Takao M. Phototransduction by retinal ganglion cells that set the circadian clock. Science. 2002;295(5557):1070–3.
4. Schmoll C, Lascaratos G, Dhillon B, Skene D, Riha RL. The role of retinal regulation of sleep in health and disease. Sleep Med Rev. 2011;15:107–13.
5. La Morgia C, Ross-Cisneros FN, Koronyo Y, et al. Melanopsin retinal ganglion cell loss in Alzheimer disease. Ann Neurol. 2016;79:90–109.
6. Drouyer E, Dkhissi-Benyahya O, Chiquet C, et al. Glaucoma alters the circadian timing system. PLoS One. 2008;3(12):e3931.
7. Berry RB. Circadian rhythm sleep disorders. In: Berry RB, editor. Fundamentals of sleep medicine. 1st ed. Philadelphia: Elsevier; 2012. p. 515–43.
8. Figueiro MG. Light, sleep and circadian rhythms in older adults with Alzheimer's disease and related dementias. Neurodegener Dis Manag. 2017;7(2):119–45.
9. Skene DJ, Vivien-Roels B, Sparks DL, Hunsaker JC, Pevet P, Ravid D, Swaab DF. Daily variation in the concentration of melatonin and 5-methoxytryptophol in the human pineal gland: effect of age and Alzheimer's disease. Brain Res. 1990;528:170–4.
10. Abbott SM, Zee PC. Irregular sleep-wake rhythm disorder. Sleep Med Clin. 2015;10(4):517–22.
11. Kripke DF, Youngstedt SD, Elliott JA, et al. Circadian phase in adults of contrasting ages. Chronobiol Int. 2005;22:695–709.
12. Yoon IY, Kripke DF, Elliott JA, et al. Age-related changes of circadian rhythms and sleep-wake cycles. J Am Geriatr Soc. 2003;51:1085–91.
13. Hofman MA, Swaab DF. Living by the clock: the circadian pacemaker in older people. Ageing Res Rev. 2006;5:33–51.
14. Mander BA, Winer JR, Walker MP. Sleep and human aging. Neuron. 2017;94(1):19–36.
15. Coppola G, Di Renzo A, Ziccardi L, Martelli F, Fadda A, Manni G, et al. Optical coherence tomography in Alzheimer's disease: a meta-analysis. PLoS One. 2015;10:e0134750.
16. Koronyo-Hamaoui M, Koronyo Y, Ljubimov AV, et al. Identification of amyloid plaques in retinas from Alzheimer's patients and noninvasive in vivo optical imaging of retinal plaques in a mouse model. NeuroImage. 2011;54:S204–17.
17. Rosenwasser AM, Turek FW. Neurobiology of circadian rhythm regulation. Sleep Med Clin. 2015;10(4):403–12.
18. Bunger MK, et al. Mop3 is an essential component of the master circadian pacemaker in mammals. Cell. 2000;103:1009–17.
19. Song H, et al. Aβ-induced degradation of BMAL1 and CBP leads to circadian rhythm disruption in Alzheimer's disease. Mol Neurodegener. 2015;10:13.
20. Cermakian N, Lamont EW, Boudreau P, Boivin DB. Circadian clock gene expression in brain regions of Alzheimer's disease patients and control subjects. J Biol Rhythm. 2011;26:160–70.
21. Logan RW, McClung CA. Rhythms of life: circadian disruption and brain disorders across the lifespan. Nat Rev Neurosci. 2019;20(1):49–65.

22. Van Someren EJW. Circadian rhythms and sleep in human aging. Chronobiol Int. 2000;17(3):233–43.
23. Ancoli-Israel S, Clopton P, Klauber MR, Fell R, Mason W. Use of wrist activity for monitoring sleep/wake in demented nursing-home patients. Sleep. 1997;20(1):24–7.
24. Gehrman P, Marler M, Martin JL, Shochat T, Corey-Bloom J, Ancoli-Israel S. The relationship between dementia severity and rest/activity circadian rhythms. Neuropsychiatr Dis Treat. 2005;1(2):155–63.
25. Tranah GJ, Blackwell T, Stone KL, et al. Circadian activity rhythms and risk of incident dementia and mild cognitive impairment in older women. Ann Neurol. 2011;70(5):722–32.
26. Rogers-Soeder TS, et al. Rest-activity rhythms and cognitive decline in older men: the osteoporotic fractures in men sleep study. JAGS. 2018;66:2136–43.
27. Marco EM, Velarde E, Llorente R, Laviola G. Disrupted circadian rhythm as a common player in developmental models of neuropsychiatric disorders. Curr Top Behav Neurosci. 2016;29:155–81.
28. Pat-Horenczyk R, Klauber MR, Shochat T, et al. Hourly profiles of sleep and wakefulness in severely versus mild-moderately demented nursing home patients. Aging Clin Exp Res. 1998;10:308–15.
29. ICSD-3. The international classification of sleep disorders: diagnostic and coding manual. 2nd ed. Darien (IL): American Academy of Sleep Medicine; 2014.
30. Witting W, Kwa IH, Eikelenboom P, Mirmiran M, Swaab DF. Alterations in the circadian rest-activity rhythm in aging and Alzheimer's disease. Biol Psychiatry. 1990;27(6):563–72.
31. Okawa M, Mishima K, Hishikawa Y, Hozumi S, Hori H, Takahashi K. Circadian rhythm disorders in sleep-waking and body temperature in elderly patients with dementia and their treatment. Sleep. 1991;14(6):478–85.
32. Lim AS, Yu L, Costa MD, et al. Increased fragmentation of rest activity patterns is associated with a characteristic pattern of cognitive impairment in older individuals. Sleep. 2012;35(5):633B–40B.
33. Esbensen AJ, Schwichtenberg AJ. Sleep in neurodevelopmental disorders. Int Rev Res Dev Disabil. 2016;51:153–91.
34. Takaesu Y, Komada Y, Inoue Y. Melatonin profile and its relation to circadian rhythm sleep disorders in Angelman syndrome patients. Sleep Med. 2012;13(9):1164–70.
35. Potocki L, Glaze D, Tan DX, et al. Circadian rhythm abnormalities of melatonin in Smith-Magenis syndrome. J Med Genet. 2000;37(6):428–33.
36. Heikkila E, Hatonen TH, Telakivi T, et al. Circadian rhythm studies in neuronal ceroid-lipofuscinosis (NCL). Am J Med Genet. 1995;57(2):229–34.
37. Santoro SD, Giacheti CM, Rossi NF, Campos LM, Pinato L. Correlations between behavior, memory, sleep-wake and melatonin in Williams-Beuren syndrome. Physiol Behav. 2016;159:14–9.
38. Okawa M, Nanami T, Wada S, et al. Four congenitally blind children with circadian sleep-wake rhythm disorder. Sleep. 1987;10(2):101–10.
39. Maaskant M, van de Wouw E, van Eijck R, Evenhuis HM, Echteld MA. Circadian sleep-wake rhythm of older adults with intellectual abilities. Res Dev Disabil. 2013;34:1144–51.
40. Afonso P, Brissos S, Figueira ML, et al. Schizophrenia patients with predominantly positive symptoms have more disturbed sleep-wake cycles measured by actigraphy. Psychiatry Res. 2011;189(1):62–6.
41. Bromundt V, Koster M, Georgiev-Kill A, et al. Sleepwake cycles and cognitive functioning in schizophrenia. Br J Psychiatry. 2011;198(4):269–76.
42. Vigano D, Lissoni P, Rovelli F, et al. A study of light/ dark rhythm of melatonin in relation to cortisol and prolactin secretion in schizophrenia. Neuro Endocrinol Lett. 2001;22(2):137–41.
43. Ayalon L, Borodkin K, Dishon L, Kanety H, Dagan Y. Circadian rhythm sleep disorders following mild traumatic brain injury. Neurology. 2007;68(14):1136–40.
44. Duclos C, Dumont M, Blais H, Paquet J, Laflamme E, de Beaumont L, et al. Rest-activity cycle disturbances in the acute phase of moderate to severe traumatic brain injury. Neurorehabil Neural Repair. 2014;28(5):472–82.

45. DelRosso LM, Hoque R, James S, et al. Sleep-wake pattern following gunshot suprachiasmatic damage. J Clin Sleep Med. 2014;10(4):443–5.
46. Borodkin K, Ayalon L, Kanety H, et al. Dysregulation of circadian rhythms following prolactin-secreting pituitary microadenoma. Chronobiol Int. 2005;22(1):145–56.
47. Mishima K, Okawa M, Hishikawa Y, Hozumi S, Hori H, Takahashi K. Morning bright light therapy for sleep and behavior disorders in elderly patients with dementia. Acta Psychiatr Scand. 1994;89:1–7.
48. Skjerve A, Holsten F, Aarsland D, Bjorvatn B, Nygaard HA, Johansen IM. Improvement in behavioral symptoms and advance of activity acrophase after short-term bright light treatment in severe dementia. Psychiatry Clin Neurosci. 2004;58(4):343–7.
49. Satlin A, Volicer L, Ross V, Herz L, Campbell S. Bright light treatment of behavioral and sleep disturbances in patients with Alzheimer's disease. Am J Psychiatry. 1992;149:1028–32.
50. Van Someren EJ, Kessler A, Mirmiran M, Swaab DF. Indirect bright light improves circadian rest-activity rhythm disturbances in demented patients. Biol Psychiatry. 1997;41:955–63.
51. Ancoli-Israel S, Martin JL, Kripke DF, Marler M, Klauber MR. Effect of light treatment on sleep and circadian rhythms in demented nursing home patients. J Am Geriatr Soc. 2002;50:282–9.
52. Ancoli-Israel S, Gehrman P, Martin JL, et al. Increased light exposure consolidates sleep and strengthens circadian rhythms in severe Alzheimer's disease patients. Behav Sleep Med. 2003;1:22–36.
53. Fetveit A, Skjerve A, Bjorvatn B. Bright light treatment improves sleep in institutionalised elderly—an open trial. Int J Geriatr Psychiatry. 2003;18:520–6.
54. Fetveit A, Bjorvatn B. The effects of bright-light therapy on actigraphical measured sleep last for several weeks post-treatment. A study in a nursing home population. J Sleep Res. 2004;13:153–8.
55. Dowling GA, Burr RL, Van Someren EJ, et al. Melatonin and bright-light treatment for rest-activity disruption in institutionalized patients with Alzheimer's disease. J Am Geriatr Soc. 2008;56:239–46.
56. Auger RR, Burgess HJ, Emens JS, Deriy LV, Thomas SM, Sharkey KM. Clinical practice guideline for the treatment of intrinsic circadian rhythm sleep-wake disorders: advanced sleep-wake phase disorder (ASWPD), delayed sleep-wake phase disorder (DSWPD), non-24-hour sleep-wake rhythm disorder (N24SWD), and irregular sleep-wake rhythm disorder (ISWRD). An update for 2015: an American academy of sleep medicine clinical practice guideline. J Clin Sleep Med. 2015;11(10):1199–236.
57. Singer C, Tractenberg RE, Kaye J, Schafer K, Gamst A, Grundman M, Thomas R, Thal LJ. Alzheimer's disease cooperative study. A multicenter, placebo-controlled trial of melatonin for sleep disturbance in Alzheimer's disease. Sleep. 2003;26(7):893–901.
58. Riemersma-van der Lek RF, Swaab DF, Twisk J, Hol EM, Hoogendijk WJ, Van Someren EJ. Effect of bright light and melatonin on cognitive and noncognitive function in elderly residents of group care facilities: a randomized controlled trial. JAMA. 2008;299(22):2642–55.
59. Wright B, Sims D, Smart S, et al. Melatonin versus placebo in children with autism spectrum conditions and severe sleep problems not amenable to behaviour management strategies: a randomised controlled crossover trial. J Autism Dev Disord. 2011;41:175–84.
60. Pillar G, Shahar E, Peled N, Ravid S, Lavie P, Etzioni A. Melatonin improves sleep-wake patterns in psychomotor retarded children. Pediatr Neurol. 2000;23:225–8.
61. Gringras P, Nir T, Breddy J, Frydman-Marom A, Findling RL. Efficacy and safety of pediatric prolonged-release melatonin for insomnia in children with autism spectrum disorder. J Am Acad Child Adolesc Psychiatry. 2017;56(11):948–57.
62. Braam W, van Geijlswijk I, Keijzer H, Smits MG, Didden R, Curfs LM. Loss of response to melatonin treatment is associated with slow melatonin metabolism. J Intellect Disabil Res. 2010;54(6):547–55.
63. Braam W, Keijzer H, Struijker Boudier H, Didden R, Smits M, Curfs L. CYP1A2 polymorphisms in slow melatonin metabolisers: a possible relationship with autism spectrum disorder? J Intellect Disabil Res. 2013;57(11):993–1000.

64. Kennaway DJ. Potential safety issues in the use of the hormone melatonin in paediatrics. J Paediatr Child Health. 2015;51(6):584–9.
65. De Leersnyder H, Bresson JL, de Blois MC, Souberbielle JC, Mogenet A, Delhotal-Landes B, et al. Beta 1-adrenergic antagonists and melatonin reset the clock and restore sleep in a circadian disorder, Smith-Magenis syndrome. J Med Genet. 2003;40(1):74–8.
66. Murray G, Harvey A. Circadian rhythms and sleep in bipolar disorder. Bipolar Disord. 2010;12(5):459–72.
67. Logan RW, McClung CA. Rhythms of life: circadian disruption and brain disorders across the lifespan. Nat Rev Neurosci. 2019;20(1):49–65.

Chapter 11
Shift Work Sleep Disorder

Alok Sachdeva and Cathy Goldstein

Introduction

Shift work is labor that occurs outside of traditional clock times (09:00 to 17:00). A shift may occur in the early morning (starting between 04:00 and 07:00 hours), the evening (from 14:00 hours to midnight), or the night (from 21:00 to 08:00 hours), and the schedule of one worker may include a rotation of different shifts [1, 2]. As long as there have been human civilizations with economies supporting a labor force, laborers have worked during shifts that overlapped with their habitual sleep periods. The commonality of shift work increased markedly during the Industrial Revolution (1760–1840 C.E.) and following the widespread use of electric lighting (late nineteenth century) [3]. Now more than ever, in a global economy driven by the rapid transfer of information, currency, and products across time zones, shift work is an economic necessity.

Although some shift workers are asymptomatic, a significant proportion develop difficulties sleeping outside of the work period and/or remaining alert during their shift, as a result of the mismatch between the endogenous homeostatic and circadian drives for sleep and wakefulness. If these complaints are accompanied by a reduction of total sleep time, and are present for at least 3 months, she/he may be diagnosed with shift work disorder (SWD) [1]. This condition can affect up to one-third of shift workers and is associated with adverse health consequences [4]. During the past 50 years, research has furthered our understanding of the generation of circadian rhythms and the complex interaction between the circadian timing system with

A. Sachdeva (✉) · C. Goldstein
Michael S. Aldrich Sleep Disorders Center, Department of Neurology, University of Michigan Health System, Ann Arbor, MI, USA
e-mail: aloks@med.umich.edu

R. R. Auger (ed.), *Circadian Rhythm Sleep-Wake Disorders*,
https://doi.org/10.1007/978-3-030-43803-6_11

sleep, performance, and health. In this chapter, we will review key aspects of shift work and SWD and will discuss promising directions for future research in this area.

Epidemiology

Although there is an increasing amount of information on the prevalence of shift work and SWD, more studies are needed to characterize changes in the prevalence over time, to elucidate the socioeconomic costs, and to identify populations at greatest risk. Approximately 10–20% of the workforce of industrialized nations worldwide participate in shift work [2, 5–7]. These estimates derive primarily from surveys conducted by nations' Departments of Labor; there are no studies that formally compare shift work in different nations with respect to variables such as gross domestic product (GDP), dominant industries (as a percentage of GDP), primary exports, or extent of industrialization. Therefore, we have a sense for the global prevalence of shift work without a deeper understanding of the factors driving the comparative prevalence in different economies. In the United States of America (USA), the most recent 2004 survey by the Department of Labor showed that approximately 15% of the workforce participated in shift work with 3% in night shifts. The same survey demonstrated that more men than women (17% vs. 12%) and more African-Americans than other races (21% vs. 14–16%) participated in shift work [6]. The prevalence of shift work is known to vary by industry and is highest in protective and food services (40–50% of employees) followed by the healthcare and transportation industries (25% of employees) [2, 6, 8].

Prevalence estimates for SWD decrease to approximately 1–5% with the accompanying requirements of insomnia and/or reduced alertness [1]. Based on this estimate and given a 2018 US population of 320 million, there may be up to 16 million people in the USA with SWD. Drake and colleagues found SWD in 32% of night shift workers and 26% of rotating shift workers. In that investigation, the "true prevalence" of SWD in shift workers (given insomnia and/or excessive sleepiness in 18% of day workers) was extrapolated to 10% [4]. In an Australian cohort, a questionnaire yielded a similar SWD prevalence of 32% in night shift workers and 10% in "day" workers (defined as those who worked during the morning, the afternoon, or a combination of these) [9]. Although these two studies had methodological differences, different populations, and different ways of defining SWD, both suggest that the condition may be present in up to one-third of night shift workers. Importantly, SWD that produces significant negative impact on social, family, or work relationships was estimated at 9% among night workers [9].

Shift work disorder prevalence rates are almost always higher in night shift workers than in workers of other shifts (rotating, evening, etc.), suggesting that misalignment between the endogenous circadian phase and the required rest/wake times is the primary driving force in its development. Data suggest that the prevalence of SWD increases with age and is higher in women than in men, but there are no studies that report differences in SWD prevalence based upon race or ethnicity

[9–12]. Evidence for varying SWD prevalence in different occupations has emerged (32% in nurses, 23% in oil rig workers, and 5% in police officers), but the effect of variables such as shift duration and intensity remains unknown [10, 13, 14].

Physiology and Pathophysiology

The regulation of sleep is based upon a two-process model which summarizes interactions between the homeostatic sleep drive ("Process S") and the circadian timing system ("Process C") [15, 16]. When functioning appropriately, the homeostatic sleep drive increases during wakefulness and decreases during sleep, while the circadian alerting signal increases during the day and peaks in the evening as a counteractive force, thereby maintaining a consolidated period of wakefulness. By contrast, the circadian alerting signal declines sharply in the evening and, combined with a high homeostatic sleep drive, facilitates sleep onset (Fig. 11.1) [17].

Process S is mediated largely by molecules called somnogens that increase sleep propensity; examples include (but are not limited to) adenosine, prostaglandin D2, interleukin 1-beta, and tumor necrosis factor alpha [18–22]. The concentration of sleep-promoting adenosine, for example, has been shown to increase in the basal forebrain during prolonged wakefulness and decrease during sleep [18, 19, 23].

The central circadian clock, located in the suprachiasmatic nucleus (SCN) of the hypothalamus, regulates the oscillation of multiple biological processes (including the sleep-wake cycle) which repeat, on average, every 24.2 hours (circadian period or *tau*) in the absence of environmental cues [24–30]. Circadian rhythms are essential for timing physiological functions of the organism to achieve optimal coordination with the external light-dark cycle. As such, the genetic mechanisms underlying

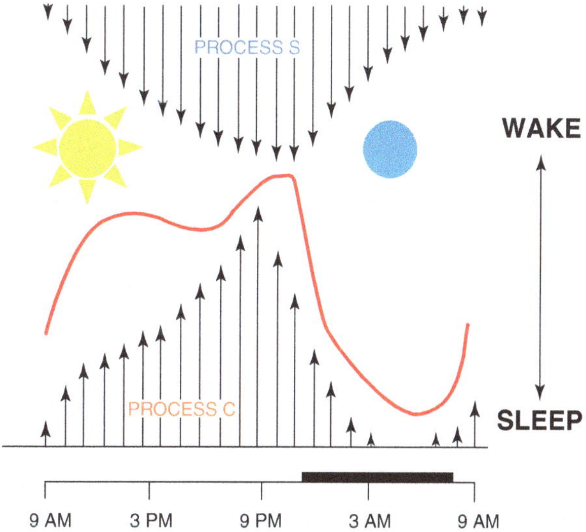

Fig. 11.1 Homeostatic sleep drive (Process S) and the circadian alerting signal (Process C)

circadian rhythmicity have been tremendously well-preserved throughout the evolution of species [31–36]. In addition to the central circadian regulation of sleep-wake, melatonin, body temperature, cortisol, and other endocrine processes, peripheral pacemakers are widespread throughout the human body and, in concert with the SCN, play an important role regulating the circadian rhythmicity of physiologic processes [37–39].

The human SCN maintains entrainment to the 24-hour day via external "time givers" or *zeitgebers*, the strongest of which is light, although exogenous melatonin, feeding, exercise, and social patterns can also influence one's circadian timing. Light is received by the melanopsin-containing intrinsically photosensitive retinal ganglion cells (iPRGCs) and afferent signals are relayed to the SCN through the retinohypothalamic tract (RHT) [40–44]. The SCN then regulates the pineal gland's secretion of melatonin and therefore, the pattern of melatonin secretion reflects the entrainment of the central clock. Additionally, melatonin provides feedback to the SCN. The mechanisms of SCN entrainment, self-sustained rhythmicity, and numerous downstream pathways are beyond the scope of this chapter; please see Reppert et al. for a detailed review [25].

The ability of light and exogenous melatonin to provoke either an advance or a delay of circadian phase is dependent upon biological time as summarized by phase response curves [45, 46]. For light exposure, the transition point from delay to advance occurs 2–3 hours prior to habitual sleep offset coinciding with the timing of the core body temperature minimum (CBTmin). Therefore, light received during the last approximately 4 hours before habitual sleep onset through the time of CBTmin produces a phase delay. Conversely, light stimuli that occur after the time of CBTmin through the first few hours following natural wake time produce a phase advance. Exogenous melatonin's phase shifting capabilities are anchored by endogenous dim light melatonin onset (DLMO), which occurs approximately 2 hours prior to habitual sleep onset. Exogenous melatonin will produce a phase advance of 0.5–1 hour when administered prior to the time of DLMO and a phase delay of similar magnitude if dosed after the occurrence of DLMO. The magnitude of phase shift produced by exogenous melatonin is greatest when endogenous melatonin secretion is lowest [26, 47–49]. The utility of the phase response curve in the treatment of SWD will be discussed later in this chapter.

The symptoms of excessive sleepiness and insomnia that are core to SWD are due to both misalignment of circadian phase with the required schedule and reduced total sleep time. Insomnia results from a high circadian alerting signal coinciding with the allotted period for sleep (e.g., during the day in a night shift worker). The inability to sleep during the available window results in sleep deprivation and increased homeostatic sleep drive. Excessive sleepiness due to sleep deprivation is further magnified by a work shift aligning with the lowest circadian alerting signal (biological nighttime).

Indeed, symptomatic night workers demonstrated a circadian phase (as measured by the pattern of melatonin secretion) that promoted daytime wakefulness and nighttime sleep, while the melatonin levels of asymptomatic individuals were high during the day and lowest at night [50]. The capacity to shift the central circadian rhythm to the desired schedule is likely multifactorial and may be behavioral (e.g., related to light exposure on work days and days off) and/or related to inherent factors such as chronotype and genotype. Chronotype, or individual diurnal preference, is known to influence tolerance to shift work and morning types may have greater difficulty adapting to a night shift work schedule [51]. Circadian phenotype is driven in part by genotype, and variable number tandem repeat (VNTR) polymorphisms of the clock gene *PER3* have been associated with chronotype [52, 53]. Additionally, the VNTR polymorphism may predict resilience to sleep loss, propensity for sleep independent from circadian phase, and whether a night shift worker's daytime sleep disturbance is driven by stress reactivity versus circadian misalignment [54–56]. Further work to understand the genetic differences that underlie the interindividual variability in tolerance to atypical schedules will likely provide greater insight into identifying vulnerabilities to SWD.

Clinical Features

Insomnia

Shift workers report 30–60 minutes less sleep than day workers, while individuals with SWD report 90 minutes less sleep [4]. Shift workers often report a sleep duration of less than 6 hours per 24-hour period, a problem that may be most pronounced in those working night or rotating shifts [4, 57, 58]. According to the Consensus Statement of the American Academy of Sleep Medicine and Sleep Research Society, regularly sleeping less than 7 hours per 24-hour period is associated with adverse health outcomes, impaired performance, and a greater risk of accidents [59].

The prevalence of insomnia in shift workers is significantly higher than the general population, ranging from 20 to 50% depending on the characteristics of the cohort and the instrument used to assess symptoms [4, 13, 60–62]. Insomnia in shift workers is associated with sleepiness, fatigue, impaired work performance, anxiety, and depressed mood [60, 61]. One investigation demonstrated an increased likelihood of insomnia among shift workers (odds ratio [OR] = 1.45, 95% CI = 1.04–2.02) even after termination of shift work [63, 64].

Interindividual variability has been observed in shift workers with insomnia, and phenotypes of alert (AI) and sleepy insomniacs (SI) have been described [65]. The AI group was found to have increased sleep onset latency and low sleep efficiency during both nocturnal and diurnal sleep and normal mean sleep latencies during the multiple sleep latency test (MSLT). Collectively, these findings suggest cortical hyperarousal. By contrast, sleepy insomniacs had very low mean sleep latencies on MSLT and an increased sleep onset latency only during diurnal and not nocturnal sleep. Thus, it has been hypothesized that the insomnia of the SI group may have a stronger association with circadian misalignment [65–67]. Of note, the AI group was found to have greater impairment on functional assessments compared to the SI phenotype. As discussed previously, impaired sleep often coexists with and can worsen excessive sleepiness, functional impairment, and increased accident rates during work.

Excessive Sleepiness, Work Dysfunction, and Accidents

Fatigue and sleepiness during shift work, most prominent during night and rotating shifts, result from a combination of factors including circadian misalignment and cumulative sleep deprivation [4]. A number of devastating industrial errors have occurred during the night and were associated with probable lapses in shift worker performance: these include the 1979 release of radioactive material from the nuclear plant at Three Mile Island in Dauphin County, Pennsylvania, USA; the 1984 leak of methyl isocyanate from the Union Carbide India pesticide plant in Bhopal, Madhya Pradesh, India; the 1986 explosion of the core of the Chernobyl nuclear power plant in Pripyat in northern Ukraine; and the 1989 *Exxon Valdez* oil spill near Tatitlek, Alaska, USA [68]. In the context of shift work, the commonly used term "excessive daytime sleepiness" must be replaced with "excessive sleepiness" for purposes of inclusivity and accuracy, given that nighttime rather than daytime sleepiness is often the primary problem.

Both excessive sleepiness and unintentional sleep are common in night shift workers and far exceed rates in day shift workers. Prior studies of sleepiness in shift workers have shown that 80–90% report feeling "sleepy" or "tired" at some point during their shift with a peak of sleepiness between 4 and 6 AM, corresponding with the nadir of one's core body temperature and circadian alerting signal [69]. Examples include a cohort of train drivers evaluated by Akerstedt and colleagues; in this group 11% of drivers reported "dozing off" on most night trips, while zero reported dozing off during day trips [69–71]. In a Massachusetts hospital, more than 30% of rotating and night shift nurses endorsed "nodding off" at work at least once per week, while this occurred in only approximately 3% of day and evening nurses [72]. A 2002 study of train drivers and traffic controllers demonstrated a 6–14-fold greater risk of severe sleepiness during the night shift compared with the day shift, and the risk increased by 15% for each additional hour worked [73].

As described by Dinges and colleagues, cumulative sleep debt and circadian misalignment increase one's sleep propensity, requiring increased efforts to remain awake and involuntary intrusions of drowsiness ("microsleeps"). These phenomena decrease work efficiency, performance quality, response time, processing speed, and short-term memory function, making shift workers more vulnerable to errors and accidents [74]. Electroencephalogram (EEG) studies of night shift workers have shown a doubling of the alpha and theta power density as well as slow roving eye movements in 50% or more of the electrooculogram (EOG) recordings [71, 75].

A decline in the performance of shift workers, especially night shift workers, has been shown in many studies across different industries. Similar to the time of peak sleepiness during a night shift, the greatest decline in performance is thought to occur between 4 and 6 AM, corresponding to the minimum core body temperature [17, 74, 76–78]. Tasks requiring sustained attention may pose the greatest challenge to sleepy shift workers [79], a problem that is especially important in high-stakes industries such as transportation and healthcare [69, 74]. Accident and error rates have been shown to be twice as high for rotating nurses than for day or evening nurses, even when rates were adjusted for age and experience [72]. Furthermore, the risk of accidents increased over successive nights of work with 6% increased risk on night two, 17% increased risk on night three, and 36% increased risk on night four. Longer work periods also correlated to a greater risk of errors, in keeping with prior studies showing more sleepiness with longer work periods [68, 73]. For example, errors were significantly more likely to occur when nurses worked longer than 12.5 hours (OR = 3.29, $p = 0.001$), worked overtime, or worked more than 40 hours per week (OR = 1.96, $p < 0.0001$) [80]. Other groups have also shown that there is an exponential increase in the likelihood of accidents as the duration of a shift increases; the risk of accidents after 12 hours of work is thought to be approximately twice the risk after 8 hours of work [81].

Accidents and errors in shift workers are not wholly explained by the low circadian alerting signal that coincides with shift timing, but also due to the cumulative sleep debt that results from insomnia; losing 2 hours of sleep per day for 1 week has been shown to cause decrements of performance comparable to those seen after 24 hours of continuous wakefulness [82, 83]. The severity of this impairment is further highlighted by the finding that task performance after 24 hours of wakefulness was comparable to that of individuals with a blood alcohol concentration (BAC) of 0.10 (above the legal limit of 0.08 in most US States) [83, 84]. While sleep deprivation causes all people to experience a decline in cognitive performance, the extent of that decline varies between subjects, and more research is needed to elucidate why some people may have a greater tolerance for sleep deprivation [85].

In closing, SWD is associated with disruptions of sleep and wakefulness that are most common in rotating and night shift workers, likely due to the greater degree of circadian misalignment. Insomnia during planned sleep periods and excessive sleepiness during work shifts promote a self-reinforcing cycle of sleep-wake impairment, often resulting in a decline in work performance and an increased susceptibility to accidents (Fig. 11.2).

Accidents that occur during shift work can cause injuries, material damage, and/or financial loss and may have negative consequences extending far beyond the boundaries of the workplace. The magnitude of these risks is rivaled by the risk of adverse health consequences that may result from shift work.

Adverse Health Consequences

The potential adverse health consequences include cardiovascular, metabolic, neoplastic, gastrointestinal, reproductive, and psychiatric disorders, among others.

Studies that evaluated the association between shift work and various chronic diseases have revealed contradictory results. The assessment and comparison of

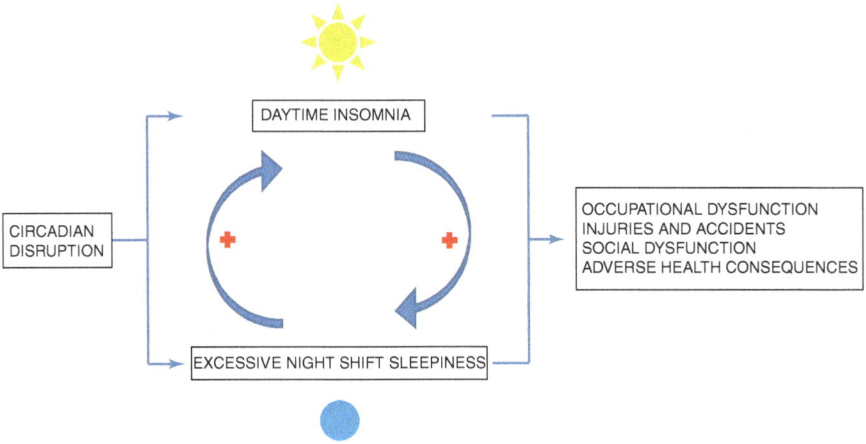

Fig. 11.2 Circadian disruption can cause a self-reinforcing cycle of sleep-wake impairment with negative medical, occupational, and social consequences in patients with shift work disorder

prior studies is limited by different definitions of shift work, heterogeneous populations, and small sample sizes that may under-power analyses. Selection bias, especially the "healthy worker" effect (healthier workers remain in their job and less healthy workers drop out of the workforce and are not studied), may also alter the perceived impact of shift work on general health [86, 87].

Cardiovascular Disease

Both circadian misalignment and sleep loss may contribute to the increased risk of cardiovascular disease in shift workers, either directly or by increasing the risk of other mediating factors such as impaired glucose metabolism or obesity. Specific mechanisms that may predispose to cardiovascular disease include a "non-dipper" blood pressure pattern in night workers [86] and altered circadian gene expression (e.g., PER and CRY) that may increase the risk of vascular endothelial damage [88].

The largest review of observational studies on shift work and cardiovascular disease by Vyas and colleagues (2,011,935 subjects) found an association between shift work and myocardial infarction (RR = 1.23, 95% CI = 1.15–1.31) despite controlling for relevant confounders such as tobacco use. Shift work was not associated with an increased mortality risk. Workers of night shifts had the greatest increased risk for "coronary events" (RR = 1.41, 95% CI = 1.13–1.76), although other investigations have demonstrated greater cardiovascular risk with rotating shifts [89, 90]. Individual studies of ischemic heart disease (IHD) in shift workers report a relative risk ranging from approximately 1.2–2.8, depending on the study and the population [89–93]. Exposure time appears relevant to increasing risk after longer duration of shift work [91, 92]. A few well-powered studies have found no difference in the risk of IHD or mortality between shift workers and day workers [94–96]. As mentioned previously, these divergent conclusions likely result from the heterogeneity of populations, study design, and bias.

Studies on the association between shift work, carotid artery disease, and ischemic stroke are more limited. A greater common carotid artery mean intima-media thickness in male shift workers compared to day workers was noted in a population of 1543 Finnish men and women ages 24–39. Male but not female shift workers also were more likely to have a carotid plaque (OR = 2.2, 95% CI = 1.2–4.0) [97]. However, in a large study of female nurses, a 4% increased risk of ischemic stroke for every 5 years of rotating night shift work was observed [98].

While there is a high likelihood that shift work increases the risk of cardiovascular disease, conflicting results necessitate more well-powered prospective studies with precisely defined exposure to test this hypothesis.

Metabolic Abnormalities

Lending further credence to the link between shift work and cardiovascular disease is the strong association between shift work and the metabolic syndrome, a clustering in one person of at least three of the following conditions: central obesity, hypertension, hyperglycemia, hypertriglyceridemia, and low high-density lipoprotein (HDL) levels. Many studies across various populations worldwide have shown an odds ratio varying from 1.1 to 2.4 for the development of the metabolic syndrome in shift workers relative to day workers [99–103]. This association remained in all cases after adjustment for other factors such as smoking, physical activity, and age. Most of these studies compared day workers to rotating shift workers, and one study focused exclusively on men [102]. The variation in the calculated odds ratio among studies may have resulted from population heterogeneity, methodological variation, and differences in the definition of "metabolic syndrome"; nonetheless, an increased risk for incident metabolic syndrome in shift workers was consistently found. Furthermore, a study of 26,382 workers in China supported the association given that longer durations of shift work progressively increased the odds of developing the metabolic syndrome, a finding that was most prominent in women [103].

Additionally, the single components of the metabolic syndrome are all independently more likely to occur in shift workers than day workers [104–108]. Obesity in particular has been highly linked to shift work and found to be increasingly likely the longer and more frequently one works a night shift with an odds ratio as high as 3.9 (95% CI = 1.5–9.9) for women that worked eight or more night shifts per month [105, 109].

Similar dose-response results have been reported regarding shift work and incident diabetes [110–112]. An analysis of thousands of nurses from the Nurses' Health Study showed an increased risk of type 2 diabetes in night shift workers. The risk of developing diabetes increased from a hazard ratio of 1.05 (95% CI = 1.00–1.11) to 1.58 (95% CI = 1.43–1.74) with night shift work of 1–2 years as opposed to

20 years in duration [110]. Similar results were obtained in a study of the Danish Nurse Cohort after a 15-year follow-up period; the greatest odds of diabetes were seen in night shift workers, though evening shift workers were also at greater risk than day workers [111]. Similar to the metabolic syndrome, a strong association between diabetes and shift work persists despite differences in population, method, and terminology [113, 114].

Metabolic abnormalities, such as decreased leptin, increased glucose and insulin, and prediabetes, have been demonstrated in controlled laboratory settings that use carefully designed protocols to induce misalignment between endogenous circadian phase, sleep-wake, and feeding [115–118]. Therefore, the association between circadian misalignment and metabolic dysfunction is particularly strong, and shift workers are at relatively high risk for medical problems such as obesity, hypertriglyceridemia, and diabetes. Frequent assessment of the nutritional status of shift workers and counseling on healthy eating habits is an essential preventive measure to ensure optimal health in this vulnerable population [119, 120].

Neoplastic

In 2007, the International Agency for Research on Cancer (IARC) asserted, "shift work that involves circadian disruption is probably carcinogenic to humans." Governments and large employers with many shift workers must take heed of this problem and generate plans for education and to increase accessibility to preventive healthcare with cancer surveillance programs. In Denmark, women with breast cancer who have worked a night shift for 20 or more years may receive governmental compensation to aid with medical expenses [121].

The association between night shift work and cancer is strongest for breast carcinoma, even after controlling for confounding factors such as age, family history of breast cancer, postmenopausal hormone use, and others. A meta-analysis by Wang and colleagues reported a pooled adjusted relative risk of 1.19 (95% CI = 1.05–1.35) for any exposure to night shift work. Furthermore, they calculated a 3% increased risk for every 5 additional years of night shift work and a 13% increased breast cancer risk for every 500 additional night shifts worked [122]. Other studies have replicated this trend for increasing breast cancer risk as the duration of night shift work increases [123–125]. Schernhammer and colleagues showed an increased risk of breast cancer among those who had worked 20+ years of night shift (RR 1.79, 95% CI = 1.06–3.01) compared to those without night shift work, but no significant increased risk among those who worked a night shift for less than 20 years [126, 127]. Not all studies have confirmed this association, however [128, 129]. A similar association has been found between night shift work and endometrial cancer (especially among obese women with 20 or more years of rotating night shift work), but not ovarian cancer [130, 131].

Men who work night shifts face an increased risk of prostate cancer, with one meta-analysis estimating relative risk at 1.24 (95% CI = 1.05–1.46) which increased with increasing duration of work [132]; however, other investigations have not confirmed this association [129, 133–135].

Shift work has also been associated with an increased risk of colorectal cancer, bladder cancer, pancreatic cancer, and non-Hodgkin's lymphoma (NHL) [135–137]. Although an association between shift work and increased lung cancer risk has also been reported, cigarette smoking is a major confounder in these studies [135, 138].

The pathogenesis of shift work and cancer is thought to be related, in part, to the oncogenic potential of melatonin suppression in the setting of nocturnal light exposure [139–142]. Melatonin suppression may be particularly relevant to the development of breast and endometrial cancer, given melatonin's peripheral role as an estrogen receptor modulator with an antiproliferative effect [143–147]. However, melatonin suppression is only one of the mechanisms leading to an increased cancer risk in shift workers; circadian desynchrony, sleep deprivation, immune impairment, and systemic inflammation may also be implicated in promoting neoplastic cellular processes [143, 148–150].

Gastrointestinal Dysfunction

Shift work also can provoke gastrointestinal dysfunction via different mechanisms, including circadian disruption, sleep deprivation, and occupational stress. The gastrointestinal tract relies upon many processes that are tied to the body's circadian rhythms. For example, apart from its pineal origin, melatonin is produced by enterochromaffin cells of the gastrointestinal tract and plays a role in regulating intestinal myoelectric rhythms as well as balancing hunger and satiety. In addition, the timing of secretion of the peptide hormones ghrelin, somatostatin, and gastrin demonstrates circadian rhythmicity [151, 152]. While many gastrointestinal symptoms and diagnoses have been studied in the context of shift work, the majority of studies have investigated the connection between rotating or night shift work and gastrointestinal ulcer formation. It is possible that circadian disruption due to rotating or night shift work increases gastrin and pepsinogen secretion, resulting in increased ulcer risk [4].

A meta-analysis by Knutsson and colleagues in 2010 concluded that there is an increased risk of peptic ulcer disease (PUD) in shift workers, but also found inadequate control for confounders such as age, smoking, and socioeconomic status in some studies [153]. Interestingly, symptomatic SWD may be more relevant than exposure to shift work alone; Drake and colleagues reported that ulcers of the gastrointestinal tract were more common in rotating and night shift workers if insomnia or excessive sleepiness were present [4].

Associations between shift work (especially evening and rotating shifts) and a higher likelihood of irritable bowel syndrome (IBS), constipation, diarrhea, bowel movement irregularity, and abdominal pain also has been found [154–157].

Reproduction

Shift work during pregnancy seems to be a risk factor for low birth weight (accounting for gestational age) and preterm (less than 37 weeks) delivery with one study estimating an approximately twofold increased odds [158–161]. The risk of miscarriage also is likely higher in pregnant women who work irregular hours or rotating shifts compared to day workers, with large effect sizes (up to fourfold increased risk) [160, 162, 163]. The impact of shift work on pregnant women deserves further study, particularly because temporary scheduling changes may be feasible and have significant benefits.

Psychiatric Disorders

Sleep disturbance and circadian disruption are strongly and bidirectionally associated with increased rates of anxiety and depression [164–167]. It is not surprising, then, that shift work and SWD have been associated with the same, which can contribute to social and occupational dysfunction [4]. In a study of 98 current and former shift workers, a high prevalence of major depressive disorder was found, most pronounced in women and those with greater shift exposure [168]. Kalmbach and colleagues studied the effect of a transition to rotating shift work on 96 non-shift workers without baseline sleep or psychiatric disturbances. Individuals with greater sleep-reactivity (as measured by the Ford Insomnia Response to Stress Test) were more than five times more likely to develop SWD. In addition, SWD was associated with a greater increase in depression and anxiety symptoms [169]. Therefore, the assessment of sleep reactivity may allow us to identify shift workers at high risk for SWD and to take measures to prevent depression and anxiety.

Other

Shift work has also been implicated in increased risk of musculoskeletal and rheumatic disease [170–172] as well as multiple sclerosis [173, 174]. The effects of shift work on multiple organ systems demonstrate the profound relevance that adequate sleep duration and appropriate circadian alignment have for health and wellness.

Diagnostic Criteria and Differential Diagnosis

Diagnostic criteria for SWD have been defined by the American Academy of Sleep Medicine (AASM) in the third edition of the International Classification of Sleep Disorders (ICSD-3) and by the American Psychiatric Association (APA) in the fifth edition of the Diagnostic and Statistical Manual of Mental Disorders (DSM-5).

ICSD-3 diagnostic criteria for SWD include a 3-month history of insomnia and/or excessive daytime sleepiness with decreased sleep duration in the context of a work schedule that overlaps with the habitual sleep period (Fig. 11.3) [1].

The DSM-5 includes SWD in the category of circadian rhythm sleep disorders (CRSD), which is broadly described as clinical distress and/or functional impairment caused by excessive sleepiness or insomnia resulting from misalignment of one's intrinsic circadian rhythm with one's sleep-wake schedule (as determined by occupational, educational, social, or other obligations) [175]. Therefore, the fundamental features of SWD are similar in both the ICSD-3 and the DSM-5, but some differences are present; for example, sleep logs and actigraphy are explicitly mentioned in the ICSD-3, but not in the DSM-5. Additionally, the DSM-5 provides distinct time course designations for SWD symptoms: episodic (present for at least

Fig. 11.3 AASM ICSD-3 shift work disorder diagnostic criteria

SHIFT WORK DISORDER DIAGNOSTIC CRITERIA (AASM, ICSD-3)
1. Insomnia and/or excessive sleepiness with a reduction of total sleep time associated with a recurring work schedule that overlaps with the usual sleep period
2. Symptoms have been present (and associated with shift work) for at least 3 months
3. Symptoms cause clinically significant distress or functional impairment
4. Whenever possible, sleep logs and actigraphy for at least 14 days confirm a sleep-wake disturbance
5. Sleep and/or wake disturbances are not better explained by another sleep disorder, medical disorder, neurological disorder, mental disorder, medication use, or substance use disorder

1 month but less than 3 months), persistent (persistent for at least 3 months), or recurrent (two or more episodes within 1 year). As always, the clinician's assessment of the unique details of each patient's symptoms is an integral part of the diagnostic process.

The diagnosis of SWD is based primarily on the patient's history and establishing that the likely source of symptoms is exposure to a work schedule that interferes with the endogenous propensity for sleep and wakefulness. As indicated in both the ICSD-3 and DSM-5 diagnostic criteria, the diagnosis is predicated on the exclusion of other conditions as the primary cause of insomnia and/or excessive sleepiness. Given the strong association of shift work with metabolic and psychiatric disturbances, the exclusion of other causes of insomnia and/or excessive sleepiness (e.g., depression or sleep-disordered breathing) should be carefully pursued.

Evaluation

History

A thorough history is the cornerstone of a complete evaluation of a patient for SWD. This should include a detailed work history, including hours, location and commute, job responsibilities, occupational hazards/exposures, light-dark environment, and work-related stress. A safety assessment should be completed, including an assessment of sleepiness while driving a motor vehicle or operating heavy machinery and documentation of work-related errors or injuries that occurred or nearly occurred due to sleepiness.

An effort should be made to determine if the patient's symptoms are associated with the atypical shift schedule and resultant misalignment with the endogenous circadian phase. Therefore, the clinician must determine how the patient's habitual sleep period overlaps with the shift work schedule and screen for sleepiness at work as well as difficulty sleeping during the time allotted between shifts. The timing and quality of sleep on non-work days may reveal the individual's naturally preferred sleep-wake times. Because light exposure is the primary mechanism that entrains the central circadian clock, light exposure must be determined on both work and off days.

Sleep hygiene must be assessed with special attention given to light and noise exposure during sleep periods, planned or unplanned naps at work, and the use of stimulating substances (such as caffeine) near the end of a work shift. Once these fundamental historical features are clear, the clinician should screen for the presence of other sleep disorders, such as sleep-disordered breathing, restless legs syndrome, and chronic insomnia (not related to shift work). A comprehensive past medical and social history as well as review of systems may reveal medical and psychiatric causes of sleepiness and/or insomnia.

QUESTIONNAIRE	DESCRIPTION	NUMBER OF ITEMS	USE IN SHIFT WORKERS
Shiftwork Disorder Screening Questionnaire	Brief questionnaire identifies patients at high risk for SWD (PPV = 89%, NPV = 62%)[176]	4	• To identify possible SWD in the primary care setting • Validated by Barger et al.[176]
Munich Chronotype Questionnaire	Self-rated scale assesses circadian phase on work and work-free days	13	• Identification of chronotype may help guide shift selection and treatment of SWD • A version specific to shift workers has been developed[177]
Horne-Östberg Morningness-Eveningness Qusetionnaire	Self-rated scale classifies individuals as morning, evening or intermediate types	19	• Identification of morningness/eveningness may help guide shift selection and treatment of SWD • Not validated in shift workers[179]
Epworth Sleepiness Scale	The likelihood of dozing off or falling asleep in different situations generates a measure of one's overall sleep propensity	8	• To assess sleepiness Has been used extensively in studies of shift workers[4,10,13,63] • Lack of excessive sleepiness does not rule out SWD
Insomina Severity Index	Questions about insomnia symptoms and their impact quantify insomnia severity	7	• To assess insomina • Has been used extensively in studies of shift workers[60,65,190] • Lack of insomnia does not rule out SWD

Fig. 11.4 Questionnaires in the evaluation of insomnia and/or excessive sleepiness in shift workers

Questionnaires

Subjective historical information may be supplemented with the use of validated SWD questionnaires and chronotype assessments. Although a helpful adjunct, these tools are not systematically integrated into clinical practice. Figure 11.4 summarizes available questionnaires that may provide beneficial information in the evaluation of a shift worker (Fig. 11.4) [176–189].

Sleep Log and Actigraphy

In all cases of suspected shift work disorder, the patient should complete a sleep log for at least 2 weeks [190, 191]. Given that shift work disorder includes circadian disruption with insomnia and/or excessive sleepiness, both the 24-hour sleep log (which provides a helpful graphical representation of circadian sleep-wake patterns) and the consensus sleep diary (which adds important information about awake

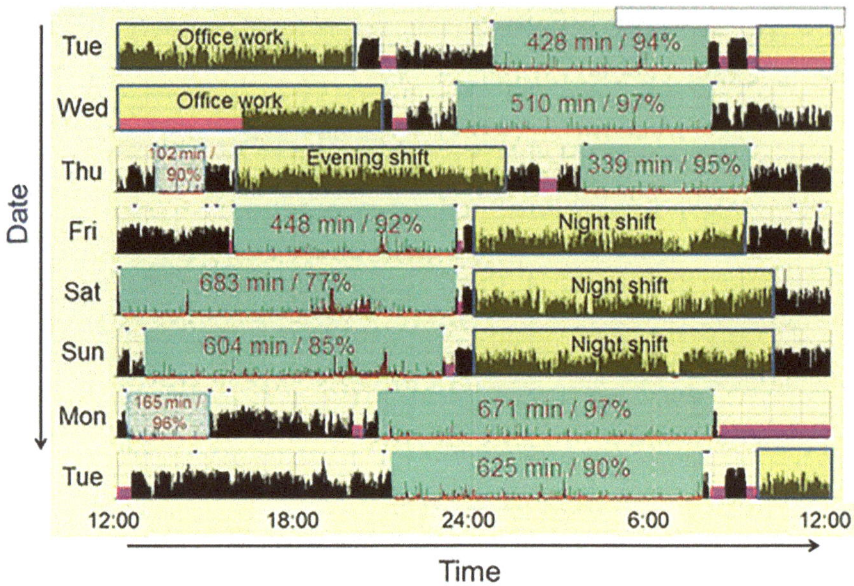

Fig. 11.5 Sample actigraphy of a patient with shift work disorder. Reprinted from Mizuno et al. [194] under the terms of the Creative Commons Attribution 4.0 International License (http://creativecommons.org/licenses/by/4.0/)

behaviors and sleep hygiene) are beneficial [192]. In addition, 2 weeks of concurrent actigraphy should be compared to the patient's sleep diary whenever possible [193]. Multiple important estimated sleep metrics can be derived from the sleep log and actigraphy data; these include bedtime, sleep onset latency, wake after sleep onset time, sleep efficiency, total sleep time, and wake-up time (Fig. 11.5).

Data from off days and work days may provide an enhanced degree of insight into the patient's symptoms and behavior. Certain actigraphy devices also can record light exposure [195]. Actigraphy is also useful to determine the frequency and duration of planned and unplanned naps that occur during a work shift and may be underreported in the sleep diary [196, 197].

Dim Light Melatonin Onset

Dim light melatonin onset (DLMO), often measured in saliva, can be used to determine one's circadian phase [47]. The DLMO time is less likely than core body temperature or serum cortisol to be "masked" or altered by other variables, but, as the name suggests, must be collected in dim light [198]. However, due to the cost and complexity of measurement and interpretation, the determination of the DLMO time is not routinely used in clinical practice [48]. Some have argued that circadian rhythm sleep-wake disorders such as SWD should not be treated until the objective circadian phase has been determined [199]. Current ongoing work pursues the development of a biomarker that accurately estimates circadian phase with low complexity and cost [200].

Other Laboratory Studies

Laboratory studies are not required to diagnose SWD, but may be helpful to rule out other medical causes of excessive sleepiness and/or insomnia. Clinical laboratory tests of a patient's blood and/or urine, including a urine drug screen, should be obtained judiciously and based upon the provider's working differential diagnosis. In addition, given shift workers' increased risk for medical problems such as hyperglycemia and hypertriglyceridemia, the clinician should consider screening for conditions such as diabetes that can decrease a patient's overall health and quality of life.

Polysomnography and Multiple Sleep Latency Test (MSLT)

Polysomnography (PSG) in shift workers may reveal a sleep architecture consistent with sleep deprivation, but PSG is not required to diagnose SWD [201]. Nonetheless, sleep laboratory testing should be pursued when appropriate to assess for other possible sleep disorders (such as obstructive sleep apnea and/or narcolepsy). Diagnosing central disorders of hypersomnolence such as narcolepsy can be difficult in shift workers given the nearly eightfold increased odds of positive in-laboratory testing (two or more sleep onset rapid-eye-movement (REM) periods during PSG and MSLT with a mean sleep latency of 8 minutes or less) [202].

Treatment: Shift Work Disorder and Symptoms Associated with Shift Work

Overview

In recent years, many published reviews have summarized the evidence supporting different treatment options for SWD and symptoms associated with shift work [2, 5, 190, 191, 203–205]. In 2007, the AASM published practice parameters for the evaluation and treatment of circadian rhythm sleep-wake disorders, including SWD. This report ranked the strength of treatment options for SWD as "standard," "guideline," or "option," in order of decreasing clinical certainty and level of evidence supporting each intervention (Fig. 11.6) [190].

SHIFT WORK DISORDER TREATMENT RECOMMENDATIONS (AASM Practice Parameters, Morgenthaler et al, 2007)[190]	
THERAPY	**LEVEL OF RECOMMENDATION**
Planned Sleep Schedules	Standard
Timed Light Exposure	Guideline
Timed Melatonin Exposure	Guideline
Hypnotic Medications	Guideline
Alerting Medications	Guideline
Stimulant Medications	Option

Standard:	A generally accepted patient-care strategy that reflects a high degree of clinical certainty. Level 1 Evidence, which directly addressed the clinical issue, or overwhelming Level 2 Evidence.
Guideline:	A patient-care strategy that reflects a moderate degree of clinical certainty. Level 2 Evidence or a consensus of Level 3 Evidence.
Option:	A patient-care strategy that reflects uncertain clinical use. Either inconclusive or conficiting evidence or conflicting expert opinion.

Fig. 11.6 Shift work disorder treatment recommendations. Compiled from data in Morgenthaler et al. [190]

The definitive treatment of SWD is discontinuation of shift work. If traditional work hours are not feasible, treatment should be pursued and includes four general components: prescribed sleep scheduling, circadian phase shifting, hypnotic (sleep-promoting) medications, and stimulant (wake-promoting) medications [191]. Despite a growing number of studies on this topic, there remains a paucity of randomized-controlled trials.

Sleep/Wake Scheduling and Naps

Because shift work can result in decreased total sleep time, sleep deprivation, and sleepiness, shift workers may benefit from sleep schedule interventions that aim to increase total sleep time and reduce sleepiness during awake periods. Shift workers need to allow for a sleep opportunity of sufficient duration prior to a work shift. Yeung and colleagues showed that a bedtime 10 hours before the start of a scheduled early-morning shift (between 0400 and 0730 hours) increased total sleep time by approximately 1 hour without detrimental effects on sleep onset latency, sleep efficiency, or sleep quality [206]. Well-timed naps that are short in duration (20–60 minutes) both before [207] and during the earlier parts of a shift [208] also can help ease sleepiness in shift workers. In addition to reduced sleepiness, naps can result in improved performance on vigilance testing [208–211]. The benefits of naps in shift workers were still observed despite low sleep efficiency and lack of stage REM and slow wave sleep during PSG recordings [211]. The impact of a nap on the main sleep bout should be assessed; one study reported impaired daytime sleep in night shift workers who took a 50-minute nap at either 0100 hours or 0400 hours [210]. It is important to note that appropriately scheduled naps were the only intervention recommended by the AASM at the "standard" level for the treatment of SWD [190].

Therefore, creating a greater sleep opportunity (advancing the bedtime) and scheduling naps can help to mitigate the excessive sleepiness that often results from circadian misalignment and sleep deprivation in shift workers. Naps during a shift should occur in the first half of a shift and should not exceed 45 minutes' duration, and patients should be cautioned about possible brief periods of sleep inertia and decreased alertness upon awakening from a nap. More studies are needed to assess the impact of naps on safety outcomes in the workplace and to study the relative benefit of prescribed sleep scheduling in patients with SWD [212].

Circadian Phase Shifting

Because the primary pathophysiology underlying SWD is the misalignment of the required schedule for sleep and wakefulness with the endogenous circadian rhythm, circadian phase shifting is a rational treatment approach. As previously discussed, although the central circadian clock oscillates in a self-sustaining manner that is independent of time cues, light and non-photic stimuli such as melatonin can entrain the endogenous circadian rhythm to external time.

Entraining a shift worker to a permanent night-work/day-sleep schedule is typically problematic given that most workers desire some degree of social and family engagement on days off. As a solution to this problem, a series of studies [213–217] defined a "compromise phase position" such that the nadir of circadian alertness occurs during the sleep period on both work days and days off. To reach this position, subjects in the experimental group were exposed to bright light pulses during a simulated night shift. They had a scheduled sleep period (in darkness) of 8:30–15:30 after the first two consecutive simulated night shifts and 8:30–13:30 after the third (last) simulated night shift. On the four simulated days off, subjects were instructed to sleep from 03:00 to 12:00. With the exception of light exposure within the first 2 hours of awakening during work days, subjects in the experimental group avoided daylight with sunglasses. Individuals fell into one of three groups based on estimated nadir of circadian alertness which was extrapolated from DLMO: completely re-entrained (circadian nadir the second half of daytime sleep following night shifts), partially re-entrained (circadian nadir during the first half of daytime sleep and towards the end of nighttime sleep on days off), and not re-entrained (circadian nadir during the night shift or commute home). Those with partial circadian re-entrainment experienced an improvement of performance, sleepiness, and mood that was comparable to those with complete re-entrainment.

Subjects without re-entrainment showed worse neurobehavioral performance despite adequate total sleep time during the day [218]. Therefore, partial re-entrainment to the "compromise phase position" is likely a sustainable method to promote alignment to night work while allowing for some degree of diurnal interaction on days off [216].

Many studies have shown that 4–6 hours of bright light exposure (2500–12,000 lux) either sustained or occurring intermittently throughout a night shift along with decreased light exposure during the day can re-entrain the circadian rhythm, improve alertness and cognitive performance, improve the quality of daytime sleep, and increase total sleep time in night shift workers [219–226]. Although most investigations have employed light exposure to produce a phase delay and enable night shift workers to sleep in the morning (0800–1600 hours) that follows their shift, others have demonstrated that bright light timed to promote a phase advance can result in high-quality evening sleep (1400–2200 hours) prior to the work period and improve performance during the night shift [227].

Night shift workers, especially those with SWD, should expose themselves to bright light (at least 2500 lux for 4–6 hours) throughout the night shift and minimize light exposure during the day (with sunglasses, blackout shades, etc.) to reduce circadian misalignment and improve daytime sleep quality. Light exposure interventions are also expected to benefit work performance.

Both the circadian phase shifting and soporific properties of exogenous melatonin have been studied extensively in night shift workers and during simulated shift work protocols, but not specifically among subjects with SWD. Placebo-controlled trials of varying doses (1.8–10 mg) and formulations of melatonin given to night shift workers prior to daytime sleep have shown improved sleep quality and increased total sleep time, though these changes were not always statistically significant [228–231]. Patients can develop tolerance to the sleep-promoting effects of melatonin and report decreasing effectiveness over time [230]. Studies of melatonin prior to daytime sleep have not shown a consistent improvement of night shift alertness or performance [230, 231].

The circadian phase shifting properties of melatonin are well-established; the greatest phase advance occurs when melatonin is timed 2–4 hours prior to the DLMO, while a phase delay is most pronounced when melatonin is timed 12–14 hours after DLMO [232]. Therefore, in addition to promoting sleep, melatonin dosed prior to the daytime sleep period could elicit a phase delay to help individuals acclimate to night shift work.

In summary, melatonin before daytime sleep may improve sleep quality and total sleep time for night shift workers, though these benefits could decrease over time due to tolerance. Unfortunately, melatonin appears unlikely to provide measurable benefit with respect to night shift alertness or performance.

Ramelteon and tasimelteon, melatonin receptor agonists, also may be effective for phase shifting and daytime insomnia in night shift workers, but the relative efficacy of these medications compared to melatonin has not been established and may not justify their increased cost [233, 234].

Hypnotic (Sleep-Promoting) Medications

Benzodiazepine receptor agonist (BZRA) medications exert their sleep-promoting effects by way of receptors for gamma-aminobutyric acid (GABA), the chief inhibitory neurotransmitter of the central nervous system. The usefulness of multiple benzodiazepine (triazolam, temazepam) and non-benzodiazepine (zopiclone, zolpidem) BZRA medications have been formally evaluated in shift workers and in simulated shift work conditions. Although all of these studies showed increased total sleep time, improved sleep efficiency, decreased sleep onset latency, and/or increased sleep quality in subjects during the designated sleep period, improvement of alertness and psychomotor performance during work hours was insignificant or absent in all but one study [235–239]. "Hangover effect" and reduced work performance were not seen in hypnotic-treated subjects during these studies, but one study reported worsened next-day mood in subjects treated with zolpidem [237, 239]. To assess combination therapy in simulated shift work conditions, zolpidem 1 hour before bedtime was given with 10 mg of oral methamphetamine 1 hour after awakening, and these medications together did not significantly improve shift-change mood or performance [240].

Despite the lack of negative effects of hypnotic medications on work performance in the above studies, brief interventions often in simulated shift work conditions may lack external validity and generalizability. Medications such as zolpidem and triazolam can impair memory, learning, and psychomotor performance in some patients; as such, clinicians should individualize care and counsel shift workers (with or without SWD) about possible adverse effects of hypnotic medications that could compromise work performance and increase the likelihood of accidents and injuries [190, 241].

Stimulant (Wake-Promoting) Medications

As discussed previously, excessive sleepiness during shift work can have serious implications for performance and safety. Therefore, stimulant medications may be required to treat excessive sleepiness and reduce the likelihood of work-related error and injury. Stimulants such as caffeine and amphetamine-based medications have been tested in both real and simulated shift work conditions. Modafinil and armodafinil have been tested extensively for sleepiness in SWD and are approved by the US Food and Drug Administration (FDA) for this purpose.

Caffeine, a methylxanthine stimulant that acts primarily by way of adenosine receptor antagonism, has been shown in doses equivalent to 2–4 cups of coffee to decrease sleepiness during a work shift and reduce the number of errors made by shift workers compared to placebo [242–244]. In addition, napping and caffeine have been shown to improve alertness more so than either intervention alone [245].

Analyses of the effect of caffeine on measures of performance have yielded mixed results, though a meta-analysis showed that caffeine likely causes some degree of improvement in reasoning, memory, and attention during shift work [243–245]. Whether or not caffeine can reduce injury rates in shift workers is unknown. Although caffeine is relatively safe, it can provoke symptoms such as insomnia or palpitations and should be used cautiously, especially in patients with cardiovascular disease.

Formal studies of the use of dextroamphetamine and/or methylphenidate for shift work or SWD are sparse, and there are no randomized-controlled trials. Methamphetamine doses of 5–10 mg, taken prior to a simulated night shift, were found to reverse the decline in performance and mood related to circadian disruption [246]. While these results are encouraging, clinicians and patients must be cautious about possible central nervous system (insomnia, headache, anxiety, emotional lability) and cardiovascular (hypertension, tachycardia, palpitations) side effects of amphetamines, as well as the potential for abuse. More studies are needed to assess the safety and efficacy of long-term amphetamine use in patients with SWD.

Modafinil and armodafinil increase dopamine levels in the brain by inhibiting its reuptake; however, the complete mechanism of action by which these medications promote wakefulness is unknown. Modafinil's efficacy in patients with SWD has been shown in two randomized, double-blind, placebo-controlled trials. Modafinil increased mean sleep latency on MSLT and reduced both the frequency and duration of lapses of attention as measured by the psychomotor vigilance test (PVT). Patients taking modafinil reported less accidents or near accidents while commuting home, but excessive subjective sleepiness persisted [247]. Workers with SWD who received modafinil doses of 200 mg or 300 mg 30–60 minutes before the night shift also displayed improved measures of health-related quality of life without significant daytime insomnia resulting from treatment [248].

Randomized, double-blind trials of armodafinil 150 mg 30–60 minutes before a night shift versus placebo in hundreds of subjects with SWD have also shown reduced objective sleepiness, reduced subjective sleepiness, and improved performance on standardized memory and attention tasks [249–252]. In addition to the reduction of objective sleepiness as measured by the multiple sleep latency test (MSLT), Drake and colleagues showed that armodafinil 150 mg improved driving performance relative to placebo during a simulated driving test after night shift work [251]. Armodafinil was well-tolerated in all studies and, similar to modafinil, the most common side effect was headache [250, 253]. A minority of patients taking modafinil or armodafinil also reported daytime insomnia [248, 254].

In summary, the optimal treatment of SWD requires a careful, individualized strategy that integrates some or all of the following treatment modalities: prescribed sleep/wake scheduling, circadian phase shifting, hypnotic (sleep-promoting) medications, and stimulant (wake-promoting) medications.

Conclusion

Shift work is very common, especially in certain industries such as healthcare and transportation. As many as one-third of shift workers may develop SWD, a condition characterized by insomnia and/or excessive sleepiness with reduced total sleep time. SWD can impair job performance, provoke work-related injuries, reduce one's overall quality of life, and promote a decline in general systemic health. Since the middle of the twentieth century, new diagnostic technologies (e.g., actigraphy), medications (e.g., modafinil), and a tremendous body of knowledge can be used to improve the well-being of shift workers, prevent injury and disease, and decrease healthcare costs. While SWD is still under-recognized, its importance to health, safety, and productivity is more fully appreciated now than it was 50 years ago, and organizations such as the National Sleep Foundation are raising awareness of SWD among the general public. More studies are needed to describe the global, national, and industry-specific prevalence of SWD so that we may better understand the personal and socioeconomic costs associated with this condition. In addition, we must strive to discover more individual and organizational risk factors for SWD so that we can effectively identify and treat the most vulnerable populations.

Educational interventions should be delivered so that workers may improve circadian alignment, reduce shift sleepiness, decrease insomnia, and mitigate modifiable risk factors of disorders linked to shift work. Referrals to board-certified sleep physicians may be indicated if symptoms persist despite self-care strategies and treatment by the primary care physician. If there is concern about excessive sleepiness while operating a motor vehicle during the commute to and from work, alternatives such as a car pool or ride share program may be utilized.

Ultimately, different phenotypes of shift work disorder may emerge, allowing clinicians to further individualize treatment, refine prognosis, and guide shift workers with critical employment decisions such as shift selection. The care of shift workers with and without SWD is an international priority that requires thoughtful cooperation between shift workers, employers, and healthcare providers.

References

1. Medicine AAoS. International classification of sleep disorders–third edition (ICSD-3); 2014.
2. Wright KP Jr, Bogan RK, Wyatt JK. Shift work and the assessment and management of shift work disorder (SWD). Sleep Med Rev. 2013;17(1):41–54.
3. Harrington JM, Great Britain. Employment Medical Advisory Service. Shift work and health: a critical review of the literature. London: H.M.S.O.; 1978. p. 28.
4. Drake CL, Roehrs T, Richardson G, Walsh JK, Roth T. Shift work sleep disorder: prevalence and consequences beyond that of symptomatic day workers. Sleep. 2004;27(8):1453–62.
5. Wickwire EM, Geiger-Brown J, Scharf SM, Drake CL. Shift work and shift work sleep disorder: clinical and organizational perspectives. Chest. 2017;151(5):1156–72.
6. Labor USDo. Workers on flexible and shift schedules in 2004. US: Bureau of Labor Statistics; 2004.

7. Lee S, McCann DM, Messenger JC. Working time around the world : trends in working hours, laws and policies in a global comparative perspective. New York: Routledge; 2007. p. xviii, 220.
8. Presser HB. Job, family, and gender: determinants of nonstandard work schedules among employed Americans in 1991. Demography. 1995;32(4):577–98.
9. Di Milia L, Waage S, Pallesen S, Bjorvatn B. Shift work disorder in a random population sample—prevalence and comorbidities. PLoS One. 2013;8(1):e55306.
10. Flo E, Pallesen S, Mageroy N, Moen BE, Gronli J, Hilde Nordhus I, et al. Shift work disorder in nurses—assessment, prevalence and related health problems. PLoS One. 2012;7(4):e33981.
11. Saksvik IB, Bjorvatn B, Hetland H, Sandal GM, Pallesen S. Individual differences in tolerance to shift work—a systematic review. Sleep Med Rev. 2011;15(4):221–35.
12. Oginska H, Pokorski J, Oginski A. Gender, ageing, and shiftwork intolerance. Ergonomics. 1993;36(1–3):161–8.
13. Waage S, Moen BE, Pallesen S, Eriksen HR, Ursin H, Akerstedt T, et al. Shift work disorder among oil rig workers in the North Sea. Sleep. 2009;32(4):558–65.
14. Rajaratnam SM, Barger LK, Lockley SW, Shea SA, Wang W, Landrigan CP, et al. Sleep disorders, health, and safety in police officers. JAMA. 2011;306(23):2567–78.
15. Saper CB, Cano G, Scammell TE. Homeostatic, circadian, and emotional regulation of sleep. J Comp Neurol. 2005;493(1):92–8.
16. Borbely AA, Achermann P, Trachsel L, Tobler I. Sleep initiation and initial sleep intensity: interactions of homeostatic and circadian mechanisms. J Biol Rhythm. 1989;4(2):149–60.
17. Dijk DJ, Czeisler CA. Contribution of the circadian pacemaker and the sleep homeostat to sleep propensity, sleep structure, electroencephalographic slow waves, and sleep spindle activity in humans. J Neurosci. 1995;15(5 Pt 1):3526–38.
18. Radulovacki M, Virus RM, Djuricic-Nedelson M, Green RD. Adenosine analogs and sleep in rats. J Pharmacol Exp Ther. 1984;228(2):268–74.
19. Strecker RE, Morairty S, Thakkar MM, Porkka-Heiskanen T, Basheer R, Dauphin LJ, et al. Adenosinergic modulation of basal forebrain and preoptic/anterior hypothalamic neuronal activity in the control of behavioral state. Behav Brain Res. 2000;115(2):183–204.
20. Urade Y, Hayaishi O. Prostaglandin D2 and sleep/wake regulation. Sleep Med Rev. 2011;15(6):411–8.
21. Krueger JM, Fang J, Taishi P, Chen Z, Kushikata T, Gardi J. Sleep. A physiologic role for IL-1 beta and TNF-alpha. Ann N Y Acad Sci. 1998;856:148–59.
22. Yasuda K, Churchill L, Yasuda T, Blindheim K, Falter M, Krueger JM. Unilateral cortical application of interleukin-1beta (IL1beta) induces asymmetry in fos, IL1beta and nerve growth factor immunoreactivity: implications for sleep regulation. Brain Res. 2007;1131(1): 44–59.
23. Porkka-Heiskanen T, Strecker RE, Thakkar M, Bjorkum AA, Greene RW, McCarley RW. Adenosine: a mediator of the sleep-inducing effects of prolonged wakefulness. Science. 1997;276(5316):1265–8.
24. Jin X, Shearman LP, Weaver DR, Zylka MJ, de Vries GJ, Reppert SM. A molecular mechanism regulating rhythmic output from the suprachiasmatic circadian clock. Cell. 1999;96(1):57–68.
25. Reppert SM, Weaver DR. Coordination of circadian timing in mammals. Nature. 2002;418(6901):935–41.
26. Czeisler CA, Duffy JF, Shanahan TL, Brown EN, Mitchell JF, Rimmer DW, et al. Stability, precision, and near-24-hour period of the human circadian pacemaker. Science. 1999;284(5423):2177–81.
27. Wyatt JK, Ritz-De Cecco A, Czeisler CA, Dijk DJ. Circadian temperature and melatonin rhythms, sleep, and neurobehavioral function in humans living on a 20-h day. Am J Phys. 1999;277(4 Pt 2):R1152–63.
28. Carskadon MA, Labyak SE, Acebo C, Seifer R. Intrinsic circadian period of adolescent humans measured in conditions of forced desynchrony. Neurosci Lett. 1999;260(2):129–32.
29. Menaker M, Moreira LF, Tosini G. Evolution of circadian organization in vertebrates. Braz J Med Biol Res. 1997;30(3):305–13.

30. Moore RY, Eichler VB. Loss of a circadian adrenal corticosterone rhythm following suprachiasmatic lesions in the rat. Brain Res. 1972;42(1):201–6.
31. Peschel N, Helfrich-Forster C. Setting the clock—by nature: circadian rhythm in the fruitfly Drosophila melanogaster. FEBS Lett. 2011;585(10):1435–42.
32. O'Neill JS, van Ooijen G, Dixon LE, Troein C, Corellou F, Bouget FY, et al. Circadian rhythms persist without transcription in a eukaryote. Nature. 2011;469(7331):554–8.
33. Konopka RJ, Benzer S. Clock mutants of Drosophila melanogaster. Proc Natl Acad Sci U S A. 1971;68(9):2112–6.
34. Zehring WA, Wheeler DA, Reddy P, Konopka RJ, Kyriacou CP, Rosbash M, et al. P-element transformation with period locus DNA restores rhythmicity to mutant, arrhythmic Drosophila melanogaster. Cell. 1984;39(2 Pt 1):369–76.
35. Bargiello TA, Jackson FR, Young MW. Restoration of circadian behavioural rhythms by gene transfer in Drosophila. Nature. 1984;312(5996):752–4.
36. Vosshall LB, Price JL, Sehgal A, Saez L, Young MW. Block in nuclear localization of period protein by a second clock mutation, timeless. Science. 1994;263(5153):1606–9.
37. Dibner C, Schibler U, Albrecht U. The mammalian circadian timing system: organization and coordination of central and peripheral clocks. Annu Rev Physiol. 2010;72:517–49.
38. Dardente H, Cermakian N. Molecular circadian rhythms in central and peripheral clocks in mammals. Chronobiol Int. 2007;24(2):195–213.
39. Cermakian N, Boivin DB. The regulation of central and peripheral circadian clocks in humans. Obes Rev. 2009;10(Suppl 2):25–36.
40. Wurtman RJ, Axelrod J, Fischer JE. Melatonin synthesis in the pineal gland: effect of light mediated by the sympathetic nervous system. Science. 1964;143(3612):1328–9.
41. Moore RY. Neural control of the pineal gland. Behav Brain Res. 1996;73(1–2):125–30.
42. Moore RY, Lenn NJ. A retinohypothalamic projection in the rat. J Comp Neurol. 1972;146(1):1–14.
43. Moore RY, Speh JC, Card JP. The retinohypothalamic tract originates from a distinct subset of retinal ganglion cells. J Comp Neurol. 1995;352(3):351–66.
44. Gooley JJ, Lu J, Fischer D, Saper CB. A broad role for melanopsin in nonvisual photoreception. J Neurosci. 2003;23(18):7093–106.
45. Golombek DA, Rosenstein RE. Physiology of circadian entrainment. Physiol Rev. 2010;90(3):1063–102.
46. Lewy AJ, Sack RL, Miller LS, Hoban TM. Antidepressant and circadian phase-shifting effects of light. Science. 1987;235(4786):352–4.
47. Lewy AJ, Sack RL. The dim light melatonin onset as a marker for circadian phase position. Chronobiol Int. 1989;6(1):93–102.
48. Pandi-Perumal SR, Smits M, Spence W, Srinivasan V, Cardinali DP, Lowe AD, et al. Dim light melatonin onset (DLMO): a tool for the analysis of circadian phase in human sleep and chronobiological disorders. Prog Neuro-Psychopharmacol Biol Psychiatry. 2007;31(1):1–11.
49. Van Cauter E, Leproult R, Kupfer DJ. Effects of gender and age on the levels and circadian rhythmicity of plasma cortisol. J Clin Endocrinol Metab. 1996;81(7):2468–73.
50. Gumenyuk V, Howard R, Roth T, Korzyukov O, Drake CL. Sleep loss, circadian mismatch, and abnormalities in reorienting of attention in night workers with shift work disorder. Sleep. 2014;37(3):545–56.
51. Hilliker NA, Muehlbach MJ, Schweitzer PK, Walsh JK. Sleepiness/alertness on a simulated night shift schedule and morningness-eveningness tendency. Sleep. 1992;15(5):430–3.
52. Archer SN, Robilliard DL, Skene DJ, Smits M, Williams A, Arendt J, et al. A length polymorphism in the circadian clock gene Per3 is linked to delayed sleep phase syndrome and extreme diurnal preference. Sleep. 2003;26(4):413–5.
53. Jones KH, Ellis J, von Schantz M, Skene DJ, Dijk DJ, Archer SN. Age-related change in the association between a polymorphism in the PER3 gene and preferred timing of sleep and waking activities. J Sleep Res. 2007;16(1):12–6.
54. Archer SN, Viola AU, Kyriakopoulou V, von Schantz M, Dijk DJ. Inter-individual differences in habitual sleep timing and entrained phase of endogenous circadian rhythms of BMAL1, PER2 and PER3 mRNA in human leukocytes. Sleep. 2008;31(5):608–17.

55. Viola AU, Archer SN, James LM, Groeger JA, Lo JC, Skene DJ, et al. PER3 polymorphism predicts sleep structure and waking performance. Curr Biol. 2007;17(7):613–8.
56. Cheng P, Tallent G, Burgess HJ, Tran KM, Roth T, Drake CL. Daytime sleep disturbance in night shift work and the role of PERIOD3. J Clin Sleep Med. 2018;14(3):393–400.
57. Pilcher JJ, Lambert BJ, Huffcutt AI. Differential effects of permanent and rotating shifts on self-report sleep length: a meta-analytic review. Sleep. 2000;23(2):155–63.
58. Mitler MM, Miller JC, Lipsitz JJ, Walsh JK, Wylie CD. The sleep of long-haul truck drivers. N Engl J Med. 1997;337(11):755–61.
59. Watson NF, Badr MS, Belenky G, Bliwise DL, Buxton OM, Buysse D, et al. Recommended amount of sleep for a healthy adult: a joint consensus statement of the American academy of sleep medicine and sleep research society. Sleep. 2015;38(6):843–4.
60. Vallieres A, Azaiez A, Moreau V, LeBlanc M, Morin CM. Insomnia in shift work. Sleep Med. 2014;15(12):1440–8.
61. Harma M, Tenkanen L, Sjoblom T, Alikoski T, Heinsalmi P. Combined effects of shift work and life-style on the prevalence of insomnia, sleep deprivation and daytime sleepiness. Scand J Work Environ Health. 1998;24(4):300–7.
62. Nakata A, Haratani T, Takahashi M, Kawakami N, Arito H, Fujioka Y, et al. Job stress, social support at work, and insomnia in Japanese shift workers. J Hum Ergol (Tokyo). 2001;30(1–2):203–9.
63. Garbarino S, De Carli F, Nobili L, Mascialino B, Squarcia S, Penco MA, et al. Sleepiness and sleep disorders in shift workers: a study on a group of italian police officers. Sleep. 2002;25(6):648–53.
64. Oyane NM, Pallesen S, Moen BE, Akerstedt T, Bjorvatn B. Associations between night work and anxiety, depression, insomnia, sleepiness and fatigue in a sample of Norwegian nurses. PLoS One. 2013;8(8):e70228.
65. Gumenyuk V, Belcher R, Drake CL, Roth T. Differential sleep, sleepiness, and neurophysiology in the insomnia phenotypes of shift work disorder. Sleep. 2015;38(1):119–26.
66. Belcher R, Gumenyuk V, Roth T. Insomnia in shift work disorder relates to occupational and neurophysiological impairment. J Clin Sleep Med. 2015;11(4):457–65.
67. Gumenyuk V, Roth T, Korzyukov O, Jefferson C, Kick A, Spear L, et al. Shift work sleep disorder is associated with an attenuated brain response of sensory memory and an increased brain response to novelty: an ERP study. Sleep. 2010;33(5):703–13.
68. Folkard S, Tucker P. Shift work, safety and productivity. Occup Med (Lond). 2003;53(2):95–101.
69. Akerstedt T. Sleepiness as a consequence of shift work. Sleep. 1988;11(1):17–34.
70. Akerstedt T. Psychological and psychophysiological effects of shift work. Scand J Work Environ Health. 1990;16(Suppl 1):67–73.
71. Akerstedt T, Kecklund G, Knutsson A. Manifest sleepiness and the spectral content of the EEG during shift work. Sleep. 1991;14(3):221–5.
72. Gold DR, Rogacz S, Bock N, Tosteson TD, Baum TM, Speizer FE, et al. Rotating shift work, sleep, and accidents related to sleepiness in hospital nurses. Am J Public Health. 1992;82(7):1011–4.
73. Harma M, Sallinen M, Ranta R, Mutanen P, Muller K. The effect of an irregular shift system on sleepiness at work in train drivers and railway traffic controllers. J Sleep Res. 2002;11(2):141–51.
74. Dinges DF. An overview of sleepiness and accidents. J Sleep Res. 1995;4(S2):4–14.
75. Akerstedt T. Work hours, sleepiness and the underlying mechanisms. J Sleep Res. 1995;4(S2):15–22.
76. Dijk DJ, Duffy JF, Czeisler CA. Circadian and sleep/wake dependent aspects of subjective alertness and cognitive performance. J Sleep Res. 1992;1(2):112–7.
77. Berger AM, Hobbs BB. Impact of shift work on the health and safety of nurses and patients. Clin J Oncol Nurs. 2006;10(4):465–71.
78. Rajaratnam SM, Arendt J. Health in a 24-h society. Lancet. 2001;358(9286):999–1005.

79. Carskadon MA, Dement WC. Daytime sleepiness: quantification of a behavioral state. Neurosci Biobehav Rev. 1987;11(3):307–17.
80. Rogers AE, Hwang WT, Scott LD, Aiken LH, Dinges DF. The working hours of hospital staff nurses and patient safety. Health Aff (Millwood). 2004;23(4):202–12.
81. Wagstaff AS, Sigstad Lie JA. Shift and night work and long working hours—a systematic review of safety implications. Scand J Work Environ Health. 2011;37(3):173–85.
82. Van Dongen HP, Maislin G, Mullington JM, Dinges DF. The cumulative cost of additional wakefulness: dose-response effects on neurobehavioral functions and sleep physiology from chronic sleep restriction and total sleep deprivation. Sleep. 2003;26(2):117–26.
83. Barger LK, Lockley SW, Rajaratnam SM, Landrigan CP. Neurobehavioral, health, and safety consequences associated with shift work in safety-sensitive professions. Curr Neurol Neurosci Rep. 2009;9(2):155–64.
84. Dawson D, Reid K. Fatigue, alcohol and performance impairment. Nature. 1997;388(6639):235.
85. Van Dongen HP. Shift work and inter-individual differences in sleep and sleepiness. Chronobiol Int. 2006;23(6):1139–47.
86. Esquirol Y, Perret B, Ruidavets JB, Marquie JC, Dienne E, Niezborala M, et al. Shift work and cardiovascular risk factors: new knowledge from the past decade. Arch Cardiovasc Dis. 2011;104(12):636–68.
87. Leclerc A. Shift-work and cardiovascular disease. Eur J Epidemiol. 2010;25(5):285–6.
88. Mosendane T, Mosendane T, Raal FJ. Shift work and its effects on the cardiovascular system. Cardiovasc J Afr. 2008;19(4):210–5.
89. Tenkanen L, Sjoblom T, Kalimo R, Alikoski T, Harma M. Shift work, occupation and coronary heart disease over 6 years of follow-up in the Helsinki Heart Study. Scand J Work Environ Health. 1997;23(4):257–65.
90. Fujino Y, Iso H, Tamakoshi A, Inaba Y, Koizumi A, Kubo T, et al. A prospective cohort study of shift work and risk of ischemic heart disease in Japanese male workers. Am J Epidemiol. 2006;164(2):128–35.
91. Kawachi I, Colditz GA, Stampfer MJ, Willett WC, Manson JE, Speizer FE, et al. Prospective study of shift work and risk of coronary heart disease in women. Circulation. 1995;92(11):3178–82.
92. Knutsson A, Akerstedt T, Jonsson BG, Orth-Gomer K. Increased risk of ischaemic heart disease in shift workers. Lancet. 1986;2(8498):89–92.
93. Akerstedt T, Knutsson A, Alfredsson L, Theorell T. Shift work and cardiovascular disease. Scand J Work Environ Health. 1984;10(6 Spec No):409–14.
94. Hublin C, Partinen M, Koskenvuo K, Silventoinen K, Koskenvuo M, Kaprio J. Shift-work and cardiovascular disease: a population-based 22-year follow-up study. Eur J Epidemiol. 2010;25(5):315–23.
95. Boggild H, Suadicani P, Hein HO, Gyntelberg F. Shift work, social class, and ischaemic heart disease in middle aged and elderly men; a 22 year follow up in the Copenhagen male study. Occup Environ Med. 1999;56(9):640–5.
96. Frost P, Kolstad HA, Bonde JP. Shift work and the risk of ischemic heart disease - a systematic review of the epidemiologic evidence. Scand J Work Environ Health. 2009;35(3):163–79.
97. Puttonen S, Kivimaki M, Elovainio M, Pulkki-Raback L, Hintsanen M, Vahtera J, et al. Shift work in young adults and carotid artery intima-media thickness: The Cardiovascular Risk in Young Finns study. Atherosclerosis. 2009;205(2):608–13.
98. Brown DL, Feskanich D, Sanchez BN, Rexrode KM, Schernhammer ES, Lisabeth LD. Rotating night shift work and the risk of ischemic stroke. Am J Epidemiol. 2009;169(11):1370–7.
99. De Bacquer D, Van Risseghem M, Clays E, Kittel F, De Backer G, Braeckman L. Rotating shift work and the metabolic syndrome: a prospective study. Int J Epidemiol. 2009;38(3):848–54.
100. Puttonen S, Viitasalo K, Harma M. The relationship between current and former shift work and the metabolic syndrome. Scand J Work Environ Health. 2012;38(4):343–8.

101. Esquirol Y, Bongard V, Mabile L, Jonnier B, Soulat JM, Perret B. Shift work and metabolic syndrome: respective impacts of job strain, physical activity, and dietary rhythms. Chronobiol Int. 2009;26(3):544–59.
102. Sookoian S, Gemma C, Fernandez Gianotti T, Burgueno A, Alvarez A, Gonzalez CD, et al. Effects of rotating shift work on biomarkers of metabolic syndrome and inflammation. J Intern Med. 2007;261(3):285–92.
103. Guo Y, Rong Y, Huang X, Lai H, Luo X, Zhang Z, et al. Shift work and the relationship with metabolic syndrome in Chinese aged workers. PLoS One. 2015;10(3):e0120632.
104. Karlsson B, Knutsson A, Lindahl B. Is there an association between shift work and having a metabolic syndrome? Results from a population based study of 27,485 people. Occup Environ Med. 2001;58(11):747–52.
105. Kim MJ, Son KH, Park HY, Choi DJ, Yoon CH, Lee HY, et al. Association between shift work and obesity among female nurses: Korean Nurses' survey. BMC Public Health. 2013;13:1204.
106. Karlsson BH, Knutsson AK, Lindahl BO, Alfredsson LS. Metabolic disturbances in male workers with rotating three-shift work. Results of the WOLF study. Int Arch Occup Environ Health. 2003;76(6):424–30.
107. Suwazono Y, Dochi M, Sakata K, Okubo Y, Oishi M, Tanaka K, et al. A longitudinal study on the effect of shift work on weight gain in male Japanese workers. Obesity (Silver Spring). 2008;16(8):1887–93.
108. Di Lorenzo L, De Pergola G, Zocchetti C, L'Abbate N, Basso A, Pannacciulli N, et al. Effect of shift work on body mass index: results of a study performed in 319 glucose-tolerant men working in a Southern Italian industry. Int J Obes Relat Metab Disord. 2003;27(11):1353–8.
109. Peplonska B, Bukowska A, Sobala W. Association of rotating night shift work with BMI and abdominal obesity among nurses and midwives. PLoS One. 2015;10(7):e0133761.
110. Pan A, Schernhammer ES, Sun Q, Hu FB. Rotating night shift work and risk of type 2 diabetes: two prospective cohort studies in women. PLoS Med. 2011;8(12):e1001141.
111. Hansen AB, Stayner L, Hansen J, Andersen ZJ. Night shift work and incidence of diabetes in the Danish Nurse Cohort. Occup Environ Med. 2016;73(4):262–8.
112. Morikawa Y, Nakagawa H, Miura K, Soyama Y, Ishizaki M, Kido T, et al. Shift work and the risk of diabetes mellitus among Japanese male factory workers. Scand J Work Environ Health. 2005;31(3):179–83.
113. Knutsson A, Kempe A. Shift work and diabetes—a systematic review. Chronobiol Int. 2014;31(10):1146–51.
114. Kivimaki M, Batty GD, Hublin C. Shift work as a risk factor for future type 2 diabetes: evidence, mechanisms, implications, and future research directions. PLoS Med. 2011;8(12):e1001138.
115. Fonken LK, Workman JL, Walton JC, Weil ZM, Morris JS, Haim A, et al. Light at night increases body mass by shifting the time of food intake. Proc Natl Acad Sci U S A. 2010;107(43):18664–9.
116. Opperhuizen AL, van Kerkhof LW, Proper KI, Rodenburg W, Kalsbeek A. Rodent models to study the metabolic effects of shiftwork in humans. Front Pharmacol. 2015;6:50.
117. Scheer FA, Hilton MF, Mantzoros CS, Shea SA. Adverse metabolic and cardiovascular consequences of circadian misalignment. Proc Natl Acad Sci U S A. 2009;106(11):4453–8.
118. Buxton OM, Cain SW, O'Connor SP, Porter JH, Duffy JF, Wang W, et al. Adverse metabolic consequences in humans of prolonged sleep restriction combined with circadian disruption. Sci Transl Med. 2012;4(129):129ra43.
119. Lowden A, Moreno C, Holmback U, Lennernas M, Tucker P. Eating and shift work - effects on habits, metabolism and performance. Scand J Work Environ Health. 2010;36(2):150–62.
120. Antunes LC, Levandovski R, Dantas G, Caumo W, Hidalgo MP. Obesity and shift work: chronobiological aspects. Nutr Res Rev. 2010;23(1):155–68.
121. Fritschi L. Shift work and cancer. BMJ. 2009;339:b2653.
122. Wang F, Yeung KL, Chan WC, Kwok CC, Leung SL, Wu C, et al. A meta-analysis on dose-response relationship between night shift work and the risk of breast cancer. Ann Oncol. 2013;24(11):2724–32.

123. Davis S, Mirick DK, Stevens RG. Night shift work, light at night, and risk of breast cancer. J Natl Cancer Inst. 2001;93(20):1557–62.
124. Hansen J. Increased breast cancer risk among women who work predominantly at night. Epidemiology. 2001;12(1):74–7.
125. Hansen J, Lassen CF. Nested case-control study of night shift work and breast cancer risk among women in the Danish military. Occup Environ Med. 2012;69(8):551–6.
126. Schernhammer ES, Kroenke CH, Laden F, Hankinson SE. Night work and risk of breast cancer. Epidemiology. 2006;17(1):108–11.
127. Schernhammer ES, Laden F, Speizer FE, Willett WC, Hunter DJ, Kawachi I, et al. Rotating night shifts and risk of breast cancer in women participating in the nurses' health study. J Natl Cancer Inst. 2001;93(20):1563–8.
128. O'Leary ES, Schoenfeld ER, Stevens RG, Kabat GC, Henderson K, Grimson R, et al. Shift work, light at night, and breast cancer on Long Island, New York. Am J Epidemiol. 2006;164(4):358–66.
129. Schwartzbaum J, Ahlbom A, Feychting M. Cohort study of cancer risk among male and female shift workers. Scand J Work Environ Health. 2007;33(5):336–43.
130. Viswanathan AN, Hankinson SE, Schernhammer ES. Night shift work and the risk of endometrial cancer. Cancer Res. 2007;67(21):10618–22.
131. Poole EM, Schernhammer ES, Tworoger SS. Rotating night shift work and risk of ovarian cancer. Cancer Epidemiol Biomark Prev. 2011;20(5):934–8.
132. Rao D, Yu H, Bai Y, Zheng X, Xie L. Does night-shift work increase the risk of prostate cancer? A systematic review and meta-analysis. Onco Targets Ther. 2015;8:2817–26.
133. Papantoniou K, Castano-Vinyals G, Espinosa A, Aragones N, Perez-Gomez B, Burgos J, et al. Night shift work, chronotype and prostate cancer risk in the MCC-Spain case-control study. Int J Cancer. 2015;137(5):1147–57.
134. Kubo T, Oyama I, Nakamura T, Kunimoto M, Kadowaki K, Otomo H, et al. Industry-based retrospective cohort study of the risk of prostate cancer among rotating-shift workers. Int J Urol. 2011;18(3):206–11.
135. Parent ME, El-Zein M, Rousseau MC, Pintos J, Siemiatycki J. Night work and the risk of cancer among men. Am J Epidemiol. 2012;176(9):751–9.
136. Schernhammer ES, Laden F, Speizer FE, Willett WC, Hunter DJ, Kawachi I, et al. Night-shift work and risk of colorectal cancer in the nurses' health study. J Natl Cancer Inst. 2003;95(11):825–8.
137. Lahti TA, Partonen T, Kyyronen P, Kauppinen T, Pukkala E. Night-time work predisposes to non Hodgkin lymphoma. Int J Cancer. 2008;123(9):2148–51.
138. Schernhammer ES, Feskanich D, Liang G, Han J. Rotating night-shift work and lung cancer risk among female nurses in the United States. Am J Epidemiol. 2013;178(9):1434–41.
139. Vinogradova IA, Anisimov VN, Bukalev AV, Semenchenko AV, Zabezhinski MA. Circadian disruption induced by light-at-night accelerates aging and promotes tumorigenesis in rats. Aging (Albany NY). 2009;1(10):855–65.
140. Anisimov VN. The light-dark regimen and cancer development. Neuro Endocrinol Lett. 2002;23(Suppl 2):28–36.
141. Anisimov VN, Hansen J. Light, endocrine systems and cancer—a meeting report. Neuro Endocrinol Lett. 2002;23(Suppl 2):84–7.
142. Straif K, Baan R, Grosse Y, Secretan B, El Ghissassi F, Bouvard V, et al. Carcinogenicity of shift-work, painting, and fire-fighting. Lancet Oncol. 2007;8(12):1065–6.
143. Haus EL, Smolensky MH. Shift work and cancer risk: potential mechanistic roles of circadian disruption, light at night, and sleep deprivation. Sleep Med Rev. 2013;17(4):273–84.
144. Cohen M, Lippman M, Chabner B. Pineal gland and breast cancer. Lancet. 1978;2(8104–5):1381–2.
145. Stevens RG. Electric power use and breast cancer: a hypothesis. Am J Epidemiol. 1987;125(4):556–61.
146. Stevens RG. Light-at-night, circadian disruption and breast cancer: assessment of existing evidence. Int J Epidemiol. 2009;38(4):963–70.

147. Hill SM, Spriggs LL, Simon MA, Muraoka H, Blask DE. The growth inhibitory action of melatonin on human breast cancer cells is linked to the estrogen response system. Cancer Lett. 1992;64(3):249–56.
148. Blask DE. Melatonin, sleep disturbance and cancer risk. Sleep Med Rev. 2009;13(4): 257–64.
149. Costa G. Shift work and occupational medicine: an overview. Occup Med (Lond). 2003;53(2):83–8.
150. Costa G, Haus E, Stevens R. Shift work and cancer - considerations on rationale, mechanisms, and epidemiology. Scand J Work Environ Health. 2010;36(2):163–79.
151. Konturek PC, Brzozowski T, Konturek SJ. Gut clock: implication of circadian rhythms in the gastrointestinal tract. J Physiol Pharmacol. 2011;62(2):139–50.
152. Vener KJ, Szabo S, Moore JG. The effect of shift work on gastrointestinal (GI) function: a review. Chronobiologia. 1989;16(4):421–39.
153. Knutsson A, Boggild H. Gastrointestinal disorders among shift workers. Scand J Work Environ Health. 2010;36(2):85–95.
154. Knutsson A. Health disorders of shift workers. Occup Med (Lond). 2003;53(2):103–8.
155. Nojkov B, Rubenstein JH, Chey WD, Hoogerwerf WA. The impact of rotating shift work on the prevalence of irritable bowel syndrome in nurses. Am J Gastroenterol. 2010;105(4):842–7.
156. Caruso CC, Lusk SL, Gillespie BW. Relationship of work schedules to gastrointestinal diagnoses, symptoms, and medication use in auto factory workers. Am J Ind Med. 2004;46(6):586–98.
157. Saberi HR, Moravveji AR. Gastrointestinal complaints in shift-working and day-working nurses in Iran. J Circadian Rhythms. 2010;8:9.
158. Armstrong BG, Nolin AD, McDonald AD. Work in pregnancy and birth weight for gestational age. Br J Ind Med. 1989;46(3):196–9.
159. McDonald AD, McDonald JC, Armstrong B, Cherry NM, Nolin AD, Robert D. Prematurity and work in pregnancy. Br J Ind Med. 1988;45(1):56–62.
160. Axelsson G, Rylander R, Molin I. Outcome of pregnancy in relation to irregular and inconvenient work schedules. Br J Ind Med. 1989;46(6):393–8.
161. Xu X, Ding M, Li B, Christiani DC. Association of rotating shiftwork with preterm births and low birth weight among never smoking women textile workers in China. Occup Environ Med. 1994;51(7):470–4.
162. Uehata T, Sasakawa N. The fatigue and maternity disturbances of night workwomen. J Hum Ergol (Tokyo). 1982;11(Suppl):465–74.
163. Infante-Rivard C, David M, Gauthier R, Rivard GE. Pregnancy loss and work schedule during pregnancy. Epidemiology. 1993;4(1):73–5.
164. Boivin DB, Czeisler CA, Dijk DJ, Duffy JF, Folkard S, Minors DS, et al. Complex interaction of the sleep-wake cycle and circadian phase modulates mood in healthy subjects. Arch Gen Psychiatry. 1997;54(2):145–52.
165. Tsuno N, Besset A, Ritchie K. Sleep and depression. J Clin Psychiatry. 2005;66(10):1254–69.
166. Alvaro PK, Roberts RM, Harris JK. A systematic review assessing bidirectionality between sleep disturbances, anxiety, and depression. Sleep. 2013;36(7):1059–68.
167. Mellman TA. Sleep and anxiety disorders. Psychiatr Clin North Am. 2006;29(4):1047–58; abstract x.
168. Scott AJ, Monk TH, Brink LL. Shiftwork as a risk factor for depression: a pilot study. Int J Occup Environ Health. 1997;3(Supplement 2):S2–9.
169. Kalmbach DA, Pillai V, Cheng P, Arnedt JT, Drake CL. Shift work disorder, depression, and anxiety in the transition to rotating shifts: the role of sleep reactivity. Sleep Med. 2015;16(12):1532–8.
170. Puttonen S, Oksanen T, Vahtera J, Pentti J, Virtanen M, Salo P, et al. Is shift work a risk factor for rheumatoid arthritis? The Finnish public sector study. Ann Rheum Dis. 2010;69(4): 779–80.
171. Cooper GS, Parks CG, Treadwell EL, St Clair EW, Gilkeson GS, Dooley MA. Occupational risk factors for the development of systemic lupus erythematosus. J Rheumatol. 2004;31(10):1928–33.

172. Trinkoff AM, Le R, Geiger-Brown J, Lipscomb J, Lang G. Longitudinal relationship of work hours, mandatory overtime, and on-call to musculoskeletal problems in nurses. Am J Ind Med. 2006;49(11):964–71.
173. Hedstrom AK, Akerstedt T, Hillert J, Olsson T, Alfredsson L. Shift work at young age is associated with increased risk for multiple sclerosis. Ann Neurol. 2011;70(5):733–41.
174. Hedstrom AK, Akerstedt T, Olsson T, Alfredsson L. Shift work influences multiple sclerosis risk. Mult Scler. 2015;21(9):1195–9.
175. American Psychiatric Association: Diagnostic and Statistical Manual of Mental Disorders: Diagnostic and Statistical Manual of Mental Disorders, Fifth Edition. Arlington, VA: American Psychiatric Association, 2013.
176. Barger LK, Ogeil RP, Drake CL, O'Brien CS, Ng KT, Rajaratnam SM. Validation of a questionnaire to screen for shift work disorder. Sleep. 2012;35(12):1693–703.
177. Juda M, Vetter C, Roenneberg T. The Munich ChronoType Questionnaire for Shift-Workers (MCTQShift). J Biol Rhythm. 2013;28(2):130–40.
178. Roenneberg T, Wirz-Justice A, Merrow M. Life between clocks: daily temporal patterns of human chronotypes. J Biol Rhythm. 2003;18(1):80–90.
179. Horne JA, Ostberg O. A self-assessment questionnaire to determine morningness-eveningness in human circadian rhythms. Int J Chronobiol. 1976;4(2):97–110.
180. Kantermann T, Sung H, Burgess HJ. Comparing the morningness-eveningness questionnaire and Munich ChronoType questionnaire to the dim light melatonin onset. J Biol Rhythm. 2015;30(5):449–53.
181. Zavada A, Gordijn MC, Beersma DG, Daan S, Roenneberg T. Comparison of the Munich Chronotype questionnaire with the Horne-Ostberg's morningness-eveningness score. Chronobiol Int. 2005;22(2):267–78.
182. Gamble KL, Motsinger-Reif AA, Hida A, Borsetti HM, Servick SV, Ciarleglio CM, et al. Shift work in nurses: contribution of phenotypes and genotypes to adaptation. PLoS One. 2011;6(4):e18395.
183. Boivin DB, Boudreau P. Impacts of shift work on sleep and circadian rhythms. Pathol Biol (Paris). 2014;62(5):292–301.
184. Vetter C, Fischer D, Matera JL, Roenneberg T. Aligning work and circadian time in shift workers improves sleep and reduces circadian disruption. Curr Biol. 2015;25(7):907–11.
185. Johns MW. A new method for measuring daytime sleepiness: the Epworth sleepiness scale. Sleep. 1991;14(6):540–5.
186. Bastien CH, Vallieres A, Morin CM. Validation of the insomnia severity index as an outcome measure for insomnia research. Sleep Med. 2001;2(4):297–307.
187. Spitzer RL, Kroenke K, Williams JB, Lowe B. A brief measure for assessing generalized anxiety disorder: the GAD-7. Arch Intern Med. 2006;166(10):1092–7.
188. Kroenke K, Spitzer RL, Williams JB. The PHQ-9: validity of a brief depression severity measure. J Gen Intern Med. 2001;16(9):606–13.
189. Martin A, Rief W, Klaiberg A, Braehler E. Validity of the Brief Patient Health Questionnaire Mood Scale (PHQ-9) in the general population. Gen Hosp Psychiatry. 2006;28(1):71–7.
190. Morgenthaler TI, Lee-Chiong T, Alessi C, Friedman L, Aurora RN, Boehlecke B, et al. Practice parameters for the clinical evaluation and treatment of circadian rhythm sleep disorders. An American Academy of Sleep Medicine report. Sleep. 2007;30(11):1445–59.
191. Sack RL, Auckley D, Auger RR, Carskadon MA, Wright KP Jr, Vitiello MV, et al. Circadian rhythm sleep disorders: part I, basic principles, shift work and jet lag disorders. An American Academy of Sleep Medicine review. Sleep. 2007;30(11):1460–83.
192. Carney CE, Buysse DJ, Ancoli-Israel S, Edinger JD, Krystal AD, Lichstein KL, et al. The consensus sleep diary: standardizing prospective sleep self-monitoring. Sleep. 2012;35(2):287–302.
193. Morgenthaler T, Alessi C, Friedman L, Owens J, Kapur V, Boehlecke B, et al. Practice parameters for the use of actigraphy in the assessment of sleep and sleep disorders: an update for 2007. Sleep. 2007;30(4):519–29.

194. Mizuno K, Matsumoto A, Aiba T, Abe T, Ohshima H, Takahashi M, et al. Sleep patterns among shift-working flight controllers of the International Space Station: an observational study on the JAXA Flight Control Team. J Physiol Anthropol. 2016;35(1):19.
195. Marino M, Li Y, Rueschman MN, Winkelman JW, Ellenbogen JM, Solet JM, et al. Measuring sleep: accuracy, sensitivity, and specificity of wrist actigraphy compared to polysomnography. Sleep. 2013;36(11):1747–55.
196. Borges FN, Fischer FM. Twelve-hour night shifts of healthcare workers: a risk to the patients? Chronobiol Int. 2003;20(2):351–60.
197. Daurat A, Foret J. Sleep strategies of 12-hour shift nurses with emphasis on night sleep episodes. Scand J Work Environ Health. 2004;30(4):299–305.
198. Lewy AJ. The dim light melatonin onset, melatonin assays and biological rhythm research in humans. Biol Signals Recept. 1999;8(1–2):79–83.
199. Keijzer H, Smits MG, Duffy JF, Curfs LM. Why the dim light melatonin onset (DLMO) should be measured before treatment of patients with circadian rhythm sleep disorders. Sleep Med Rev. 2014;18(4):333–9.
200. Mullington JM, Abbott SM, Carroll JE, Davis CJ, Dijk DJ, Dinges DF, et al. Developing biomarker arrays predicting sleep and circadian-coupled risks to health. Sleep. 2016;39(4):727–36.
201. Tepas DI, Carvalhais AB. Sleep patterns of shiftworkers. Occup Med. 1990;5(2):199–208.
202. Goldbart A, Peppard P, Finn L, Ruoff CM, Barnet J, Young T, et al. Narcolepsy and predictors of positive MSLTs in the Wisconsin Sleep Cohort. Sleep. 2014;37(6):1043–51.
203. Zee PC, Goldstein CA. Treatment of shift work disorder and jet lag. Curr Treat Options Neurol. 2010;12(5):396–411.
204. Roth T. Appropriate therapeutic selection for patients with shift work disorder. Sleep Med. 2012;13(4):335–41.
205. Morrissette DA. Twisting the night away: a review of the neurobiology, genetics, diagnosis, and treatment of shift work disorder. CNS Spectr. 2013;18(Suppl 1):45–53. quiz 4.
206. Yeung J, Sletten TL, Rajaratnam SM. A phase-advanced, extended sleep paradigm to increase sleep duration among early-morning shift workers: a preliminary investigation. Scand J Work Environ Health. 2011;37(1):62–9.
207. Harma M, Knauth P, Ilmarinen J. Daytime napping and its effects on alertness and short-term memory performance in shiftworkers. Int Arch Occup Environ Health. 1989;61(5):341–5.
208. Purnell MT, Feyer AM, Herbison GP. The impact of a nap opportunity during the night shift on the performance and alertness of 12-h shift workers. J Sleep Res. 2002;11(3):219–27.
209. Bonnefond A, Muzet A, Winter-Dill AS, Bailloeuil C, Bitouze F, Bonneau A. Innovative working schedule: introducing one short nap during the night shift. Ergonomics. 2001;44(10):937–45.
210. Sallinen M, Harma M, Akerstedt T, Rosa R, Lillqvist O. Promoting alertness with a short nap during a night shift. J Sleep Res. 1998;7(4):240–7.
211. Signal TL, Gander PH, Anderson H, Brash S. Scheduled napping as a countermeasure to sleepiness in air traffic controllers. J Sleep Res. 2009;18(1):11–9.
212. Ruggiero JS, Redeker NS. Effects of napping on sleepiness and sleep-related performance deficits in night-shift workers: a systematic review. Biol Res Nurs. 2014;16(2):134–42.
213. Lee C, Smith MR, Eastman CI. A compromise phase position for permanent night shift workers: circadian phase after two night shifts with scheduled sleep and light/dark exposure. Chronobiol Int. 2006;23(4):859–75.
214. Smith MR, Cullnan EE, Eastman CI. Shaping the light/dark pattern for circadian adaptation to night shift work. Physiol Behav. 2008;95(3):449–56.
215. Smith MR, Eastman CI. Night shift performance is improved by a compromise circadian phase position: study 3. Circadian phase after 7 night shifts with an intervening weekend off. Sleep. 2008;31(12):1639–45.
216. Smith MR, Fogg LF, Eastman CI. A compromise circadian phase position for permanent night work improves mood, fatigue, and performance. Sleep. 2009;32(11):1481–9.
217. Smith MR, Fogg LF, Eastman CI. Practical interventions to promote circadian adaptation to permanent night shift work: study 4. J Biol Rhythm. 2009;24(2):161–72.

218. Crowley SJ, Lee C, Tseng CY, Fogg LF, Eastman CI. Complete or partial circadian re-entrainment improves performance, alertness, and mood during night-shift work. Sleep. 2004;27(6):1077–87.
219. Czeisler CA, Johnson MP, Duffy JF, Brown EN, Ronda JM, Kronauer RE. Exposure to bright light and darkness to treat physiologic maladaptation to night work. N Engl J Med. 1990;322(18):1253–9.
220. Dawson D, Campbell SS. Timed exposure to bright light improves sleep and alertness during simulated night shifts. Sleep. 1991;14(6):511–6.
221. Eastman CI, Martin SK. How to use light and dark to produce circadian adaptation to night shift work. Ann Med. 1999;31(2):87–98.
222. Boivin DB, Boudreau P, James FO, Kin NM. Photic resetting in night-shift work: impact on nurses' sleep. Chronobiol Int. 2012;29(5):619–28.
223. Boivin DB, Boudreau P, Tremblay GM. Phototherapy and orange-tinted goggles for night-shift adaptation of police officers on patrol. Chronobiol Int. 2012;29(5):629–40.
224. Boivin DB, James FO. Circadian adaptation to night-shift work by judicious light and darkness exposure. J Biol Rhythm. 2002;17(6):556–67.
225. Boivin DB, James FO. Light treatment and circadian adaptation to shift work. Ind Health. 2005;43(1):34–48.
226. Lowden A, Akerstedt T, Wibom R. Suppression of sleepiness and melatonin by bright light exposure during breaks in night work. J Sleep Res. 2004;13(1):37–43.
227. Santhi N, Aeschbach D, Horowitz TS, Czeisler CA. The impact of sleep timing and bright light exposure on attentional impairment during night work. J Biol Rhythm. 2008;23(4):341–52.
228. Folkard S, Arendt J, Clark M. Can melatonin improve shift workers' tolerance of the night shift? Some preliminary findings. Chronobiol Int. 1993;10(5):315–20.
229. Jorgensen KM, Witting MD. Does exogenous melatonin improve day sleep or night alertness in emergency physicians working night shifts? Ann Emerg Med. 1998;31(6):699–704.
230. Sharkey KM, Fogg LF, Eastman CI. Effects of melatonin administration on daytime sleep after simulated night shift work. J Sleep Res. 2001;10(3):181–92.
231. Yoon IY, Song BG. Role of morning melatonin administration and attenuation of sunlight exposure in improving adaptation of night-shift workers. Chronobiol Int. 2002;19(5):903–13.
232. Burgess HJ, Revell VL, Molina TA, Eastman CI. Human phase response curves to three days of daily melatonin: 0.5 mg versus 3.0 mg. J Clin Endocrinol Metab. 2010;95(7):3325–31.
233. Richardson GS, Zee PC, Wang-Weigand S, Rodriguez L, Peng X. Circadian phase-shifting effects of repeated ramelteon administration in healthy adults. J Clin Sleep Med. 2008;4(5):456–61.
234. Rajaratnam SM, Polymeropoulos MH, Fisher DM, Roth T, Scott C, Birznieks G, et al. Melatonin agonist tasimelteon (VEC-162) for transient insomnia after sleep-time shift: two randomised controlled multicentre trials. Lancet. 2009;373(9662):482–91.
235. Walsh JK, Muehlbach MJ, Schweitzer PK. Acute administration of triazolam for the daytime sleep of rotating shift workers. Sleep. 1984;7(3):223–9.
236. Walsh JK, Schweitzer PK, Anch AM, Muehlbach MJ, Jenkins NA, Dickins QS. Sleepiness/alertness on a simulated night shift following sleep at home with triazolam. Sleep. 1991;14(2):140–6.
237. Monchesky TC, Billings BJ, Phillips R, Bourgouin J. Zopiclone in insomniac shiftworkers. Evaluation of its hypnotic properties and its effects on mood and work performance. Int Arch Occup Environ Health. 1989;61(4):255–9.
238. Moon CA, Hindmarch I, Holland RL. The effect of zopiclone 7.5 mg on the sleep, mood and performance of shift workers. Int Clin Psychopharmacol. 1990;5(Suppl 2):79–83.
239. Hart CL, Ward AS, Haney M, Foltin RW. Zolpidem-related effects on performance and mood during simulated night-shift work. Exp Clin Psychopharmacol. 2003;11(4):259–68.
240. Hart CL, Haney M, Nasser J, Foltin RW. Combined effects of methamphetamine and zolpidem on performance and mood during simulated night shift work. Pharmacol Biochem Behav. 2005;81(3):559–68.

241. Troy SM, Lucki I, Unruh MA, Cevallos WH, Leister CA, Martin PT, et al. Comparison of the effects of zaleplon, zolpidem, and triazolam on memory, learning, and psychomotor performance. J Clin Psychopharmacol. 2000;20(3):328–37.
242. Walsh JK, Muehlbach MJ, Humm TM, Dickins QS, Sugerman JL, Schweitzer PK. Effect of caffeine on physiological sleep tendency and ability to sustain wakefulness at night. Psychopharmacology. 1990;101(2):271–3.
243. Muehlbach MJ, Walsh JK. The effects of caffeine on simulated night-shift work and subsequent daytime sleep. Sleep. 1995;18(1):22–9.
244. Ker K, Edwards PJ, Felix LM, Blackhall K, Roberts I. Caffeine for the prevention of injuries and errors in shift workers. Cochrane Database Syst Rev. 2010;5:CD008508.
245. Schweitzer PK, Randazzo AC, Stone K, Erman M, Walsh JK. Laboratory and field studies of naps and caffeine as practical countermeasures for sleep-wake problems associated with night work. Sleep. 2006;29(1):39–50.
246. Hart CL, Ward AS, Haney M, Nasser J, Foltin RW. Methamphetamine attenuates disruptions in performance and mood during simulated night-shift work. Psychopharmacology. 2003;169(1):42–51.
247. Czeisler CA, Walsh JK, Roth T, Hughes RJ, Wright KP, Kingsbury L, et al. Modafinil for excessive sleepiness associated with shift-work sleep disorder. N Engl J Med. 2005;353(5):476–86.
248. Erman MK, Rosenberg R, The USMSWSDSG. Modafinil for excessive sleepiness associated with chronic shift work sleep disorder: effects on patient functioning and health-related quality of life. Prim Care Companion J Clin Psychiatry. 2007;9(3):188–94.
249. Czeisler CA, Walsh JK, Wesnes KA, Arora S, Roth T. Armodafinil for treatment of excessive sleepiness associated with shift work disorder: a randomized controlled study. Mayo Clin Proc. 2009;84(11):958–72.
250. Erman MK, Seiden DJ, Yang R, Dammerman R. Efficacy and tolerability of armodafinil: effect on clinical condition late in the shift and overall functioning of patients with excessive sleepiness associated with shift work disorder. J Occup Environ Med. 2011;53(12):1460–5.
251. Drake C, Gumenyuk V, Roth T, Howard R. Effects of armodafinil on simulated driving and alertness in shift work disorder. Sleep. 2014;37(12):1987–94.
252. Howard R, Roth T, Drake CL. The effects of armodafinil on objective sleepiness and performance in a shift work disorder sample unselected for objective sleepiness. J Clin Psychopharmacol. 2014;34(3):369–73.
253. Schwartz JR, Khan A, McCall WV, Weintraub J, Tiller J. Tolerability and efficacy of armodafinil in naive patients with excessive sleepiness associated with obstructive sleep apnea, shift work disorder, or narcolepsy: a 12-month, open-label, flexible-dose study with an extension period. J Clin Sleep Med. 2010;6(5):450–7.
254. Black JE, Hull SG, Tiller J, Yang R, Harsh JR. The long-term tolerability and efficacy of armodafinil in patients with excessive sleepiness associated with treated obstructive sleep apnea, shift work disorder, or narcolepsy: an open-label extension study. J Clin Sleep Med. 2010;6(5):458–66.

Chapter 12
Duty Hour Regulations of Physicians in Training and Circadian Considerations

Eric J. Olson

Definitions

Accreditation Council for Graduate Medical Education (ACGME) Organization responsible for setting the standards for graduate medical education and for the accreditation of most training programs (residencies and fellowships) in the United States. In the 2017–2018 academic year, the ACGME accredited approximately 11,200 residency and fellowship programs in 180 specialties sponsored by approximately 830 institutions. The accreditation process is managed by specialty-specific Review Committees which consist of leaders from specific fields who set standards and provide peer evaluations of sponsoring institutions and their residencies and fellowships.

Fellow A physician in training who has completed medical school and one or more residencies and is electively undergoing further specialty training. The fellow's period of advanced training is called a fellowship which typically lasts 1–3 years. Fellows may provide clinical and educational support to interns and residents.

Institute of Medicine Component of the National Academies of Science which seeks to provide comprehensive information on health policy issues. This organization was recently renamed the National Academy of Medicine.

Harvard Work Hours Health and Safety Group Multidisciplinary group that studies occupational sleep and circadian issues and seeks strategies to improve the health and safety of workers in safety-sensitive jobs.

E. J. Olson (✉)
Department of Medicine, Division of Pulmonary and Critical Care Medicine, and Center for Sleep Medicine, Mayo Clinic, Rochester, MN, USA
e-mail: olson.eric@mayo.edu

R. R. Auger (ed.), *Circadian Rhythm Sleep-Wake Disorders*,
https://doi.org/10.1007/978-3-030-43803-6_12

Intern A physician in training who has completed medical school and is in his/her first year of postgraduate medical training (i.e., first-year resident). The intern's period of training is called an internship. Interns may be referred to as a "PGY-1" which stands for postgraduate year 1.

Night Float A patient care process in which one or more residents work only over the nighttime hours to provide relief for on-call residents by evaluating new hospital admissions and/or providing coverage for previously admitted inpatients. The night float transfers care responsibilities back to their on-call colleagues in the morning.

Resident A physician in training who has completed medical school and internship but has an additional 2–7 years (or more) of education in a medical or surgical discipline before independent practice. The term originates from the time when these junior doctors resided in hospitals to provide care around the clock. The resident's period of training is called a residency. This training typically occurs in hospital and clinic settings under supervision from institutional faculty (senior clinicians at the sponsoring facility) in that specialty. Residents oversee interns and assume greater autonomy as they gain experience and demonstrate competence. Residents may be referred to as a "PGY-x" with "x" referring to their current year of postgraduate medical training. Residents may also be referred to as "house staff" or "house officers" in recognition of their in-hospital (house) duties.

Introduction

The education pathway from medical student to independent medical practitioner goes through residency and, in some cases, specialty fellowship. Residencies and fellowships are physically and emotionally demanding experiential training periods through which providers gain the skills to enter unsupervised practice through direct care of patients supplemented by clinical teaching from supervisors and structured educational events, such as conferences, lectures, journal clubs, and simulation sessions. The first year of residency is termed an internship and is traditionally the most grueling year of graduate medical education. Extended work hours, overnight call, and rotating work shifts are characteristic features of residencies and fellowships. Historically, these challenging schedules were felt necessary for comprehensive trainee learning, respect for the physician-patient relationship, and to satisfy the clinical and economic realities of providing care for patients around the clock. However, there has been greater awareness of the impact of long work hours on trainee wellbeing, education, and patient safety, leading to efforts to curb resident and fellow duty hours. This chapter will review how medical training may disrupt wake/sleep regulation, summarize literature on the impact of sleep loss and circadian disruption on trainees and patients, provide a history of duty hour restrictions for medical trainees, and examine the consequences of work hour limits on trainees and patients.

Sleep/Wake Regulation and Medical Trainees

Sleep and wakefulness are primarily determined by the interplay of the homeostatic drive to sleep and the circadian system [1]. The homeostatic drive to sleep strengthens as time from the prior sleep bout elapses, which results in progressive sleep propensity and degradation in neurobehavioral performance. By 17 hours of continued wakefulness, the performance impairment is equivalent to that observed with alcohol intoxication in healthy subjects [2]. The circadian system, centered in the suprachiasmatic nucleus of the hypothalamus and entrained by the environmental light/dark cycle, oscillates in such a way to provide a counterbalancing drive for wakefulness during the day and reinforcing stimulus for sleep at night. Both the obtainment of adequate sleep time to satiate the homeostatic drive and the proper alignment of this drive with the circadian system are important to produce consolidated and behaviorally optimized periods of wakefulness and sleep. This interaction is routinely threatened by the prolonged and fluid work schedules typical of medical training. Before the implementation of work hour limits for residents approximately 20 years ago, it was not unusual for trainees to work 80–90-hour weeks with each week consisting of 36-hour duty shifts in hospital broken up by 12 hours or less of rest between shifts [3].

The homeostatic drive to sleep may not be quenched in residents and fellows due to acute or chronic partial sleep deprivation, which in turn leads to increased sleep propensity and impaired alertness. The on-call shifts ubiquitous in training programs are often marked by acute sleep restriction. Using ambulatory electroencephalographic (EEG) monitoring, Richardson [4] documented that internal medicine residents averaged just 3.6 hours of sleep when on call. Even if afforded an opportunity to sleep on call, pager interruptions are commonly followed by sustained wakefulness [4]. Efforts to balance other life demands make it challenging for trainees to obtain adequate recovery sleep during time away from work, leading to chronic partial sleep deprivation. Nationwide surveys of residents before [3, 5] the imposition of work hour limits revealed reported average nightly sleep durations of less than 6 hours, well below the 7 hours recommended for optimal health [6]. In experimental subjects, curtailing sleep to 6 hours per night results in vigilance deficits equivalent to two nights of total sleep deprivation and an accompanying alarming inability to perceive progressive impairments [7]. From a circadian standpoint, night shifts force trainees to work at the peak of sleep propensity which may jeopardize waking performance. While the homeostatic drive to sleep is strong at the end of such a shift, ensuing daytime recovery sleep quality is jeopardized by the building circadian drive for wakefulness. Rotating work shifts (as opposed to "straight" night shifts), common in specialties such as emergency medicine and critical care medicine, may further exacerbate circadian misalignment.

Sleep deprivation and/or circadian misalignment may also increase vulnerability to sleep inertia in trainees. Sleep inertia is characterized by the desire to return to sleep upon awakening and can be accompanied by varying degrees of grogginess, disorientation, slurred speech, impaired cognition, and automatic behaviors. This phenomenon is most prominent within the first minutes of awakening. The cognitive performance of individuals immediately upon awakening may be just 65% of their peak cognitive performance and is worse than their cognitive performance after 24 hours without sleep [8]. Sleep inertia may pose an important safety concern when trainees are awakened to provide patient care.

Impact of Sleep Loss and Circadian Disruptions on the Personal and Professional Lives of Trainees

The first publication on the impact of sleep loss in residents appeared in 1971 when Friedman [9] reported that post-call interns made nearly twice as many errors when reading electrocardiograms in comparison to rested colleagues. What followed was a vast literature exploring the impact of long work hours and sleep loss on trainee wellbeing and performance and patient safety. Studies reported associations of sleep deprivation with diminished cognitive function, memory, fine motor skills, motivation, and mood, errors in patient care, and serious conflicts with other members of the healthcare team [10, 11]. Surveyed internal medicine residents admitted to falling asleep while performing clinical duties, such as writing notes in the chart (69%), reviewing medication lists (61%), interpreting labs (51%), writing orders (46%), placing central lines, drawing blood cultures, and running codes [12]! Philibert [13] compiled a list of the 30 most influential and widely cited articles from this body of literature, culled from over 1000 investigations. Works highlighted included:

- The report of Reuben [14], which was the first to associate depressive symptoms to residency, with interns exhibiting the highest prevalence. The training context was relevant, as symptoms were greatest among interns during rotations on inpatient care wards and intensive care units. Rosen [15] subsequently first reported temporally linked trends for ratings of chronic sleep deprivation, depression, and burnout in internal medicine interns.
- Jacques' [16] initial demonstration that sleep loss diminished resident performance on standardized tests. There was a significant decline in the in-training test scores among family medicine residents with decreasing sleep the night before the test.
- Hillson's [17] analysis of 22,000 internal medicine admissions that revealed patients admitted at night under the care of residents experienced an increased relative risk for inpatient mortality.
- Marcus and Laughlin's [18] survey to first report that post-call pediatric residents were significantly more likely to fall asleep while driving and to be involved in motor vehicle crashes. A disturbing 49% of surveyed residents reported falling asleep at the wheel with 90% of the events occurring post call. Steele [19] later surveyed emergency medicine residents with 8% reporting crashes and 58% reporting near miss crashes with the overwhelming majority occurring after a night shift.

- Parks' [20] finding of a 1.5-fold increased risk of experiencing a blood-borne pathogen exposure (needle stick, laceration, or splash) for residents and medical students working during night versus daytime hours.
- Bellini's [21] report of decreased trainee caring and enthusiasm after the transition from medical school to residency. Five months into training, residents showed increased scores of depression/dejection, anger/hostility, fatigue/inertia, distress, and decreased empathy.
- Barger's [22] results from a national survey of approximately 2700 interns who completed online monthly reports detailing their work schedules, work activities, and sleep. The odds of a motor vehicle crash increased twofold and of a near miss accident fivefold during the trip home after an extended work shift compared to nonextended shifts. Each extended work shift increased the risk of crash on the home commute by 16%. Although not cited by Philibert [13], these investigators used the same survey process to also reveal an increased risk for sustaining a percutaneous injury the day after working overnight compared to the same time period on the previous day [23]. Self-reported fatigue-related medical errors climbed from 4% during months with no extended shifts to 10% during months with one to four extended shifts to 16% during months with five or more extended shifts [24]. Self-reported falling asleep during educational and patient care activities also increased in proportion to the number of extended shifts worked.

Although compelling, this literature suffers from limitations [25], as studies often involve only one institution or a single specialty, which limits generalizability. The studies also fail to account for circadian factors, baseline chronic partial sleep deprivation in control residents, practice differences, and trainee motivation. Moreover, some results were contradictory, which may have been a result of small sample sizes, leaving studies open to confounding from interindividual differences in resilience to sleep loss [26]. It is challenging to distinguish the impact of sleep loss per se from that of long hours and other job-related stressors characteristic of residency training. Finally, study designs have been heterogeneous, and outcomes measures may not be sensitive to more subtle levels of impairment or may lack relevance to actual work performed by residents. Despite these drawbacks, a meta-analysis of the literature concluded that sleep loss has significant effects on resident clinical performance, cognitive function, memory, and vigilance [27].

Other challenges with this literature include a paucity of analyses performed in true-to-life circumstances and a careful accounting of whether resident impairment actually led to patient harm. Landrigan [28] and colleagues of the Harvard Work Hours Health and Safety Group endeavored to address these shortcomings by performing a carefully designed prospective trial comparing a modified work schedule without shifts longer than 16 hours to a traditional every third night 24-hour call schedule among interns working in the intensive care unit. Greater than 2200 patient-days were monitored in a comprehensive multidisciplinary manner. Interns in the traditional schedule worked 19.5 hours more per week, slept 6 less hours per week, and committed 36% more serious errors (defined by the investigators as "a medical error that causes harm or has the potential to cause harm" [28]), including

21% more serious medication errors and six times more serious diagnostic errors in comparison to their peers on the modified schedule. This group also experienced twice the rate of ambulatory-EEG captured attentional failures during on-call nights [28, 29]. Despite these differences, the adverse event rates for patients cared for by each group were similar, highlighting that the support on-call residents receive from supervisory staff and other members of the multidisciplinary care team (e.g., nurses, pharmacists, therapists) makes it very challenging to determine the extent to which resident sleep deprivation and circadian misalignment result in adverse patient outcomes.

History of Resident Work Hour Reforms (Table 12.1)

The death of 18-year-old Libby Zion on March 5, 1984 in a New York hospital while under the care of residents is widely considered a seminal catalyst with respect to residency training reforms in the United States. Resident fatigue was determined to be a contributing factor to Ms. Zion's death. Her father, Sidney Zion, a journalist and former federal prosecutor, pushed for an investigation and mounted a crusade publicizing the inadequacies of the residency training system. In a *New York Times* op-ed, he wrote, "You don't need kindergarten to know that a resident working a 36-hour shift is in no condition to make any kind of judgment call—forget about life and death" [30]. A grand jury decided not to indict Ms. Zion's doctors, but the case prompted the creation of the Bell Commission, which issued a report recommending greater resident supervision and limits on residency work hours. In 1989, the State of New York became the first jurisdiction to mandate that residents work no more than 80 hours per week. In 1989, the Internal Medicine Residency Review Committee endorsed the same.

In 1998, the European Working Time Directive became law for all workers in the European Union. This statute called for a maximum work week of 48 hours, a minimum rest period of 11 consecutive hours per 24-hour duty, and a minimum rest period of 24 hours per 7-day duty. The rollout for medical trainees across the continent has been uneven with full implementation in the United Kingdom delayed until 2009 [31].

Under pressure from various advocacy groups and the threat of federal intervention, the Accreditation Council for Graduate Medical Education (ACGME), the organization responsible for overseeing residency and fellowship training programs in the United States, instituted compulsory nationwide work limits for residents and fellows in July 2003. The requirements imposed a maximum 80-hour work week averaged over 4 weeks (with an allowance of up to 88 hours in programs with a compelling educational rationale), capped duty shift length to a maximum of 24 hours with an additional 6 hours allowed to maintain continuity of care and participate in didactic activities ("24 + 6"), limited in-hospital call frequency to no more than every third night, stipulated a minimum time between shifts of 10 hours, and mandated 1 day in seven free from all training responsibilities (the latter 3

Table 12.1 Timeline of selected events in the regulation of work hours for medical trainees

1971	Friedman [9] report of increased electrocardiogram interpretation errors in post-call interns as compared to rested colleagues
1984	Death of New York college student, Libby Zion, which was ultimately attributed to suboptimal residency supervision and resident fatigue from extended work hours
1989	New York imposes 80-hour week limit for medical trainees within the state in response to investigations following Ms. Zion's death
1989	Internal Medicine Residency Review Committee endorses 80-hour work week for its accredited training programs
1998	European Working Time Directive enacted which established a maximum work week of 48 hours for European Union workers
2003	Implementation of first ACGME duty hour limits which mandated an 80-hour work week limit, maximum shift length of 24 + 6 hours, maximum on-call frequency of every third night, and 1 day off in 7
2006	Harvard Work Hours Health and Safety Group [28] report demonstrating less sleep and more errors in intensive care unit interns working on traditional call schedule with 24-hour shifts compared with interns on a schedule with a maximum shift length of 16 hours
2008	Release of Institute of Medicine's report, *Resident Duty Hours: Enhancing Sleep, Supervision, and Safety* [33], which advocated for more stringent resident duty hour limits
2011	Implementation of second ACGME duty hour limits [34] which capped the maximum intern work shift at 16 hours and 24 + 4 hours for all other residents. The 80-hour work week limit, maximum on-call frequency of every third night, and 1 day off in 7 continued
2016	FIRST [35] results published demonstrating equivalent patient outcomes from surgical programs managed under the 2011 ACGME work limits versus those with flexibility with respect to maximum shift length and time off between shifts. Residents expressed a strong preference for training under more flexible conditions [37]
2017	Implementation of third ACGME duty hour limits [38] which reversed the 16-hour maximal shift length cap for interns and subjected all residents to the 24 + 4 hour shift limit. The 80-hour work week limit, maximum shift length of 24 + 6 hours, maximum on-call frequency of every third night, and 1 day off in 7 continued
2018	iCOMPARE [36] results published demonstrating more resident dissatisfaction yet greater residency program director satisfaction in internal medicine residencies that allowed flexibilities with respect to maximum shift length and time off between shifts versus programs managed under the 2011 ACGME work limits
2019	Internal medicine programs with flexible duty hours participating in iCOMPARE report noninferior patient outcomes compared to standard duty hour programs [38]

averaged over 4 weeks). Internal moonlighting performed by residents (covering extra shifts within their teaching institution) counted toward the 80-hour weekly limit. The requirements also obligated training programs to adhere to principles of prioritizing resident learning over service duties, to provide appropriate supervision of resident patient care activities, and to educate faculty and residents on recognizing and mitigating sleep deprivation.

The duty hour limits were immediately controversial and have remained so to the present day. Critics voiced concern that increased rest for residents and the accompanying personal wellbeing benefits would be offset by increased transitions of patient care, decreased professionalism (clock-based versus patient-based ethic), reduced access to procedures and educational activities, and greater differences between training and real-life practice encountered after training completion [32]. There was also the realization that work hour limits didn't necessarily translate into dramatically longer recovery sleep. A nationwide survey of trainee self-reported sleep revealed a modest increase in mean nightly sleep duration of 22 minutes (from 5.91 hours to 6.27 hours) [5]. Proponents argued that the work limits were not stringent enough, as aggregate data suggested residents working within the 30-hour maximal allowable shift may function at a level comparable to the 15th percentile of rested colleagues [27]. Beyond the debate was the practical challenge of replacing care coverage previously provided by residents. On the eve of the 2003 work hour limits' implementation, first-year obstetrics/gynecology and general surgery trainees provided an average of 91 and 102 hours of patient care per week, respectively [3]. Hospitals responded by hiring more advanced practice providers and hospitalists, by implementing night-only call teams ("night floats") to relieve on-call residents, and by shifting duties to supervising faculty physicians.

The Landrigan modified intern schedule study [28] and recommendations from the Institute of Medicine's (IOM) 2008 report, *Resident Duty Hours: Enhancing Sleep, Supervision, and Safety* [33], significantly influenced the ACGME's next version of its duty hour regulations. The IOM advocated for further work hour limits, improved patient handoffs, and closer resident supervision. Specific recommendations included no change in the 80-hour work week limit, but a reduction in the maximum allowable shift length for all residents to either 16 or 30 hours (the latter if there were 5 hours protected time for sleep between the circadian favorable periods of 10 PM–8 AM); a minimum of 10 and 12 hours off after a day and night shift, respectively; 14 hours off after an extended duty period with no return to the hospital before 6 AM the next day; a mandatory 5 days off per month with at least 1 day off per week (not averaged); a maximum of four in-hospital night shifts per month with 48 hours off after three or four consecutive night shifts to minimize circadian disruption; and inclusion of internal and external moonlighting hours toward the 80-hour weekly limit. The economic reality of hiring the personnel to implement these changes was estimated at $1.7 billion.

The second version of the ACGME revised common program requirements, enacted in 2011, further limited the maximum duty period for interns to 16 hours [34]. The maximum shift length for intermediate and senior residents remained at 24 hours, but the extra time allowed for transitional and educational activities was cut to 4 hours ("24 + 4"). The rules granted residents the flexibility to remain beyond their maximal allowable duty period to continue to provide care to a single patient under special clinical, academic, or humanistic circumstances. Strategic napping between 10 PM and 8 AM was strongly suggested for intermediate or senior residents working beyond 16 hours without mandating protected time for this rest. The mandatory minimum time off between 24-hour shifts was increased to 14 hours. Night duty was limited to six consecutive nights, interns were now prohibited from moonlighting, and internal *and* external moonlighting now counted against the 80-hour work week limit. The foundational requirements of the 80-hour work week limit, maximum on-call frequency of every third night, and 1 day off in 7 were preserved. The revised rules encouraged residents to prioritize sleep over other discretionary activities when away from work (so as to arrive on duty appropriately rested), required training programs to adjust schedules to alleviate burdensome service demands while minimizing patient handoffs, and mandated the provision of adequate sleep facilities and/or transportation options for residents too sleepy to return home, with an overall cultural emphasis of safety, quality improvement, trainee wellbeing, and team-based patient care.

Controversy continued after the implementation of the 2011 duty hour requirements, especially surrounding the 16-hour shift limit for interns. There was concern, especially from surgical disciplines, that this limit was diminishing continuity of care and imperiling trainee opportunities to witness the trajectory of acute clinical conditions. The ACGME heeded the call of the IOM [33] to foster multicenter research on the impacts of duty hour limits by providing seed funding and waivers for several duty hour requirements for programs involved in two national trials that examined the impact of work hour limits: the Flexibility in Duty Hour Requirements for Surgical Trainees (FIRST) trial [35] and the Individualized Comparative Effectiveness of Models Optimizing Patient Safety and Resident Education (iCOMPARE) trial [36, 38]. FIRST was a noninferiority trial that randomized 117 general surgery programs during the 2014–2015 academic year to the 2011 ACGME duty hour policies (standard policy group) or to more flexible policies that waived limits on maximum shift lengths and time off between shifts (flexible policy group) while still abiding by the 80-hour work week limit, maximum on-call frequency of every third night, and an average of 1 day off in every 7 days. There was no difference between groups in the primary composite outcome measure of 30-day rate of postoperative death or serious complications or of secondary postoperative outcomes determined through the American College of Surgeons' National Surgical Quality Improvement Program. Residents in the flexible policy group did not report greater dissatisfaction with their education quality and were less likely than standard group residents to perceive negative impacts of their duty hours on patient safety, care continuity, clinical skills acquisition, operative skills acquisition, autonomy, operation volumes and case completion, conference attendance, and professionalism, but they were more likely to notice negative effects on case preparation, research work, family time, extracurricular activity participation, and rest. Only 14% of FIRST residents surveyed expressed a preference for training under standard ACGME policies [37].

The iCOMPARE trial similarly randomized 63 internal medicine programs to the 2011 ACGME duty hour policies or to more flexible policies that waived limits on maximum shift lengths and time off between shifts. Trainees in the two groups had similar average percentages of directly observed shift time spent in direct patient care, in-training examination scores, perception of appropriate balance of service demands and education in their clinical rotations, and burnout scores. The flexible group residents were more likely to report dissatisfaction with the overall quality of their education and impact of work on their personal lives but less likely to report negative impacts of duty hours on continuity of care [36].

Faculty perceived the balance of resident workload to capacity equivalent in both groups, while program directors in flexible duty hour programs were less likely to report dissatisfaction with many aspects of the training environment, including intern responsibility for patient care, intern morale, frequency of handoffs, continuity of care, and time for teaching [36]. Programs with flexible duty hours had noninferior differences in pre-trial year to trial year 30-day mortality, 7-day readmission rates, and patient safety indicators, compared to programs adhering to standard ACGME duty hours [38].

The ACGME's third version of its program requirements for duty hours was implemented during the 2017–2018 academic year (Table 12.2) [39]. Based on review of published research, including FIRST, and broad input from multiple stakeholders, the ACGME reversed its decision on the intern 16-hour cap and restored the maximum work shift duration for interns to that of all other trainees (the aforementioned "24 + 4"). The 80-hour work week, maximum on-call frequency of every third night, and 1 day off in 7 limits endure. The requirements add to the 2011 rules by expanding the focus on trainee and faculty wellbeing, by mandating that faculty and residents engage in patient safety and quality improvement activities, and by requiring programs to design systems to optimize transitions in patient care and to maintain an environment that promotes the joy of professional and intellectual development.

Table 12.2 Requirements pertaining to duty hours from Section VI of the 2017 ACGME Common Program Requirements [39]

Maximum hours of clinical and education work per week
Clinical and educational work must be limited to no more than 80 hours per week, averaged over a 4-week period, inclusive of all in-house clinical and educational activities, work done from home, and all moonlighting
Mandatory time free of clinical work and education
The program must design an effective program structure that is configured to provide residents with educational opportunities, as well as reasonable opportunities for rest and personal wellbeing
Residents should have 8 hours off between scheduled clinical work and educational periods. There may be circumstances when residents choose to stay to care further for patients or to return to the hospital with fewer than 8 hours free of clinical experience and education. This must occur within the context of the 80-hour and the 1 day off in 7 requirements
Residents must have at least 14 hours free of clinical and educational work after 24 hours of in-house call
Residents must be scheduled for a minimum of 1 day off in 7 free of clinical work and required education (when averaged over 4 weeks). At-home call cannot be assigned on these free days.
Maximal clinical work and education period length
Clinical and educational work periods for residents must not exceed 24 hours of continuous scheduled clinical assignments
Up to 4 hours of additional time may be used for activities related to patient safety, such as providing effective transitions of care, and/or resident education. Additional patient care responsibilities must not be assigned to resident during this time

Table 12.2 (continued)

Clinical and educational work hour exceptions
In rare circumstances, after handing off all other responsibilities, a resident, on their own initiative, may elect to remain or return to the clinical site in the following circumstances: to continue to provide care to a single severely ill or unstable patient, to provide humanistic attention to the needs of the patient or family, or to attend unique educational events. These additional hours of care or education will be counted toward the 80-hour weekly limit
A Review Committee may grant rotation-specific exceptions for up to 10% or a maximum of 88 clinical and educational work hours to individual programs based on a sound educational rationale
Moonlighting
Moonlighting most not interfere with the ability of the resident to achieve the goals and objectives of the educational program and must not interfere with the resident's fitness for work nor compromise patients' safety
Time spent by residents in internal and external moonlighting must be counted toward the 80-hour maximum weekly limit
Postgraduate year 1 residents are not permitted to moonlight
In-house night float
Night float must occur within the context of the 80-hour and 1 day off in 7 requirements
Maximum in-house on-call frequency
Residents must be scheduled for in-house call no more frequently than every third night (when averaged over a 4-week period)
At-home call
Time spent on patient care activities by residents on at-home call must count toward the 80-hour maximum weekly limit. The frequency of at-home call is not subject to the every-third-night limitation, but must satisfy the requirement of 1 day off in 7 free of clinical work and education, when averaged over 4 weeks
At-home call must not be so frequent or taxing so as to preclude rest or reasonable personal time for each resident
Reasons are permitted to return to the hospital while on at-home call to provide direct care for new or established patients. These hours of inpatient patient care must be included in the 80-hour maximum weekly limit

Effect of Resident Duty Hour Limitations on Resident Wellbeing, Resident Education, and Patient Safety

Despite an ever-expanding number of studies, a definitive sense of the impact of the ACGME's efforts to limit resident work hours on trainees and the patients they care for has not fully emerged. The varied findings are evident in the following high-impact studies selected by Philibert [13]:

- Poulose [40] reported that the frequency of adverse patient safety indicators for surgical residents increased after implementation of the New York state work hour limits.
- Gopal [41] found that the original 2003 work hour limits were associated with decreased internal medicine resident exhaustion, depersonalization, and depression rates at a large academic medical institution, although resident educational conference attendance dropped and overall residency satisfaction fell.

- Examining on-call residents assigned to a schedule that provided protection from clinical duties from midnight to 7 AM versus a standard 24-hour call schedule, Arora [42] observed that many residents on the protected schedule chose not to use the coverage because of their desire to avoid discontinuity of care.
- Volpp [43] revealed that the 2003 ACGME work limits were not associated with statistically significant improvements or declines in mortality of Medicare medical or surgical patients when comparing the 3 years prior to the 2003 work hour limits with the two subsequent years.
- Jagannathan [44] reported that work hour limits were associated with reduced performance on the American Board of Neurological Surgery written exam and that while resident registration to national meetings increased, the number of resident abstracts dropped 7% when comparing 2002 with 2007.
- McCoy [45] observed that internal medicine residents on a rotation with a 16-hour shift limit felt less prepared to manage cross coverage of patients, although patient care metrics were unchanged.
- Desai [46] randomized internal medicine interns to schedules compliant with the 2003 or 2011 work hour limits and showed that the 2011-compliant schedules increased intern sleep but decreased intern participation in didactic sessions and increased patient handoffs. The study was terminated early because of concerns of reduced quality of care delivered by the night float.
- In a cohort of 2300 interns who self-reported information on work and wellbeing quarterly over the course of a year, Sen [47] learned that the 2011 work hour limits reduced weekly work hours but did not increase sleep. Depressive symptoms and wellbeing did not change and self-reported medical errors increased.

These conflicting findings explain why systematic reviews of this literature have come to different conclusions. In Bolster and Rourke's [48] analysis of the systematic reviews of the 2003 work hour limits' literature, five reviews examined the impact of duty hour limits on resident wellbeing with four reviews [33, 49–51] concluding the impact was favorable and one [52] determining the literature was inconclusive; five reviews examined the impact on resident education with two [51, 53] concluding there was no impact, two [49, 50] concluding the impact was inconclusive, and one [52] concluding the impact was unfavorable; and eight reviews examined the impact of duty hour limits on patient safety with two [33, 52] concluding there was a positive impact, two [53, 54] finding no impact, and four [50, 51, 55, 56] concluding the impact was inconclusive. Bolster and Rourke then reviewed literature regarding the 2011 duty hours and similarly found mixed results for resident- and patient-focused outcomes. Night float in particular, with its attendant circadian disruption from consecutive nights worked, was found in the majority of studies to have a negative impact on residents, including decreased sleep, more stress and fatigue, decreased conference attendance, and less exposure to attending physicians [48].

The complexity of studying duty hour reforms exposes the many methodologic limitations of this vast literature [57]. As mentioned above, studies often suffer from small sample sizes and may only describe a single institution's experience.

A glaring and persistent shortcoming has been the very limited attention paid to circadian factors. Investigations employing large national administrative databases for patient outcomes may lack the fidelity to account for the impact of local training processes. Follow-up of residents or patients involved in studies may be too short. Some studies have mixed and others segregated medical and surgical residents and patients. Conclusions from studies only involving interns may not generalize to residents nearing graduation. Many studies employ surveys which may not be validated and/or have low response rates. These surveys gather stakeholder perceptions of the impact of duty hour reforms but lack objective data for corroboration of their impressions. A large array of endpoints of varying rigor and appropriateness has been used to capture facets of resident wellbeing, resident education, and patient safety. Finally, investigators struggle to account for the many other factors that can influence the resident learning experience, such as care process changes that have accompanied duty hour limits or advances in medical knowledge.

Amidst all mixed findings, the ACGME drew conclusions from its literature review [51] that informed creation of their 2017 duty hour regulations. The duty hour limits have increased resident sleep, albeit modestly, and objective measures of alertness. The majority of studies have indicated a reduction in resident burnout with work hour limits, yet how this impacts the physician-patient relationship is understudied. The majority of studies have not shown a negative impact of duty hour limits on procedural volumes or in-training exam scores. Trainee work has been compressed into fewer hours which may result in a negative impact on the resident-patient relationship, patient outcomes, and resident participation in educational activities. The impact of duty hour limits may differ between medical and surgical trainees (as per FIRST and iCOMPARE). The impact on patient safety has been mixed and also medical specialty context dependent. Single-institution studies involving internal medicine residency programs have generally found a positive effect, whiles studies using large national datasets have found negative impacts on surgical patients.

Conclusion

The need to improve the work and learning environments of resident and fellow physicians which were so disruptive to sleep and circadian mechanisms was indisputable, yet how best to modify these conditions while simultaneously achieving the goals of training professional practitioners ready for fulfilling careers and the provision of excellent patient care remains in dispute. The primary approach has been to address sleep deprivation through work hour restrictions. The expansive literature examining how the duty hour limitations have influenced trainee wellbeing/education and patient safety has yielded decidedly mixed findings, although there is a suggestion that medical and surgical disciplines are differentially impacted, raising the possibility of more specialty tailored future work hour regulations.

Far less attention has been paid to studying and mitigating the circadian misalignments integral to the house staff training experience. The unpredictable trajectory of medical conditions and the complexity of modern-day inpatients require hospitals to maintain the capacity to deliver care around the clock which inevitably means trainees will work and sleep at times that are inopportune from the circadian standpoint and that places residents at increased risk for adverse events, such as percutaneous injuries. The commonly employed strategy of limiting resident exposure to overnight duty by providing them night float coverage has been understudied, and the limited data suggest inconsistent impacts. Grossly underexplored are strategies to facilitate resident circadian acclimatization in the setting of extended and/or rotating work shifts.

ACGME acknowledges that its duty hour policies are a "living document that will continue to evolve in response to changing medical and educational practices, cultural mores, and research findings" [58]. The continuing effort to optimize resident working conditions will result in an ongoing tension between sleep homeostatic and circadian considerations. To date, work hour restrictions have been prioritized, but they are just one component of the very complex training environment, and further attention to circadian and other contextual elements, such as faculty supervision and handoffs of information at transition points in patient care, is anticipated.

References

1. Borbely AA. A 2 process model of sleep regulation. Hum Neurobiol. 1982;1:195–204.
2. Dawson D, Reid K. Fatigue, alcohol and performance impairment. Nature. 1997;388:235.
3. Baldwin DC Jr, Daugherty SR, Tsai R, Scotti MJ Jr. A national survey of residents' self-reported work hours: thinking beyond specialty. Acad Med. 2003;78:1154–63.
4. Richardson GS, Wyatt JK, Sullivan JP, et al. Objective assessment of sleep and alertness in medical hour staff and the impact of protected time for sleep. Sleep. 1996;19:718–26.
5. Landrigan CP, Barger LK, Cade BE, Ayas NT, Czeisler CA. Interns' compliance with accreditation Council for Graduate Medical Education work-hour limits. JAMA. 2006;296:1063–70.
6. Watson NF, Badr MS, Belenky G, et al. Recommended amount of sleep for a healthy adult: a joint consensus statement of the American Academy of sleep medicine and Sleep Research Society. J Clin Sleep Med. 2015;11:591–2.
7. Van Dongen HPA, Maislin G, Mullington JM, Dinges DF. The cumulative cost of additional wakefulness: dose-response effects on neurobehavioral functions and sleep physiology from chronic sleep restriction and total sleep deprivation. Sleep. 2003;26:117–26.
8. Wertz AT. Effects of sleep inertia on cognition. JAMA. 2006;295:163–4.
9. Friedman RC, Bigger T, Kornfeld DS. The intern and sleep loss. N Engl J Med. 1971;285:201–3.
10. Weinger MB, Ancoli-Isreal S. Sleep deprivation and clinical performance. JAMA. 2002;287:955–7.
11. Veasey S, Rosen R, Barzansky B, Rosen I, Owens J. Sleep loss and fatigue in residency training: a reappraisal. JAMA. 2002;288:1116–24.
12. Rosen IM, Bellini LM, Shea JA. Sleep behaviors and attitudes among internal medicine housestaff in a U.S. university-based residency program. Acad Med. 2004;79:407–16.
13. Philibert I. What is known: examining the empirical literature in resident work hours using 30 influential articles. J Grad Med Edu. 2016;8:795–804.

14. Reuben DB. Depressive symptoms in medical house officers: effects of training and work rotation. Arch Intern Med. 1985;145:286–8.
15. Rosen IM, Gimotty PA, Shea JA, Bellini LM. Evolution of sleep quantity, sleep deprivation, and mood disturbances, empathy, and burnout among interns. Acad Med. 2006;81:82–5.
16. Jacques CH, Lynch JC, Samkoff JS. The effects of sleep loss on cognitive performance of resident physicians. J Fam Pract. 1990;30:223–9.
17. Hillson SD, Dowd B, Rich EC, Luxenberg MG. Call nights and patients care: effects on inpatients at one teaching hospital. J Gen Intern Med. 1992;7:405–10.
18. Marcus CL, Loughlin GM. Effect of sleep deprivation on driving in housestaff. Sleep. 1996;19:763–6.
19. Steele MT, Ma OJ, Watson WA, Thomas HA, Muelleman RL. The occupational risk of motor vehicle collisions for emergency medicine residents. Acad Emerg Med. 1999;6:1050–3.
20. Parks DK, Yetman RJ, McNeese MC, Burau K, Smolensky MH. Day-night pattern in accidental exposures to blood-borne pathogens among medical students and residents. Chronobiol Int. 2000;17:61–70.
21. Bellini LM, Baime M, Shea JA. Variation of mood and empathy during internship. JAMA. 2002;282:3143–6.
22. Barger LK, Cade BE, Ayas NT, et al. Extended work shifts and the risk of motor vehicle crashes among interns. N Engl J Med. 2005;352:125–34.
23. Ayas NT, Barger LK, Cade BE, et al. Extended work shift duration and the risk of self-reported percutaneous injuries in interns. JAMA. 2006;296:1055–62.
24. Barger LK, Ayas NT, Cade BE, et al. Impact of extended-duration shifts on medical errors, adverse events, and attentional failure. PLoS Med. 2006;3:e487.
25. Owens JA. Sleep loss and fatigue in medical training. Curr Opin Pulm Med. 2001;7:411–8.
26. Van Dongen HPA, Baynard MD, Maislin G, Dinges DF. Systematic interindividual differences in neurobehavioral impairment from sleep loss: evidence of a trait-like differential vulnerability. Sleep. 2004;27:423–33.
27. Philibert I. Sleep loss and performance in residents and nonphysicians: a meta-analytic examination. Sleep. 2005;28:1392–402.
28. Landrigan CP, Rothschild JM, Cronin JW, et al. Effect of reducing interns' work on serious medical errors in intensive care units. N Engl J Med. 2004;351:1838–48.
29. Lockley SW, Cronin JW, Evans EE, et al. Effect of reducing interns' work hours on sleep and attentional failures. N Engl J Med. 2004;351:1829–37.
30. Zion S. "Doctors know best?" New York times, 13 May 1989, https://www.nytimes.com/1989/05/13/opinion/doctors-know-best.html. Accessed 29 Jan 2019.
31. Temple J. Resident duty hours around the globe: where are we now? BMC Med Edu. 2014;14(Suppl 1):S8.
32. Pastores SM, O'Connor MF, Kleinpell RM, et al. The accreditation Council for Graduate Medical Education resident duty hour new standards: history, changes, and impact on staffing intensive care units. Crit Care Med. 2011;39:2540–9.
33. Ulmer C, Wolman D, Johns M, editors. Resident duty hours: enhancing sleep, supervision, and safety. Washington DC: National Academies Press; 2008.
34. Accreditation Council for Graduate Medical Education. Common program requirements. 2011. https://www.acgme.org/Portals/0/PFAssets/ProgramResources/Common_Program_Requirements_07012011.pdf?ver=2015-11-06-120650-183. Accessed 28 Jan 2019.
35. Bilimoria KY, Chunng JW, Hedges LV, et al. National cluster-randomized trial of duty hour flexibility in surgical training. N Engl J Med. 2016;374:713–27.
36. Desai SV, Asch LM, Bellini KH, et al. Education outcomes in a duty hour flexibility trial in internal medicine. N Engl J Med. 2018;378:1494–508.
37. Yang AD, Chung JW, Dahlke AR, et al. Differences in resident perceptions by postgraduate year of duty hour policies: an analysis for the flexibility in duty hour requirements for surgical trainees (FIRST) trial. J Am Coll Surg. 2017;224:103–12.
38. Silber JH, Bellini LM, Shea JA, et al. Patient safety outcomes under flexible and standard resident duty-hour rules. N Engl J Med. 2019;380:905–14.

39. Accreditation Council for Graduate Medical Education. Common program requirements section VI with background and intent. 2017. https://www.acgme.org/Portals/0/PFAssets/ProgramRequirements/CPRs_Section%20VI_with-Background-and-Intent_2017-01.pdf. Accessed 28 Jan 2019.
40. Poulose BK, Ray WA, Arbogast PG, et al. Resident work hour limits and patient safety. Ann Surg. 2005;241:847–56; discussion 856–60.
41. Gopal R, Glasheen JJ, Miyoshi TJ, et al. Burnout and internal medicine resident work-hour restrictions. Arch Intern Med. 2005;165:2595–600.
42. Arora V, Dunphy C, Chang VY, et al. The effects of on-duty napping on intern sleep time and fatigue. Ann Intern Med. 2006;144:792–8.
43. Volpp KG, Rosen AK, Rosenbaum PR, et al. Mortality among hospitalized Medicare beneficiaries in the first 2 years following ACGME resident duty hour reforms. JAMA. 2007;298:973–83.
44. Jagannathan J, Vates GE, Pouratian N, et al. Impact of the accreditation Council for Graduate Medical Education wotk-hour regulations on neurosurgical resident education and productivity. J Neurosurg. 2009;110:820–7.
45. McCoy CP, Halvorsen AJ, Loftus CG, McDonald FS, Oxentenko AS. Effect of 16-hour duty periods on patient care and resident education. Mayo Clin Proc. 2011;86:192–6.
46. Desai SV, Feldman L, Brown L, et al. Effect of the 2011 vs 2003 duty hour regulation-compliant models on sleep duration, trainee education, and continuity of patient care among internal medicine hour staff: a randomized trial. JAMA Intern Med. 2013;173:649–55.
47. Sen S, Kranzler HR, Didwania AK, et al. Effects of the 2011 duty hour reforms on interns and their patients: a prospective longitudinal cohort study. JAMA Intern Med. 2013;173:657–62; discussion 663.
48. Bolster L, Rourke L. The effect of restricting residents' duty hours on patient safety, resident well-being, and resident education: an updated systematic review. J Grad Med Educ. 2015;7:149–63.
49. Fletcher KE, Underwood W 3rd, Davis SQ, Mangrulkar RS, McMahn LF, Saint S. Effects of work hour reduction on residents' lives: a systematic review. JAMA. 2005;294:1088–100.
50. Fletcher KE, Reed DA, Arora VM. Patient safety, resident education and resident well-being following implementation of the 2003 ACGME duty hour rules. J Gen Intern Med. 2011;26:907–19.
51. Philibert I, Nasca T, Brigham T, Shapiro J. Duty hour limits and patient care and resident outcomes: can high-quality studies offer insight into complex relationships? Ann Rev Med. 2013;64:467–83.
52. Reed D, Fletcher K, Arora VM. Systematic review: association of shift length, protected sleep time, and night float with patient care, residents' health, and education. Ann Intern Med. 2010;153:829–42.
53. Moonesinghe SR, Lowery J, Shahi N, Millen A, Beard JD. Impact of reduction in working hours for doctors in training on postgraduate medical education and patients' outcomes: systematic review. BMJ. 2011;342:d1580.
54. Jamal M, Doi SA, Rousseau M, et al. Systematic review and meta-analysis of the effect of North American working hours restrictions on mortality and morbidity in surgical patients. Br J Surg. 2012;99:336–44.
55. Fletcher KE, Davis SQ, Underwood W, Mangrulkar RS, McMahon LF, Saint S. Systematic review: effects of resident work hours on patient safety. Ann Intern Med. 2004;141:851–7.
56. Baldwin K, Namdari S, Donegan D, Kamath A, Mehta S. Early effects of resident work-hour restrictions on patient safety: a systematic review and plea for improved studies. J Bone Joint Surg Am. 2011;93:e5.
57. Lin H, Lin E, Auditore S, Fanning J. A narrative review of high-quality literature on the effects of resident duty hours reforms. Acad Med. 2016;91:140–50.
58. Burchiel KJ, Zetterman RK, Ludmerer KM, et al. The 2017 common work hour standards: promoting physician learning and professional development in a safe, human environment. J Grad Med Edu. 2017;9:692–6.

Chapter 13
Jet Lag Sleep Disorder

Cinthya Pena-Orbea, Bhanu Prakash Kolla, and Meghna P. Mansukhani

Introduction

Jet lag disorder (JLD) occurs as a result of temporary misalignment between the internal circadian clock and the external light/dark environment due to a rapid shift across two or more time zones. The internal clock is relatively slow to reset and requires approximately 1 day per zone crossed to resynchronize [1]. The prevalence is unknown and there are conflicting data regarding the age predilection for JLD. In one study, middle-aged individuals (37–52 years) had more fragmented sleep on polysomnography and a greater impairment in daytime alertness than younger individuals (18–25 years) [2]. Results from other studies, however, suggest that older age can be protective against jet lag. The methods and age groups studied differed among these investigations which make direct comparisons difficult [3, 4].

Individuals suffering from JLD may experience a multitude of daytime and nighttime symptoms including impaired alertness, fatigue, insomnia, irritability, depression, reduced cognitive skills, psychomotor discoordination, and gastrointestinal issues (e.g., anorexia, nausea, and diarrhea) [5, 6]. It is generally easier to reacclimate subsequent to westward versus eastward travel, as the former typically favors a circadian phase delay, an adjustment facilitated by humans' typically greater than 24-hour innate circadian clock [7, 8]. Those engaged in westward air travel typically have difficulties with sleep maintenance and early morning awakenings as the internal clock promotes wakefulness during destination early morning hours until reacclimatization occurs. Sleepiness and fatigue during evening hours are common complaints at the new time zone. On the other hand, eastward air travelers typically

C. Pena-Orbea · M. P. Mansukhani (✉)
Center for Sleep Medicine, Mayo Clinic, Rochester, MN, USA

B. P. Kolla
Center for Sleep Medicine and Department of Psychiatry and Psychology, Mayo Clinic, Rochester, MN, USA

R. R. Auger (ed.), *Circadian Rhythm Sleep-Wake Disorders*,
https://doi.org/10.1007/978-3-030-43803-6_13

have difficulties with sleep onset at night and sleepiness during the daytime due to a misalignment between the sleep and wakefulness drive at the new time zone. The number of time zones traveled (three or more) [6], time of travel (morning travel is better for those traveling east and vice versa), ability to sleep while traveling, travel-related discomfort, alcohol and/or caffeine intake, as well as dehydration are other factors that can play an important role in the development of jet lag.

Jet lag is a benign and self-limited condition for many, although even those traveling for leisure will cite avoidance of related symptoms as an important goal. The condition can be more dangerous for military personnel, pilots, flight attendants, and athletes, in whom alertness and optimal performance are critical. The rapidity at which one reacclimates depends upon many variables, including one's innate phase tolerance.

Pathophysiology

The circadian regulation of sleep-wake cycles is driven by the biological clock located in the suprachiasmatic nucleus (SCN) which is situated in the anterior hypothalamus. The human circadian period is (on average) slightly greater than 24 hours and is synchronized (entrained) with the 24-hour day under the influence of environmental factors, with light exposure as the most important variable [8, 9]. The timing of the light exposure will determine the shift direction of the circadian clock. Neurons in the SCN project to various areas of the brain, modulating daily rhythms in sleep and alertness, in core body temperature (CBT), and the secretion of melatonin, cortisol, and other hormones [1].

There are two main processes that are thought to influence patterns of sleep-wake behavior [10]. The homeostatic sleep drive, called "Process S," refers to the process whereby sleep tendency increases proportionally to the duration of preceding wakefulness; this is maximal at about 40 hours of wakefulness. "Process C," in contrast, drives wakefulness and is controlled by the circadian pacemaker irrespective of behavioral state [11]. The homeostatic sleep drive begins to accumulate subsequent to a sleep bout but, in the entrained state, is counteracted by the circadian drive for wakefulness. At bedtime, the wakefulness drive subsides, and the homeostatic drive expresses itself fully. This cycle is repeated every 24 hours. In JLD, these two processes are misaligned with external time, resulting in sleep disturbances and impaired daytime alertness.

Diagnosis

The diagnosis of JLD can be made based upon clinical history. The International Classification of Sleep Disorders Third Edition (ICSD-3) [5] stipulates that the following three criteria are met:

1. There is a complaint of insomnia or excessive daytime sleepiness, accompanied by a reduction in total sleep time, associated with transmeridian jet travel across at least two time zones.
2. There is an associated impairment of daytime function, general malaise, or somatic symptoms such as gastrointestinal disturbance within 1 or 2 days after travel.
3. The sleep disturbance is not better explained by another current sleep disorder, medical or neurological disorder, mental disorder, medication use, or substance use disorder.

Treatment

Resynchronization of the sleep-wake rhythm with the external light/dark cycle resolves jet lag symptoms. Treatment principles are similar for eastward and westward travel, but the timing and administration of treatment depend upon the direction of travel and the number of time zones crossed. Realignment is not feasible when traveling for less than 2 days [12].

Regardless of the circadian phase, the traveler should ensure adequate sleep. This can be achieved by different strategies: keeping a regular sleep-wake schedule in the destination time zone, maintaining a good sleep environment that is dark and quiet and, if that is not possible, earplugs and eyeshades can be used. Travelers should also try to avoid excessive alcohol or caffeine usage, as these can have unfavorable effects on sleep [13, 14]. Sleep during the flight should be aimed to align with the destination nighttime to hasten adaptation to the new time zone.

Light/Dark Exposure for Jet Lag

Light is the primary factor for entertainment of circadian rhythms. The core body temperature minimum (CBT) drops to its lowest point ~3 hours before one habitually arises from sleep. Exposure to light prior to the CBT causes phase delays and, conversely, exposure after this inflection point promotes phase advances. Therefore, the amount and the timing of light exposure upon arrival at the destination is the main factor determining the speed and the direction of alignment. Controlling preflight light exposure has also been found to reduce jet lag symptoms in travelers [15].

When traveling east, entrainment generally favors a phase advance (≤8 time zones crossed), which requires a movement of the sleep-wake schedule to an earlier time. This can be achieved by avoiding light exposure approximately 3 hours prior to CBT and pursuing exposure to bright light (outdoor exposure or via a medical grade light box) for approximately 3 hours thereafter (Fig. 13.1).

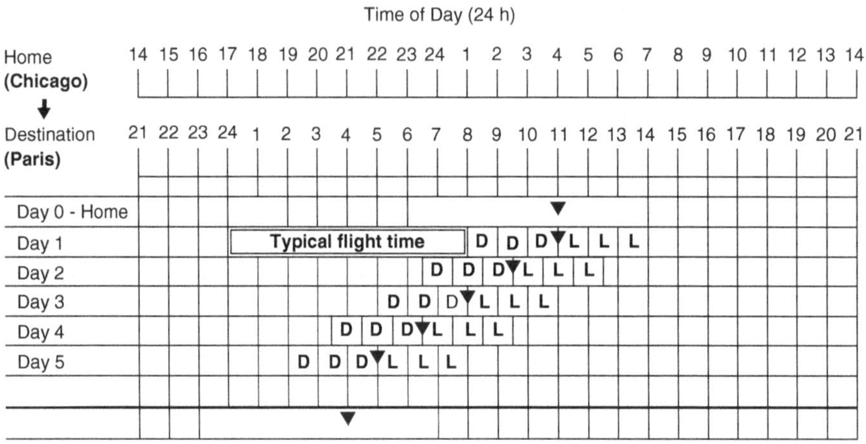

Fig. 13.1 Jet lag plan for seven time zones east. A time grid for jet travel from Chicago to Paris (7 hours eastward flight). The clear rectangles represent habitual sleep times at home (Day 0) and the desired sleep time at the travel destination (final row). The inverted triangles represent estimated CBT. D = when to seek dark, L = when to seek light. If started on arrival, jet lag symptoms should subside in 3–4 days

As an example, a flight from Chicago to Paris (7 hours eastward) will arrive early morning at destination if leaving Chicago in the evening (Fig. 13.1). Thus travelers should avoid bright light exposure (denoted by "D" in Fig. 13.1) by wearing dark glasses or staying indoors from early morning (upon arrival) until 11 am, assuming the CBT will occur at 11 am in the new destination (Day 1 in Fig. 13.1). Light exposure should be maximized for 3 hours after CBT is reached ("L" in Fig. 13.1). For each subsequent day, the CBT is moved 1.5 hours earlier until a clock time within 1 hour of the desired destination CBT time is reached (or sooner if JLD symptoms subside).

When traveling west, a phase delay is instead generally favored. The sleep-wake schedule needs to be moved to a later time and this is sought by the opposite light/dark pattern. The CBT symbol is drawn 2 hours later on day 1 than on day 0 (to account for the generally greater ease to achieve phase delays). As an example, if traveling from New York to Los Angeles, the CBT minimum will occur 2 hours later at destination. Light exposure should be avoided for 3 hours after CBT minimum is reached and sought for 3 hours before, with subsequent daily shifts in 2-hour increments until a clock time within 1 hour of the desired destination CBT minimum is reached (or sooner if JLD symptoms subside). Light exposure can be achieved using sunlight or portable artificial light.

Shifts greater than 8 hours require contemplation as to the most strategic direction to shift. For example, travel from Los Angeles to Rome favors a phase delay (Fig. 13.2). These recommendations should serve as guidelines only, and one need not pursue to completion if his/her sleep schedule is aligning without major difficulties.

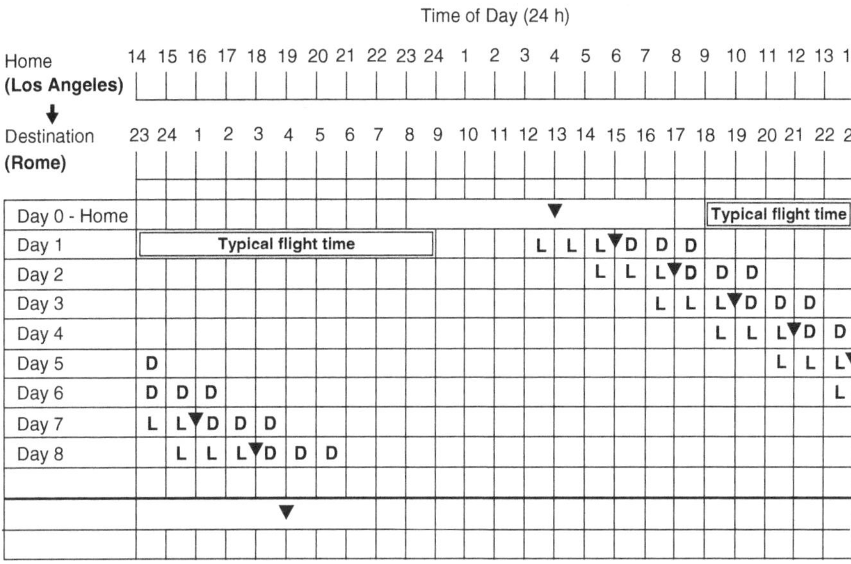

Fig. 13.2 Jet lag plan for nine time zones east. A time grid completed for jet travel from Los Angeles to Rome (9 hours eastward flight). The clear rectangles represent habitual sleep times at home (Day 0) and the desired sleep time at the travel destination (final row). The inverted triangles represent estimated CBT. D = when to seek dark, L = when to seek light. If started on arrival, jet lag symptoms should subside in 4–5 days

Pharmacologic Treatments

Melatonin

Melatonin can be taken at the destination bedtime at doses between 2 and 5 mg. This has been found to improve subjective and objective measures of night time sleep [16–18] in addition to JLD symptoms through its hypnotic and chronobiotic effects [19–21]. Different regimens have been recommended for JLD. One regimen suggests preflight dosing of melatonin at 5 mg for 3 days before the travel day at the corresponding destination bedtime with continued use for 3–4 days upon arrival [16, 20]. The other regimen is simpler and recommends 2 mg of sustained release of melatonin [18] or 5 mg of immediate release melatonin [17, 22] only upon arrival at the destination bedtime for 4–5 days.

It is important to remember that verification of purity of melatonin is difficult as it is not regulated by the U.S. Food and Drug Administration (FDA).

Sedative Hypnotics

Non-benzodiazepine benzodiazepine receptor agonists can improve nighttime sleep; however, their impact on the remainder of JLD symptoms is unclear. Two studies that used zolpidem at 10 mg showed positive effects on subjective sleep quality in the setting of jet lag [16, 23]. In one study that compared melatonin (5 mg) with zolpidem (10 mg) and a combination of zolpidem plus melatonin, zolpidem alone was superior in terms of self-rated sleep quality, but the groups taking zolpidem had more adverse effects than those using melatonin alone, including nausea, vomiting, amnesia, and somnambulism [16].

Stimulants

Caffeinated beverages are commonly used to combat daytime sleepiness. In one study, participants used 300 mg of slow release caffeine at 8:00 am at the destination for 5 days; circadian re-entrainment occurred more rapidly in the caffeine group than in the placebo group; however, jet lag symptoms per se were not assessed [24]. In a subsequent study, those receiving caffeine had more nocturnal sleep complaints in comparison to a non-caffeinated group [25].

Armodafinil at a dose of 150 mg was shown to have beneficial effects on subjective and objective alertness when flying eastward in one study [26].

Conclusion

In summary, JLD is usually a self-limited circadian sleep-wake rhythm disorder that occurs due to desynchronization between the internal rhythm and the external clock. Difficulties with staying awake, falling asleep, and maintaining sleep during the nighttime are the most common symptoms that travelers experience when they travel across more than two time zones. Eastward travel typically requires a phase advance, while westward travel typically requires a phase delay and it usually takes 1 day for each time zone crossed to entrain to the new time zone. Light exposure on the wrong side of the CBT minimum will delay resynchronization of the sleep-wake rhythm. Ensuring adequate sleep and the appropriate timing of light exposure, as well as supplemental use of melatonin, sedative hypnotics and/or stimulants may help with JLD symptoms.

References

1. Sack RL. The pathophysiology of jet lag. Travel Med Infect Dis. 2009;7(2):102–10.
2. Moline ML, Pollak CP, Monk TH, Lester LS, Wagner DR, Zendell SM, et al. Age-related differences in recovery from simulated jet lag. Sleep. 1992;15(1):28–40.
3. Waterhouse J, Edwards B, Nevill A, Carvalho S, Atkinson G, Buckley P, et al. Identifying some determinants of "jet lag" and its symptoms: a study of athletes and other travellers. Br J Sports Med. 2002;36(1):54–60.
4. Sack RL, Auckley D, Auger RR, Carskadon MA, Wright KP Jr, Vitiello MV, et al. Circadian rhythm sleep disorders: part I, basic principles, shift work and jet lag disorders. An American Academy of Sleep Medicine review. Sleep. 2007;30(11):1460–83.

5. Medicine AAoS. International classification of sleep disorders. 3rd ed. American Academy of Sleep Medicine: Darien; 2014.
6. Waterhouse J, Reilly T, Atkinson G, Edwards B. Jet lag: trends and coping strategies. Lancet. 2007;369(9567):1117–29.
7. Kolla BP, Auger RR. Jet lag and shift work sleep disorders: how to help reset the internal clock. Cleve Clin J Med. 2011;78(10):675–84.
8. Czeisler CA, Duffy JF, Shanahan TL, Brown EN, Mitchell JF, Rimmer DW, et al. Stability, precision, and near-24-hour period of the human circadian pacemaker. Science. 1999;284(5423):2177–81.
9. Auger RR, Burgess HJ, Emens JS, Deriy LV, Thomas SM, Sharkey KM. Clinical practice guideline for the treatment of intrinsic circadian rhythm sleep-wake disorders: advanced sleep-wake phase disorder (ASWPD), delayed sleep-wake phase disorder (DSWPD), non-24-hour sleep-wake rhythm disorder (N24SWD), and irregular sleep-wake rhythm disorder (ISWRD). An update for 2015: an American academy of sleep medicine clinical practice guideline. J Clin Sleep Med. 2015;11(10):1199–236.
10. Borbely AA, Achermann P. Concepts and models of sleep regulation: an overview. J Sleep Res. 1992;1(2):63–79.
11. Srinivasan V, Spence DW, Pandi-Perumal SR, Trakht I, Cardinali DP. Jet lag: therapeutic use of melatonin and possible application of melatonin analogs. Travel Med Infect Dis. 2008;6(1–2):17–28.
12. Reid KJ, Abbott SM. Jet lag and shift work disorder. Sleep Med Clin. 2015;10(4):523–35.
13. Stone BM. Sleep and low doses of alcohol. Electroencephalogr Clin Neurophysiol. 1980;48(6):706–9.
14. Walsh JK, Muehlbach MJ, Schweitzer PK. Hypnotics and caffeine as countermeasures for shiftwork-related sleepiness and sleep disturbance. J Sleep Res. 1995;4(S2):80–3.
15. Reid KJ, Burgess HJ. Circadian rhythm sleep disorders. Prim Care. 2005;32(2):449–73.
16. Suhner A, Schlagenhauf P, Hofer I, Johnson R, Tschopp A, Steffen R. Effectiveness and tolerability of melatonin and zolpidem for the alleviation of jet lag. Aviat Space Environ Med. 2001;72(7):638–46.
17. Suhner A, Schlagenhauf P, Johnson R, Tschopp A, Steffen R. Comparative study to determine the optimal melatonin dosage form for the alleviation of jet lag. Chronobiol Int. 1998;15(6):655–66.
18. Paul MA, Gray G, Sardana TM, Pigeau RA. Melatonin and zopiclone as facilitators of early circadian sleep in operational air transport crews. Aviat Space Environ Med. 2004;75(5): 439–43.
19. Edwards BJ, Atkinson G, Waterhouse J, Reilly T, Godfrey R, Budgett R. Use of melatonin in recovery from jet-lag following an eastward flight across 10 time-zones. Ergonomics. 2000;43(10):1501–13.
20. Herxheimer A, Petrie KJ. Melatonin for the prevention and treatment of jet lag. Cochrane Database Syst Rev. 2002;(2):Cd001520.
21. Petrie K, Conaglen JV, Thompson L, Chamberlain K. Effect of melatonin on jet lag after long haul flights. BMJ. 1989;298(6675):705–7.
22. Petrie K, Dawson AG, Thompson L, Brook R. A double-blind trial of melatonin as a treatment for jet lag in international cabin crew. Biol Psychiatry. 1993;33(7):526–30.
23. Jamieson AO, Zammit GK, Rosenberg RS, Davis JR, Walsh JK. Zolpidem reduces the sleep disturbance of jet lag. Sleep Med. 2001;2(5):423–30.
24. Piérard C, Beaumont M, Enslen M, Chauffard F, Tan DX, Reiter RJ, et al. Resynchronization of hormonal rhythms after an eastbound flight in humans: effects of slow-release caffeine and melatonin. Eur J Appl Physiol. 2001;85(1–2):144–50.
25. Beaumont M, Batéjat D, Piérard C, Van Beers P, Denis JB, Coste O, et al. Caffeine or melatonin effects on sleep and sleepiness after rapid eastward transmeridian travel. J Appl Physiol (1985). 2004;96(1):50–8.
26. Rosenberg RP, Bogan RK, Tiller JM, Yang R, Youakim JM, Earl CQ, et al. A phase 3, double-blind, randomized, placebo-controlled study of armodafinil for excessive sleepiness associated with jet lag disorder. Mayo Clin Proc. 2010;85(7):630–8.

Chapter 14
A Guide for Characterizing and Prescribing Light Therapy Devices

Mark S. Rea

Introduction

Light therapy is commonly advocated by clinicians as an effective, nonpharmacological palliative for a variety of maladies such as seasonal depression and circadian rhythm sleep-wake disorders (CRSWDs) [1, 2]. This advice has both empirical and, to varying extents, physiological mechanistic support [3–5]. Nevertheless, there have been skeptics who contend the efficacy of light therapy may be little better than a placebo [6, 7].

Accurate dosing is critical for any therapeutic intervention. Consistently prescribing too little or too much of an antidote can potentially compromise its effectiveness and/or create unwanted side effects. More problematically, however, is the inconsistent dosing of the antidote. It is impossible to establish a reliable and predictable outcome if the actual dose is unknown. Without reliable and predictable outcomes, skepticism will limit widespread utilization of any therapeutic intervention.

Postulated here is the notion that some of the explicit or implicit skepticism surrounding the efficacy of light therapy for treating a wide range of maladies [8–10] is rooted in the uncertainty of light dosing. The amount, spectrum, duration, and timing of the light dose all need to be specified if a causal relation can, in fact, be established between light therapy and reductions in relevant maladies. Importantly, complete specification must characterize how that light dose is, in fact, applied. If, for example, the light therapy device induces discomfort glare [11], a prescribed light dosage will not be achieved because people will not look at the device. Moreover, each malady may have a different functional relationship between light dose and amelioration. Therefore, a single light dose may not

M. S. Rea (✉)
Lighting Research Center, Rensselaer Polytechnic Institute, Troy, NY, USA
e-mail: ream@rpi.edu

R. R. Auger (ed.), *Circadian Rhythm Sleep-Wake Disorders*,
https://doi.org/10.1007/978-3-030-43803-6_14

be equally effective for all maladies. It is further postulated that the clinical uncertainty surrounding the efficacy of light therapy can be reduced if care is taken by clinicians to quantify the light dose *as applied* using the guidelines offered here.

What Is Light?

The international authority, the Commission Internationale de l'Éclairage (CIE), defines light as "any radiation capable of causing visual [and/or nonvisual] sensation directly" [12]. By this *qualitative* definition, *light* is not a characteristic of the physical world alone. Rather, light is defined in terms of the relationship between optical radiation incident on the retina and a neural response that leads to a sensation. (It should be noted that the parenthetical phrase in the above definition is not part of the CIE definition; nonvisual sensations, such as those generated by the biological clock in the suprachiasmatic nuclei (SCN), should also be part of the definition of light [13]). Light must always be conceived in terms of biophysics, that is, in terms of how the *bio*logical world, specifically retinal physiology, and the *physic*al world, specifically optical radiation, intersect. This conceptualization is critical for accurately quantifying a therapeutic dose of light as it would be applied in clinical interventions aimed at ameliorating relevant maladies.

The CIE has *quantitatively* defined *light* in terms of response to optical radiation on the retina by a single neural channel with a particular *spectral sensitivity* to optical radiation, the photopic luminous efficiency function, $V(\lambda)$. This spectral sensitivity function is illustrated in Fig. 14.1 (adapted from Rea [14]), showing the relationship between the wavelengths of optical radiation on the retina to the efficiency with which that specific neural channel converts those wavelengths into a neural signal. *All* photometric instruments are calibrated in terms of $V(\lambda)$, and *all* recommendations for light therapy devices are defined in terms of $V(\lambda)$. Specifically, the amount of optical radiation generated by a light therapy device is always specified in terms of the photopic illuminance (usually in lux) it would produce at a particular (if undefined) distance and angle. Although clinicians rarely measure the amount of optical radiation generated by a light therapy device, most recommendations prescribe 10,000 lx or 5000 lx at the patient's eyes, depending upon the duration of exposure [15].

As will be described in more detail below, however, there are multiple neural channels emanating from the retina, each with different spectral sensitivities that convert optical radiation on the retina to neural signals to the brain. The CIE quantitative definition of light will not be relevant to all of these neural channels. In particular, $V(\lambda)$ is not relevant to quantifying light therapy devices. Dysfunction of the circadian system has been strongly implicated as the cause of psychological maladies where light therapy has been successful [2, 4, 16–18]. The circadian system is much more sensitive to short wavelengths than would be measured by instruments calibrated in terms of $V(\lambda)$. Functionally then, *light* for therapeutic treatment of the circadian system should not be measured in terms of the CIE definition, but, rather, in terms of the biophysical relationship between optical radiation on the retina and the neural sensation evoked by the SCN.

Retinal Physiology

The retina is a complex structure. There are four known types of photoreceptors in the distal retina, rods and three types of cones, long-wavelength (L), middle-wavelength (M), and short-wavelength (S) sensitive cones. In addition to these well-known photoreceptors, the more proximal intrinsically photosensitive retinal ganglion cells (ipRGCs) also convert optical radiation into neural signals [19]. The five types of photoreceptors are not randomly distributed across the retina, and each of these photoreceptors has a different spectral sensitivity to optical radiation (see Fig. 14.1).

There are several neural channels, both visual and nonvisual, emanating from the retina that combine the five photoreceptor responses in different ways (Fig. 14.2). For example, the spectral sensitivity of spatial resolution by the fovea, or contrast judgments, is well represented by $V(\lambda)$, but $V(\lambda)$ is defined in terms of just two cone photoreceptors, the L- and M-cones, that dominate the central fovea (see Fig. 14.1).

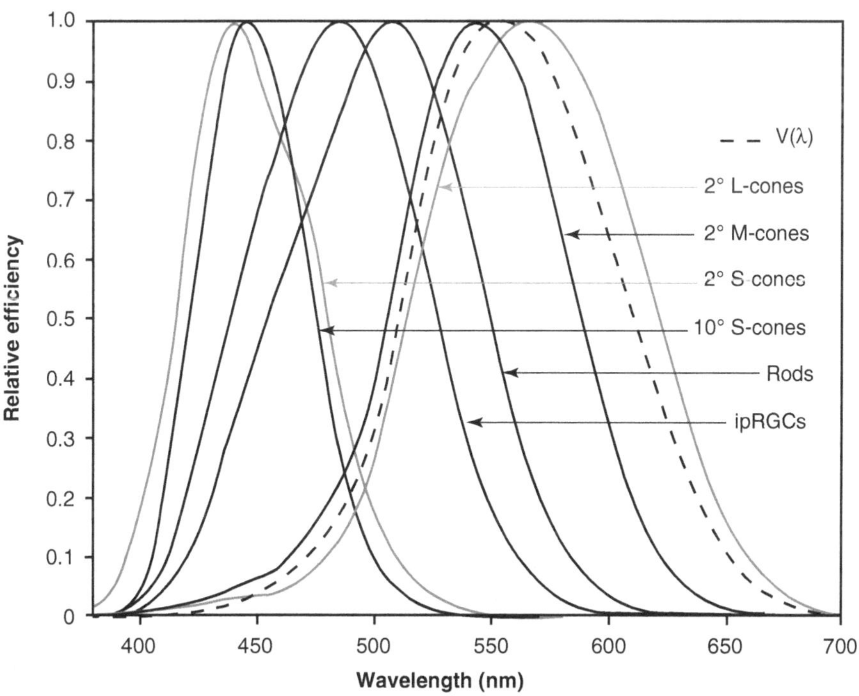

Fig. 14.1 The spectral sensitivities of the human retinal photoreceptors and the photopic luminous efficiency function [$V(\lambda)$]. (Adapted with permission from Rea [14])

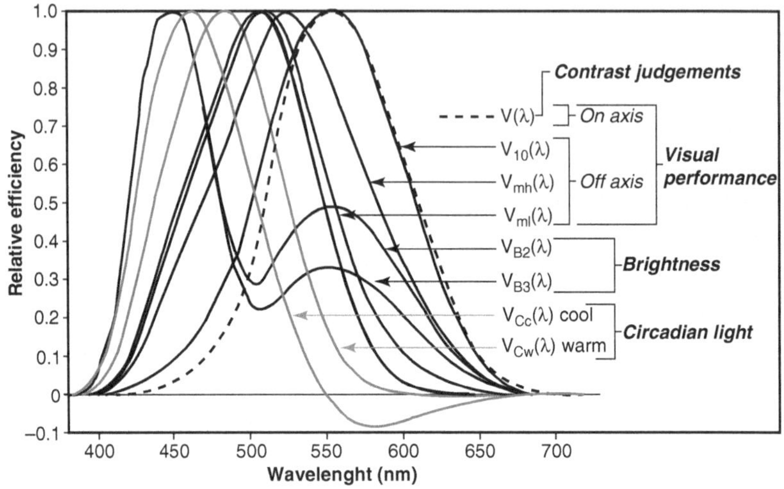

Fig. 14.2 The spectral sensitivities of different neural channels emanating from the human retina to the brain. Illustrated are four efficiency functions for visual performance, on-axis [$V(\lambda)$] and off-axis, the latter characterized by the CIE 10° photopic luminous efficiency function [$V_{10}(\lambda)$] and two off-axis mesopic (rod and cone) efficiency functions, one for high-mesopic [$V_{mh}(\lambda)$] and one for low-mesopic [$V_{ml}(\lambda)$] light levels. The on-axis contrast judgment efficiency function is also characterized by $V(\lambda)$. Two brightness efficiency functions are shown, one for high indoor light levels [$V_{B3}(\lambda)$] and one for low outdoor light levels [$V_{B2}(\lambda)$]. Also shown are two efficiency functions for circadian regulation, one for "warm" spectra [$V_{Cw}(\lambda)$] and one for "cool" spectra [$V_{Cc}(\lambda)$]. (Adapted with permission from Rea MS [14])

In contrast, brightness perception relies on all three cone photoreceptors distributed across the retina. In fact, the L- and M-cone responses play a minor role in brightness perception relative to the S-cone response (see Figs. 14.1, 14.2). Thus, contrast judgments and brightness perception are not perfectly correlated for different light sources. Two light sources that would generate the same L- and M-cone responses, and thus produce equal photopic illuminance levels on sheets of white paper, can differ in terms of their generated S-cone response. These two light sources would yield the same contrast judgments for black targets printed on sheets of white paper (e.g., printed text), but the sheets of paper would not appear equally bright; the sheet of white paper irradiated by the light source producing relatively more S-cone response would appear brighter.

Presumed Underlying Mechanism

Among clinicians and researchers, there appears to be consensus that light therapy positively affects the circadian system [18, 20]. As shown in Fig. 14.3 (also see Fig. 14.2), unlike $V(\lambda)$, the spectral sensitivity of the circadian system peaks at short wavelengths. Thus, *light* for the circadian system cannot be accurately measured in terms of the L- and M-cone response (i.e., $V(\lambda)$). In fact, the L- and M-cones have no significant impact on the circadian system response to optical radiation on the retina. Rather, the ipRGCs and, for some spectra, the S-cones dominate the spectral sensitivity of circadian phototransduction. This being the case, and similar to brightness perception, light therapy devices that produce optical radiation dominated by short wavelengths will be more effective than ones dominated by long wavelengths at exactly the same photopic illuminance at the eye. To quantify light for the circadian system, a different approach must be taken.

One approach to quantifying light for the circadian system utilizes a model of circadian phototransduction based upon well-established retinal physiology and empirical data relating optical radiation of specific spectra and amounts to the suppression of the hormone melatonin at night [21, 22]. (Nocturnal melatonin suppression is an orthodox measure of the circadian system's response to optical radiation.) The spectral sensitivity of the circadian system to monochromatic [23, 24] and polychromatic light sources is shown in Fig. 14.3, as depicted by Rea and colleagues [21, 22]. As can be inferred from this figure, the circadian system is

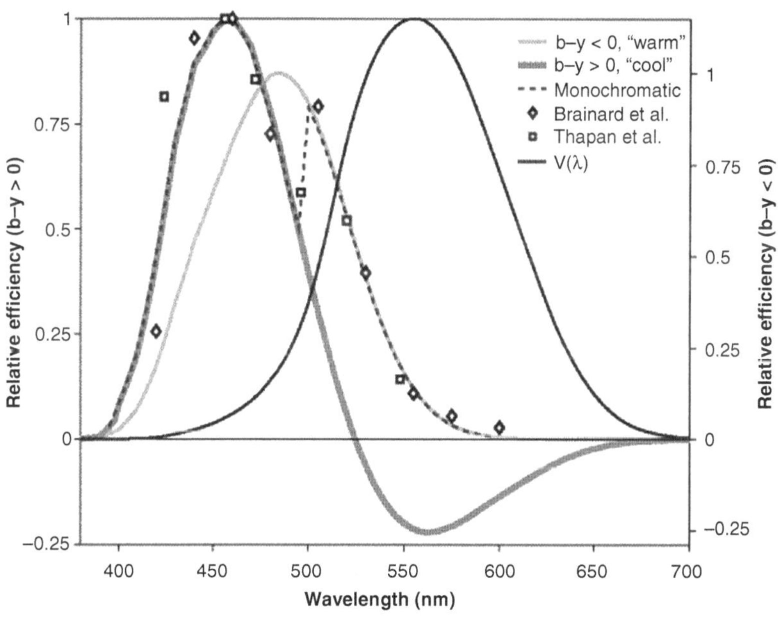

Fig. 14.3 According to the model by Rea and colleagues [21, 22], for "cool" sources where the blue vs. yellow (*b–y*) color-opponent mechanism signals "blue" ($b–y > 0$), the spectral sensitivity of the circadian system is represented by the thick, dark gray line, including a subadditive region of the spectrum where the spectral power from a light source subtracts from the overall response of the system. For "warm" sources where the *b–y* mechanism signals "yellow" ($b–y < 0$), the spectral sensitivity is represented by the thin, light gray line, the preretinal filtered spectral sensitivity of melanopsin. The measured spectral sensitivities for each monochromatic light, indicated by the open symbols [23, 24], are also well described by the two-state model (Eq. 14.1) at 300 scotopic lux, as shown here by the dashed line. At 300 scotopic lux, the circadian system is operating midway between threshold and saturation, consistent with the operating range of the circadian system where a constant criterion method can be successfully applied for determining spectral sensitivity to monochromatic lights

maximally sensitive to optical radiation at or near 460 nm; this single wavelength also evokes the color blue sensation. These data along with data from a variety of studies using polychromatic lights [25–28] have led to a definition of circadian light (CL_A), expressed in Eq. 14.1.

In addition to spectral sensitivity, the absolute sensitivity of the circadian system is also important for prescribing a dose of light. Equation 14.2 relates CL_A to circadian stimulus (CS), the *effectiveness* of circadian light, which is equivalent to nocturnal melatonin suppression in percent. Figure 14.4 shows this relationship from threshold (CS ≈ 0.1) to saturation (CS ≈ 0.6), along with fixed amounts of photopic illuminance for daylight and incandescent light sources needed to reach different CS levels.

$$CL_A = \begin{cases} 1548\left[\int \mathrm{Mc}_\lambda E_\lambda \mathrm{d}\lambda + \left(a_{b-y}\left(\int \frac{S_\lambda}{\mathrm{mp}_\lambda} E_\lambda \mathrm{d}\lambda - k\int \frac{V_\lambda}{\mathrm{mp}_\lambda} E_\lambda \mathrm{d}\lambda\right) - a_{\mathrm{rod}}\left(1 - e^{\frac{-\int V'_\lambda E_\lambda \mathrm{d}\lambda}{\mathrm{RodSat}}}\right)\right)\right] \\ \qquad if \int \frac{S_\lambda}{\mathrm{mp}_\lambda} E_\lambda \mathrm{d}\lambda - k\int \frac{V_\lambda}{\mathrm{mp}_\lambda} E_\lambda \mathrm{d}\lambda > 0 \\ 1548\int \mathrm{Mc}_\lambda E_\lambda \mathrm{d}\lambda \qquad if \int \frac{S_\lambda}{\mathrm{mp}_\lambda} E_\lambda \mathrm{d}\lambda - k\int \frac{V_\lambda}{\mathrm{mp}_\lambda} E_\lambda \mathrm{d}\lambda \le 0 \end{cases} \tag{14.1}$$

$$CS = 0.7 - \frac{0.7}{1 + \left(\frac{CL_A}{355.7}\right)^{1.1026}} \tag{14.2}$$

where:

CL_A: circadian light; the constant, 1548, sets the normalization of CL_A so that 2856 K blackbody radiation at 1000 lx has a CL_A value of 1000.

E_λ: light source spectral irradiance distribution.

Mc_λ: melanopsin (corrected for crystalline lens transmittance).

S_λ: S-cone fundamental.

mp_λ: macular pigment transmittance.

V_λ: photopic luminous efficiency function.

V'_λ: scotopic luminous efficiency function.

RodSat: half-saturation constant for bleaching rods = 6.5 W/m^2.

$k = 0.2616$.

$a_{b-y} = 0.700$.

$a_{\mathrm{rod}} = 3.300$.

From a clinical perspective, assuming that light therapy should activate the circadian system and that the model is predictive of circadian-effective light, both in terms of spectral and absolute sensitivities, one can quantitatively prescribe a light

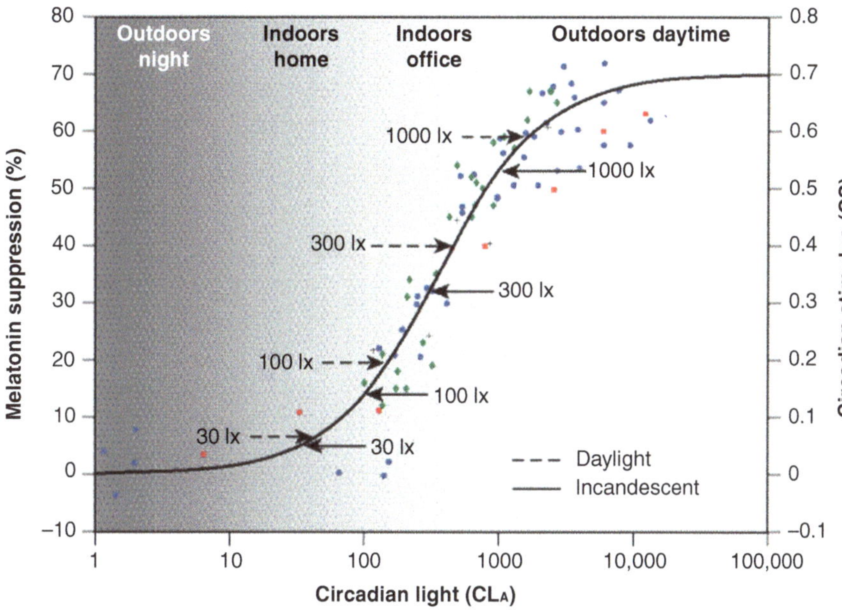

Fig. 14.4 Absolute sensitivity of the human circadian system [21, 22] to light measured in units of CS, as a function of circadian light, measured in terms of CL_A; CS, the effectiveness of circadian light, is equivalent to the percent of light-induced nocturnal melatonin suppression after 1-hour exposure. The shaded portions of the figure illustrate typical levels of illumination found outdoors at night, indoors in homes at night, indoors in offices, and outdoors during the daytime. Also shown is the relationship between photopic illuminance for daylight and incandescent spectra at different levels of CL_A and thus CS. As with the examples offered in Table 14.1, a criterion level of CS can typically be achieved at a lower photopic illuminance at the eye with a "cool" light source (e.g., daylight) than with a "warm" light source (e.g., incandescent), or, conversely, for the same photopic illuminance (e.g., 300 lx), a "cool" daylight source is more effective than a "warm" incandescent source at stimulating the human circadian system. (Data points are from Brainard et al. [23] and Thapan et al. [24])

dose for any light source (Eqs. 14.1, 14.2). The rows in Table 14.1 are for three different types of light sources, a "warm" incandescent light source, "cool" natural daylight, and a blue LED with peak emission at 470 nm. The columns in Table 14.1 represent two constant amounts of circadian-effective light, CS = 0.3 and CS = 0.6. The cell values in Table 14.1 are the levels of photopic illuminance (in lux) needed for each light source to reach the criterion amounts of circadian-effective light. As can be seen from the cell values in Table 14.1, the levels of photopic illuminance needed to reach a given CS criterion below saturation (i.e., CS ≤ 0.6) vary considerably. This wide variation is a direct result of the CIE definition of light that underlies photopic illuminance measurements. In other words, because photopic illuminance poorly represents the spectral sensitivity of the circadian system, different light sources will require different levels of photopic illuminance to be equally effective for the circadian system.

Table 14.1 Photopic illuminance levels (in lux) at the eyes required for three selected sources to provide a criterion level of circadian-effective light shown to improve sleep and reduce depression [29–33] (CS = 0.3) compared to those required to achieve near saturation of the human circadian system (CS = 0.6). All of these light levels are well below those currently prescribed for light therapy

Source	CS = 0.3	CS = 0.6
Incandescent (A lamp)	275 lx	1800 lx
Daylight (D65)	180 lx	1100 lx
Blue LED (470 nm)	18 lx	115 lx

Table 14.2 The CS levels achieved by selected sources delivering 5000 photopic lx and 10,000 photopic lx at the eyes. Both photopic light levels achieve saturation of the human circadian system for the three selected light sources

Source	5000 lx	10,000 lx
Incandescent (A lamp)	0.664	0.683
Daylight (D65)	0.682	0.693
Blue LED (470 nm)	0.699	0.699

It has been shown in nonclinical applications that white light providing a CS = 0.3 is effective for improving sleep and reducing depression [29–33]. Arguably then, much more modest levels of white light than currently prescribed (e.g., 5000–10,000 lx) [15] would be needed to reach CS = 0.3, as shown in Table 14.1. Even if one assumes that for clinical applications a patient needs more white light to ameliorate depressive symptoms of circadian dysregulation than that obtained in nonclinical applications, a near-saturating criterion of CS = 0.6 might be used to set white light levels at the patient's eyes. Indeed, no more than 20% of the light levels currently recommended would be needed to reach this criterion for any white light source. With blue light, even lower photopic illuminance levels are required to reach a criterion CS = 0.6, recognizing, again, that $V(\lambda)$ is a poor representation of the spectral sensitivity of the circadian system.

Also, it is clear from Table 14.2 that light doses of 5000 lx and 10,000 lx result in circadian system saturation for *any* light source spectral power distribution. Perhaps the past successes of light therapy lie in overprescribing the light dose needed to be clinically relevant. This over-specifying may, however, have led to patient noncompliance simply because the light therapy device is too bright, thus adding to the skepticism surrounding light therapy.

Discussion

There remains some uncertainty regarding the effectiveness of light therapy for ameliorating seasonal depression and even less certainty with respect to its effectiveness in treating CRSWDs [18]. That uncertainty may indeed be associated with "methodological problems" [10], one of which is the accurate measurements of the actual light doses (amount, spectrum, duration, and timing) provided to patients.

The uncertainty associated with light measurements has probably led to overprescribing light dosages, perhaps reducing compliance among patients. Empirically, providing a CS = 0.3 during the day has been shown to improve sleep and reduce depression [29–33]. Based upon this evidence, one could arguably reduce the amount of photopic illuminance to 3% of that currently recommended for any light source (see Table 14.1) and still obtain satisfactory results if these light doses were indeed delivered. Even if a more cautious clinician wanted to prescribe light doses near saturation (CS = 0.6), an 80% reduction in photopic illuminance levels relative to those currently prescribed could be implemented. This reduction could perhaps increase compliance without reducing clinical efficacy and thereby increase the collective clinical efficacy of light therapy. Moreover, lower levels of photopic illuminance, particularly with blue LEDs, would help minimize the risk of blue light hazard [34].

The major point here is that without a clearly defined light dose, based upon the biophysics of circadian phototransduction, proof of efficacy will remain elusive. It is *not* argued here that a CS = 0.3 will be clinically effective. Rather, it is only by accurately specifying an actual light dose, whether it be a CS = 0.3 or a CS = 0.6, that any specific light dose efficacy hypothesis can be tested clinically. Until those tests are undertaken, however, uncertainty about the efficacy of light therapy will always remain.

Guidelines

Before prescribing a light therapy device for use by a patient, the clinician will need to perform nine initial steps. These initial steps will take some time and some modest expense, but once a light therapy device is fully characterized, it will be much easier to make confident and consistent prescriptions of light dose in the future.

A significant problem facing clinicians is the ready access of photometric instrumentation that can measure and quantify circadian-effective light. Rarely will a clinician spend several thousand dollars to acquire instrumentation and software that can (a) measure the spectral power distribution of the light therapy device directly and (b) automatically compute circadian-effective light. Consequently, less expensive methods for specifying the luminous attributes of light therapy devices were developed here to help clinicians ensure delivery of prescribed light doses.

Initial Steps

1. Define a specific level of circadian-effective light at the patient's eyes. Although other models may be available (e.g., the WELL Building Standard [35]), circadian-effective light is defined here in terms of circadian stimulus (CS). For light therapy, a CS = 0.6 is recommended here, but other criteria may also be selected based upon recursive feedback.

2. Purchase a calibrated illuminance meter. All illuminance meters should be calibrated in terms of photopic illuminance and have a cosine-corrected spatial response [36]. A good illuminance meter will cost less than $200.
3. Obtain the relative spectral power distribution (SPD) of the light source in the light therapy device from the light source manufacturer. Every reputable manufacturer can provide a "typical" SPD; if the manufacturer will not or cannot provide a typical SPD, choose a different manufacturer.
4. Purchase a candidate light therapy device for testing. Ideally, it should be possible to dim the light output from the device. Dimming will provide more flexibility in placing the light therapy device at a location acceptable to the patient. In general, do *not* purchase small light therapy devices and/or ones that have undiffused light-emitting elements (e.g., visible bare LEDs). Whereas small light sources can meet a criterion amount of circadian-effective light at the patient's eyes, they may also cause discomfort glare [37]. And although the risk may be small, undiffused light-emitting elements will increase the risk of blue light hazard [34].
5. Turn all of the room lights off, and orient the illuminance meter at a fixed, measured distance in front of and perpendicular to the main light-emitting element. The distance (*d*) from the illuminance meter to the light-emitting element should be at least five times that of the maximum linear dimension of the light-emitting element. For example, if the diagonal dimension of the rectangular, diffuse light-emitting element is 0.5 m (20 in.), the illuminance meter should be no closer than 2.5 m (8.2 ft). Record the photopic illuminance (*E*).
6. Estimate the viewing angle between the patient's line of sight and the light therapy device. The amount of light entering the eye will decrease with the cosine of the angle [38]. Placing the light therapy device along the line of sight (0°) will usually interfere with visual tasks the patient may want to perform while receiving light therapy, so it will most often be displaced laterally from the line of sight. For example, if the light therapy device is placed on a desktop, 45° from the line of sight, the effective light level at the eye should be discounted by the cosine of 45°, or approximately 30%.
7. Input the discounted photopic illuminance at the measured distance and the relative SPD into a free, web-based calculator to determine CS (http://www.lrc.rpi.edu/cscalculator/).
8. Adjust the output of the light therapy device and/or the distance between the light therapy device and the patient's eyes along the expected viewing angle until the criterion CS has been reached; record the distance. The amount of illuminance varies with the square of the distance. So, reducing the distance between the light therapy device and the patient's eyes along the expected viewing angle by, for example, half, will increase the illuminance at the eye by a factor of 4. For example, if Step 5 recorded an *E* of 288 lx at 2.5 m, then at 1 m, the E would equal 1800 lx.

9. Estimate the luminance (L, in cd/m^2) of the light therapy device from the illuminance value (E) and the distance (d) recorded in Step 5 above. Measure the area (A, in m^2) of the light-emitting element of the light therapy device, and then calculate L using Eq. 14.3:

   $L = E\,d^2/A$ (14.3)

   For example, if $d = 2.5$ m, $E = 288$ lx, $A = 0.25$ m^2 (0.5 m × 0.5 m), $L = 7200$ cd/m^2.

   The luminance should not exceed 8500 cd/m^2 for any white light source and, based upon work by Bullough [37], no more than 850 cd/m^2 for any narrowband, blue light source. Discomfort glare will be avoided if the luminance of a light source is below these limits, and because the light emitted by the light therapy device is diffuse, there will be no risk of blue light hazard [34].

Specification

Once the light therapy device has been fully characterized, prescribe a distance and angle of sight to ensure the criterion *amount* and *spectrum* of circadian light (e.g., CS = 0.6) will be provided to the patient. Prescribe a *timing* and a *duration* of therapy every day in the context of desired clinical outcomes. For example, to achieve circadian phase advances, light therapy should be prescribed for use in the morning, whereas to achieve circadian phase delays, light therapy should be prescribed for the late afternoon or evening. Because mammalian circadian systems are slow to reach steady-state equilibrium (i.e., entrainment) [39], patients should be continuously exposed to light therapy for at least 1 hour, day after day. Ask the patient to document compliance with photos and diary.

Test Clinical Outcomes

With clear documentation of the patient's actual use of the fully characterized light therapy device, conduct a clinical assessment after 3 weeks. Adjust dose (i.e., timing, duration, amount, and spectrum) according to outcomes. Document the findings. Repeat steps as necessary.

References

1. Terman M, Terman JS. Light therapy for seasonal and nonseasonal depression: efficacy, protocol, safety, and side effects. CNS Spectr. 2005;10(8):647–73.
2. Auger RR, Burgess HJ, Emens JS, Deriy LV, Thomas SM, Sharkey KM. Clinical practice guideline for the treatment of intrinsic circadian rhythm sleep-wake disorders: advanced sleep-wake phase disorder (ASWPD), delayed sleep-wake phase disorder (DSWPD), non-24-hour sleep-wake rhythm disorder (N24SWD), and irregular sleep-wake rhythm disorder (ISWRD). An update for 2015. J Clin Sleep Med. 2015;11(10):1199–236.
3. Oldham MA, Ciraulo DA. Bright light therapy for depression: a review of its effects on chronobiology and the autonomic nervous system. Chronobiol Int. 2014;31(3):305–19.
4. Lam RW, Levitan RD. Pathophysiology of seasonal affective disorder: a review. J Psychiatry Neurosci. 2000;25(5):469–80.

5. Bunney BG, Bunney WE. Mechanisms of rapid antidepressant effects of sleep deprivation therapy: clock genes and circadian rhythms. Biol Psychiatry. 2013;73(12):1164–71.
6. Eastman CI. What the placebo literature can tell us about light therapy for SAD. Psychopharmacol Bull. 1990;26(4):495–504.
7. Eastman CI, Young MA, Fogg LF, Liu L, Meaden PM. Bright light treatment of winter depression: a placebo-controlled trial. Arch Gen Psychiatry. 1998;55(10):883–9.
8. Golden RN, Gaynes BN, Ekstrom RD, Hamer RM, Jacobsen F, Suppes T, et al. The efficacy of light therapy in the treatment of mood disorders: a review and meta-analysis of the evidence. Am J Psychiatry. 2005;162(4):656–62.
9. Lam RW, Levitt AJ, Levitan RD, Enns MW, Morehouse R, Michalak EE, et al. The Can-SAD study: a randomized controlled trial of the effectiveness of light therapy and fluoxetine in patients with winter seasonal affective disorder. Am J Psychiatry. 2006;163(5):805–12.
10. Mårtensson B, Pettersson A, Berglund L, Ekselius L. Bright white light therapy in depression: a critical review of the evidence. J Affect Disord. 2015;182:1–7.
11. Glickman G, Byrne B, Pineda C, Hauck WW, Brainard GC. Light therapy for seasonal affective disorder with blue narrow-band light-emitting diodes (LEDs). Biol Psychiatry. 2006;59(6):502–7.
12. Commission Internationale de l'Éclairage. Light as a true visual quantity: principles of measurement. Paris: Commission Internationale de l'Éclairage; 1978.
13. Rea MS. The lumen seen in a new light: Making distinctions between light, lighting and neuroscience. Light Res Technol. 2015;47(3):259–80.
14. Rea MS. Value metrics for better lighting. Bellingham: SPIE; 2013.
15. Eagles JM. Light therapy and seasonal affective disorder. Psychiatry. 2009;8(4):125–9.
16. Costa IC, Carvalho HN, Fernandes L. Aging, circadian rhythms and depressive disorders: a review. Am J Neurodegener Dis. 2013;2(4):228–46.
17. Lewy AJ, Bauer VK, Cutler NL, Sack RL, Ahmed S, Thomas KH, et al. Morning vs evening light treatment of patients with winter depression. Arch Gen Psychiatry. 1998;55(10):890–6.
18. Menculini G, Verdolini N, Murru A, Pacchiarotti I, Volpe U, Cervino A, et al. Depressive mood and circadian rhythms disturbances as outcomes of seasonal affective disorder treatment: a systematic review. J Affect Disord. 2018;241:608–26.
19. Berson D, Dunn F, Takao M. Phototransduction by retinal ganglion cells that set the circadian clock. Science. 2002;295(5557):1070–3.
20. Germain A, Kupfer DJ. Circadian rhythm disturbances in depression. Hum Psychopharmacol. 2008;23(7):571–85.
21. Rea MS, Figueiro MG, Bullough JD, Bierman A. A model of phototransduction by the human circadian system. Brain Res Rev. 2005;50(2):213–28.
22. Rea MS, Figueiro MG, Bierman A, Hamner R. Modelling the spectral sensitivity of the human circadian system. Light Res Technol. 2012;44(4):386–96.
23. Brainard GC, Hanifin JP, Greeson JM, Byrne B, Glickman G, Gerner E, et al. Action spectrum for melatonin regulation in humans: evidence for a novel circadian photoreceptor. J Neurosci. 2001;21(16):6405–12.
24. Thapan K, Arendt J, Skene DJ. An action spectrum for melatonin suppression: evidence for a novel non-rod, non-cone photoreceptor system in humans. J Physiol. 2001;535:261–7.
25. Rea MS, Figueiro MG. A working threshold for acute nocturnal melatonin suppression from "white" light sources used in architectural applications. J Carcinog Mutagen. 2013;4(3):1000150.
26. Figueiro MG, Bierman A, Rea MS. Retinal mechanisms determine the subadditive response to polychromatic light by the human circadian system. Neurosci Lett. 2008;438(2):242–5.
27. Figueiro MG, Bullough JD, Parsons RH, Rea MS. Preliminary evidence for spectral opponency in the suppression of melatonin by light in humans. Neuroreport. 2004;15(2):313–6.
28. Figueiro MG, Bullough JD, Bierman A, Rea MS. Demonstration of additivity failure in human circadian phototransduction. Neuro Endocrinol Lett. 2005;26(5):493–8.

29. Figueiro MG, Steverson B, Heerwagen J, Rea MS, editors. Daylight in office buildings: impact of building design on personal light exposures, sleep and mood. 28th CIE Session; 2015 June 28 – July 4; Manchester: Commission Internationale de l'Éclairage.
30. Figueiro MG, Steverson B, Heerwagen J, Kampschroer K, Hunter CM, Gonzales K, et al. The impact of daytime light exposures on sleep and mood in office workers. Sleep Health. 2017;3(3):204–15.
31. Figueiro MG, Kalsher M, Steverson BC, Heerwagen J, Kampschroer K, Rea MS. Circadian-effective light and its impact on alertness in office workers. Light Res Technol. 2019;51(2):171–83.
32. Figueiro MG, Plitnick BA, Lok A, Jones GE, Higgins P, Hornick TR, et al. Tailored lighting intervention improves measures of sleep, depression, and agitation in persons with Alzheimer's disease and related dementia living in long-term care facilities. Clin Interv Aging. 2014;9:1527–37.
33. Figueiro MG, Hunter CM, Higgins PA, Hornick TR, Jones GE, Plitnick B, et al. Tailored lighting intervention for persons with dementia and caregivers living at home. Sleep Health. 2015;1(4):322–30.
34. Bullough JD, Bierman A, Rea MS. Evaluating the blue-light hazard from solid state lighting. Int J Occup Saf Ergonomics. 2017;25(2):311–20.
35. International WELL Building Institute. The WELL building standard, version 1. New York: International WELL Building Institute; 2014.
36. Commission International de l'Éclairage. Characterization of the performance of illuminance meters and luminance meters, ISO/CIE 19476:2014(E). Vienna: Commission International de l'Éclairage; 2014.
37. Bullough J. Spectral sensitivity for extrafoveal discomfort glare. J Mod Opt. 2009;56(13):1518–22.
38. Van Derlofske J, Bierman A, Rea MS, Ramanath J, Bullough JD. Design and optimization of a retinal flux density meter. Meas Sci Technol. 2002;13(6):821–8.
39. Pittendrigh CS. On the mechanism of the entrainment of a circadian rhythm by light cycles. In: Aschoff J, editor. Circadian clocks: proceedings of the Feldafing Summer School, 7–18 September 1964. Amsterdam: North-Holland; 1965. p. 277–97.

Chapter 15
Future Directions for Lighting Environments

Mariana G. Figueiro

Introduction

Lighting for the built environment has traditionally centered on the places that we illuminate and the various things we do in those places. As a result, lighting typically has been designed, specified, and manufactured to meet a relatively limited number of very specific objectives. Historically, the primary requirement of lighting has been to illuminate spaces for optimization of visual performance (addressing concerns such as efficiency, productivity, and safety), to provide visual comfort to occupants, and to enhance the space's appearance for aesthetic appreciation. Over time, as indoor lighting proliferated to the point that it virtually became taken for granted by many end users, increasing energy demand and costs led to the adoption of energy conservation as an additional requirement. The lighting industry has been a key driver of technological advances to address these needs, from the development of incandescent and fluorescent sources to today's rapidly evolving solid-state lighting technologies. The most recent driver, the so-called nonvisual effects of light on the circadian system, has spurred lighting in the built environment to undergo yet another transformation in response. The lighted environment will undoubtedly continue to change as we develop a deeper understanding of how light impacts human physiology and behavior.

M. G. Figueiro (✉)
Lighting Research Center, Rensselaer Polytechnic Institute, Troy, NY, USA
e-mail: figuem@rpi.edu

R. R. Auger (ed.), *Circadian Rhythm Sleep-Wake Disorders*,
https://doi.org/10.1007/978-3-030-43803-6_15

What Is Light and How Do We Characterize It?

For a more detailed definition of light, refer to Chap. 14 in this volume. Briefly though, by definition, light is optical radiation that excites photoreceptors in the human eye. The photopic luminous efficiency function, V(λ), is the primary spectral weighting function used in science and commerce to differentiate light, which is associated with a very narrow range (≈ 380–780 nanometers (nm)) of the electromagnetic spectrum (Fig. 15.1). Indeed, V(λ) underlies all quantitative descriptions of light for commercial photometry and all recommendations for lighting practice [1]. Both luminous flux (i.e., the rate of flow of light from a source, measured in lumens) and intensity (i.e., the amount of light generated by a source in a given direction, measured in candelas) are universally used by lighting manufacturers to communicate the light quantities generated by their products. For lighting applications, V(λ) is also used exclusively to weight spectral irradiance and radiance functions, so both illuminance (lumens per square meter, lm/m^2) and luminance (candelas per square meter; cd/m^2) are used for recommended practices in hospitals, schools, factories, roadways, parking lots, and homes.

Lighting standards are still set primarily in terms of illuminance and lumens per watt, both of which are based upon the implicit assumption that the value of lighting can be characterized by the lumen. However, the lumen is derived from a very narrow set of experimental conditions that are only relevant to simple visual functions and does not characterize all of the visual responses that are important to modern built environments (e.g., apparent brightness). Moreover, the lumen is only indirectly related to the provision of perceptual information about the environment (e.g., linear perspective) and is, by definition, unrelated to the nonvisual, circadian effects of lighting that help to maintain entrainment of our many biological functions to the local time on Earth. Boyce and Rea have conceptualized three domains that light can affect [2], which are shown in Fig. 15.2. Again, the lumen is relevant to only one of these three domains, which is the visual system.

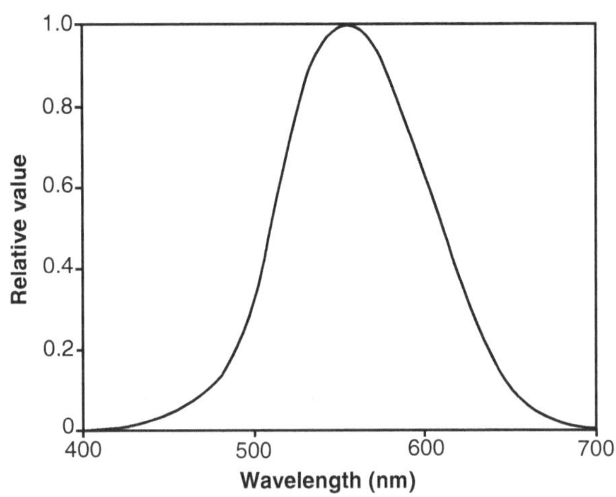

Fig. 15.1 The photopic luminous efficiency function, V(λ), adopted by the CIE in 1924 [1], is generally the only luminous efficiency function incorporated into commercially available photometric instruments and the only one used internationally for lighting application standards

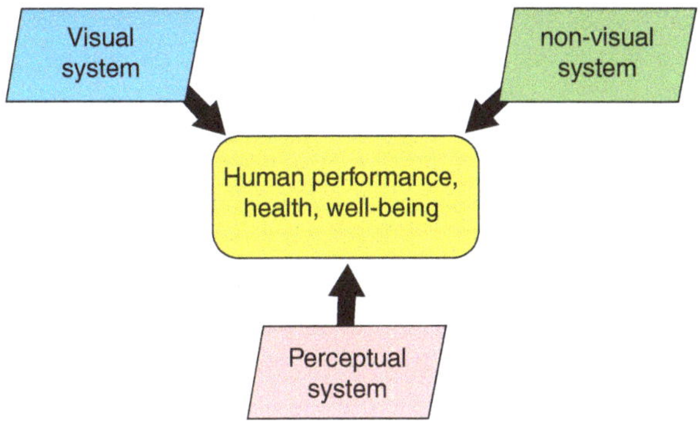

Fig. 15.2 Human performance, health, and well-being are influenced by at least three systems: visual, nonvisual, and perceptual (adapted from Boyce and Rea [2])

In addition to allowing us to see, light also impacts our nonvisual and perceptual systems. One of the most studied nonvisual responses to light is the one responsible for stimulating the circadian system. (The term *circadian* derives from Latin and is a blending of circa, or "about," and *dies*, or "day.") The 24-h light–dark patterns incident on the retina control the timing of the body's biological clock, which generates and regulates our circadian rhythms so that we are awake during the day and asleep at night. Because our circadian rhythms typically free-run on a slightly longer cycle of about 24.2 h in the absence of external cues, however, we require daily exposure to morning light to reset our biological clock and keep us synchronized (or *entrained*) with the local time on Earth. In addition to promoting entrainment, exposure to sufficient levels of light of any spectrum also exerts a direct, acute effect on people, which leads to an increase in alertness, as measured by objective (e.g., reduction in alpha power from electroencephalogram measurements) and subjective (i.e., self-reported) responses [3, 4]. Interestingly, it has been demonstrated that the spectral sensitivity of acute alertness is different than that of melatonin suppression and phase shifting of the onset of the body's secretion of melatonin in dim light (i.e., dim light melatonin onset, or DLMO) [5]. While short-wavelength ("blue") light has been shown to maximally affect the timing of the biological clock, a series of laboratory and field studies have shown that saturated red light (630 nm) at levels varying from 30 to 200 lx at the cornea exert a strong alerting effect on healthy adults [6–9]. As discussed below, when thinking about "light as a cup of coffee" and when melatonin suppression is not desired, red light may be a better choice.

Last but not least, light, especially from saturated long-wavelength "red" ($\approx$ 630 nm) and short-wavelength "blue" ($\approx$ 470 nm) sources, can affect human perceptual and/or psychological systems. As Elliot and Maier note in their review of research on the effects of color perception on human psychological functioning [10], comparatively little published research exists on this theme despite an extensive body of literature on related topics such as the physics of light color perception, the manner in which color stimuli are processed by the eyes and brain, and the linguistic categorization of color, among others. While the impacts of color and colored light on mood and behavior are still not well understood, these impacts should nevertheless be considered when thinking about the lighted environment.

Current Trends in Lighted Environments

Light–dark exposures experienced by humans have differed quantitatively between societies and through time. Prior to the widespread use of electric lighting, people of a century ago were exposed to very bright days and very dark nights, which is probably true today only for those who live in agrarian societies or perform some form of outdoor work. With a growing frequency that is projected to reach almost 70% of the world's population by 2050 (compared to 30% in 1950 and 55% today) [11], people are living in urban or suburban built environments and, in terms of circadian regulation, are thus becoming increasingly likely to experience extended dim days and light at night.

Data obtained using the Daysimeter (Fig. 15.3), an instrument designed to measure both conventional photopic illuminance and circadian light exposures at the eye throughout the waking day [12, 13], showed that people who work in the built environment experience much lower light levels compared to those who live in agrarian societies. The Daysimeter is a personal light meter that measures and is calibrated in terms of photopic illuminance (i.e., the density of light incident on a surface as perceived by the human eye, measured in lux (lx)), "circadian illuminance" (CL_A) [14], and the absolute sensitivity of the human circadian system (measured in circadian stimulus (CS)), based on the model of phototransduction proposed by Rea et al. [15–17]. The values of CL_A are scaled so that 1000 lx of a standard light source representing the spectral composition of an incandescent light source at 2856 K is equivalent to 1000 units of CL_A. Those values are in turn transformed into CS values corresponding to the relative suppression of nocturnal melatonin after a

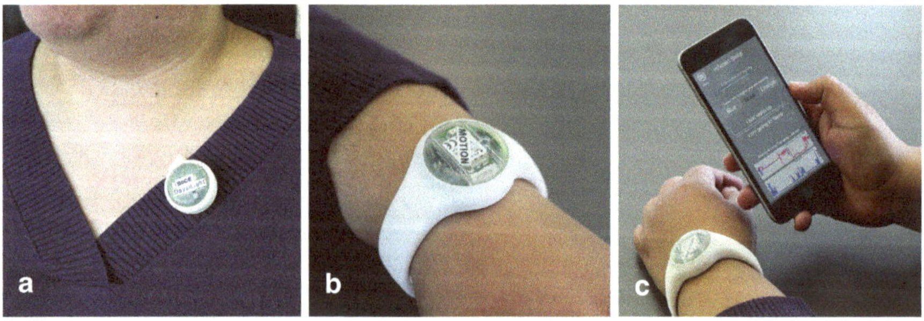

Fig. 15.3 The Daysimeter is used to measure both conventional photopic illuminance and circadian light exposures. The data presented in Table 15.1 were collected using an earlier version of the Daysimeter that was worn at eye level. For increased user compliance, newer versions of the Daysimeter are being worn as pendants or pins (**a**) to measure light exposures. The newer version of the Daysimeter also measures activity levels when worn on the wrist (**b**). A third component of the newer version of the Daysimeter uses a mobile phone app to compute light treatments (**c**)

1-h light exposure through a 2.3 mm diameter pupil during the midpoint of the body's nighttime production of that hormone [14]. Since CS is defined in terms of the circadian system's input–output relationship, from threshold (CS = 0.1) to saturation (CS ≈ 0.7), it is considered a better measure of the circadian effectiveness of light than either photopic illuminance or CL_A. The data summarized in Table 15.1 were obtained from 24 young adults aged 18–30 years [18], 22 female school teachers (mean ± standard deviation (SD) age of 36.25 ± 7.86 years) [19], 22 eighth-grade elementary school students [20], and 77 rotating- and day-shift nurses [21]. The participants in all studies wore the device for at least 5 consecutive days and were instructed to maintain their normal schedules while participating in the respective protocols.

It should be emphasized that for the rotating-shift nurses, because their morning and evening data were collected relative to bedtimes and wake times rather than actual clock time, the respective values shown in Table 15.1 could represent light exposures that occurred much earlier and/or later in the day than those of the other populations. The Daysimeter data were used to calculate mean photopic illuminance (lux), CL_A, and CS exposures. Log10 transforms of the photopic and CL_A values were also calculated because the recorded light exposures had highly skewed distributions in which brief exposures to extremely bright light (e.g., sunlight) dominated the arithmetic mean values. The log10 transforms of the photopic illuminance and CL_A values are therefore probably more representative of the light exposures' central tendency than the arithmetic means.

Several studies have postulated that exposure to electric light at night (LAN) poses health risks because it is sufficiently bright to suppress melatonin or disrupt our circadian rhythms [22–24]. However, very few data have been reported concerning actual light exposures in living and working environments during night and day. The data presented in Table 15.1 suggest that the average evening light exposures experienced by the five experimental populations were well below this level, even among the rotating-shift nurses who, research indicates, are known to be at higher risk for breast cancer [25, 26] and skin cancer [27]. The mean CS values for the five populations ranged from 0.04 to 0.07, suggesting that their evening light exposures resulted in suppression values that were below 7%.

Gooley et al. showed that an 8-h exposure to <200 lx at the cornea of a 4100 K light source resulted in significant suppression of evening melatonin in the laboratory [28]. Although no real-life light measurements were presented by the authors, and they did not utilize a photometric instrument calibrated in terms of the spectral sensitivity of the human circadian system, the group postulated that exposure to 60–180 lx at the cornea (which they referred to as the <200 lx condition) is representative of typical room lighting conditions experienced by people in their homes during the evening. Exposure to 200 lx at the cornea from a 4100 K source like the one used by Gooley et al., however, would have provided a CS value between 0.15 and 0.19, depending upon the source's specific spectral power distribution. Thus, the data in Table 15.1 underscore the fact that light in the built environment is simply too low for circadian entrainment when compared to the amount of light one would be exposed to outdoors, even on a cloudy day.

Table 15.1 Mean (±) standard error of the mean morning (4 h after rising) and evening (4 h before bed) light exposures for different populations

Population	Photopic illuminance (lx)		Log10 photopic illuminance (lx)		CL_A		Log10 CL_A		CS	
	Morning	Evening	Morning	Evening	Morning	Evening	Morning	Evening	Morning	Evening
Young adults (n = 24)	772 ± 188	38.2 ± 4.3	2.00 ± 0.06	1.22 ± 0.06	1650 ± 438	34.3 ± 3.6	2.02 ± 0.06	1.17 ± 0.05	0.193 ± 0.015	0.046 ± 0.005
Teachers (n = 22)	373 ± 80	40.4 ± 9.9	1.94 ± 0.05	1.07 ± 0.06	478 ± 105	44.1 ± 14.5	1.88 ± 0.06	0.97 ± 0.07	0.172 ± 0.013	0.036 ± 0.006
Eighth-grade students (n = 22)	268 ± 25	63.0 ± 19.6	2.04 ± 0.04	1.19 ± 0.08	305 ± 54	78.4 ± 30.7	1.96 ± 0.03	1.13 ± 0.08	0.184 ± 0.006	0.046 ± 0.008
Day-shift nurses (n = 33)	296 ± 50	73.9 ± 16.9	1.49 ± 0.04	0.94 ± 0.05	408 ± 173	35.8 ± 11.1	1.30 ± 0.06	0.79 ± 0.04	0.109 ± 0.011	0.029 ± 0.004
Rotating-shift nurses (n = 44)	277 ± 56	104.0 ± 13.7	1.37 ± 0.04	1.09 ± 0.06	414 ± 103	135 ± 22.4	1.35 ± 0.04	1.06 ± 0.05	0.114 ± 0.009	0.066 ± 0.006

Research on the Impact of Current Lighted Environments on Health and Well-being

The lack of adequate light exposure in the built environment has been associated with negative outcomes in measures of sleep, mood, and general well-being. Boubekri et al., for example, showed that people working in offices without access to windows reported poorer sleep quality, shorter sleep duration, and more-frequent sleep disturbances as assessed by self-reports and actigraphy recordings [29]. A recent study in five office buildings, on the other hand, showed that office workers who received high circadian stimulation in the morning reported better sleep and diminished depressive symptoms compared to their colleagues who received low circadian stimulation in the morning [30]. A follow-up study involving the installation of circadian-effective lighting in four other office buildings found that the intervention reduced workers' subjective sleepiness and increased their subjective feelings of alertness, vitality, and energy [31].

While a number of studies have attempted to demonstrate the benefits of lighting that provide higher circadian stimulation in the built environment during the daytime, other studies have employed a reduction in circadian stimulation during evening hours [32, 33]. Well-recognized risks for circadian disruption have been associated with evening light exposures from sources such as kitchen lighting and bathroom vanities, especially when they are furnished with high correlated color temperature (CCT) fixtures [34]. A growing risk in this regard has accompanied the burgeoning popularity of self-luminous portable electronic devices, or PEDs (e.g., cellphones, e-readers and tablets, laptop computers), whose displays are known to disrupt sleep [35] and the melatonin rhythm [36], particularly when used in the early evening (i.e., after 2000) during the onset of melatonin secretion. While at least one laboratory study has demonstrated that watching television at this time does not affect melatonin production [37], suggesting that devices' proximity to the corneas plays a role, another laboratory study demonstrated that manufacturers' solutions to avoid melatonin suppression from self-luminous devices in the evening (e.g., Apple's "Night Shift" setting for the iPad) are not necessarily effective [17].

Researchers generally agree that lighting schemes for daytime workers should be designed to promote circadian entrainment, which in turn should result in better sleep, mood, health, and perhaps some measures of performance. These recent studies emphasize not only the importance of reducing evening light exposures to maintain a regular sleep–wake cycle, but also the benefit of providing good circadian stimulation throughout the entire workday.

Adolescents present a special case, as they can be chronically sleep deprived due to their naturally later circadian phase and related inability to fall asleep at times conducive to their fixed early wake-up times on school days. Chronic insufficient sleep within this age group has been linked to depression, behavioral problems, poor performance at school, and automobile accidents [38]. This problem can be compounded by adolescents spending most of their day indoors in dimly lit classrooms, which inhibits the synchronization of their circadian systems with the solar day. Studies by the Lighting Research Center (LRC) have pointed to the importance of controlling the entirety of adolescents' 24-h light–dark patterns while they are at school and at home to promote circadian entrainment and reduce sleep restriction [39, 40], especially in view of the popular use of self-luminous electronic devices before sleep, which can delay circadian phase [41].

Most of the other published laboratory and field research to date has examined lighting for older adults living with Alzheimer's disease and related dementias (ADRD) [42–44] as well as patients in hospital rooms [45, 46]. Field research focused on older adults with ADRD has shown that a lighting intervention that provides high circadian stimulation during the day and low stimulation at night can reduce depressive symptoms among those living at home [33]. In nursing homes, where light exposures are more easily controlled, the same intervention resulted in improvements to sleep, depression, and agitation [44]. In another LRC study conducted in a nursing home, a lighting intervention with very high circadian stimulation was delivered to residents using a specially designed light-emitting diode (LED) "light table" [47]. Residents who sat around the table, as residents conventionally do in nursing home common areas, showed significantly increased sleep duration and reduced agitation and depression scores. In another non-LRC study, cancer patients who underwent myeloma transplant and received high circadian-effective light between 0700 and 1000 were less depressed than those who received a dim, control light [48]. Although more field studies are needed to confirm these initial findings, it has been postulated that lighting can improve patients' mood and health outcomes [49] and may shorten the length of hospital stays [50].

The Need for Proposing a 24-h Lighting Scheme in our Lighted Environments

The data shown in Table 15.1 clearly indicate that people working indoors are exposed to dim, extended, and aperiodic circadian-effective light, whereas those working outdoors are likely to be exposed to a robust light–dark cycle that is ideal for regulating the circadian system. These differences are probably even more pronounced during the winter, when the duration of daylight becomes shorter and those working indoors are exposed to even lesser amounts of circadian-effective light during the day. The resulting disruption of the circadian cycle that is probably now experienced by most people in modern societies may very well have a direct influence on associated health problems such as metabolic diseases, cancer, and mood disorders, among various other conditions [23, 51, 52]. Future trends in the lighted environment should therefore include 24-h lighting solutions that promote circadian entrainment.

Initial studies of light's therapeutic application focused on treating the symptoms of seasonal affective disorder (SAD) [53, 54]. While the mechanisms associated with light's beneficial effects on SAD are not yet well established, light therapy is nonetheless recognized by physicians and recommended as a sole or adjunct treatment [55]. However, conventionally prescribed light therapy boxes, which are not an integral part of the built environment and can be very bright and glaring if not equipped with a dimming control, can reduce patients' compliance with the treatment and thereby reduce its efficacy. Fortunately, newly developed, more-effective technologies such as solid-state lighting have fostered better targeted, less glaring solutions for relieving SAD symptoms. (For more information on how to measure and specify light therapy, please refer to Chap. 14 in this volume.)

A growing body of evidence indicates that any successful field application of light therapy to correct circadian sleep disorders must control the total light exposures over the course of the 24-h light–dark cycle. Using the modified model of the human circadian oscillator [56] that permits quantitative predictions of circadian phase changes resulting from light exposure, Rea et al. [57] showed that light exposures during total waking hours must be monitored and controlled in order to predict circadian phase changes resulting from a given light treatment. Several other models for predicting phase changes have been proposed, but they have met with mixed success due to their inability to account for differential light exposures from week to week. The model by Kronauer et al. [56], for example, consists of a light-stimulus phototransduction process (L) driving an oscillator-based pacemaker process (P). In the phototransduction model proposed by Rea et al. [15, 16], unlike the Kronauer et al. model, CS is used as an input to process L instead of photopic illuminance (lux). Parameters of the process P were revised based upon data from field studies where light exposures and circadian phase changes were measured. More specifically, two parameters in the process P were adjusted (k and q), and a time-dependent sensitivity modulation factor was removed. Overall, the model proposed by Rea et al. [15, 16] improves upon Kronauer's model by incorporating new knowledge of human circadian phototransduction; thus the Rea et al. model can provide more accurate quantitative predictions of circadian phase changes resulting from differential light exposures.

To illustrate the application of this model, predictions of circadian phase changes based upon continuously measured 24-h light exposures were compared to measured phase changes (DLMO) from three field studies [18, 58, 59]. Figure 15.4a shows the correlation between DLMO and the measured phase changes calculated from the Daysimeter [12, 13] data and predicted phase change using the modified Kronauer model [56, 57], which were statistically significant ($R^2 = 0.70$, $p < 0.0001$) with a prediction uncertainty of 1.75 h (95% confidence). Interestingly, and as shown in Fig. 15.4b, when only the treatment light exposures—and not the total light exposures measured over the entire waking period—are included in the model, predictions of circadian phase change are not as good, as shown by the large deviation of the best fit from the ideal fit.

Controlling light exposures for the entire 24-h day therefore requires tracking and recording—in terms of spectrum, level, timing, and history—throughout all waking and sleeping hours. It is also necessary to understand how those light exposures have interacted with a person's biological systems and to then make the necessary lighting adjustments, effectively creating a personalized light prescription that follows them everywhere they go, including while they are at work and at home.

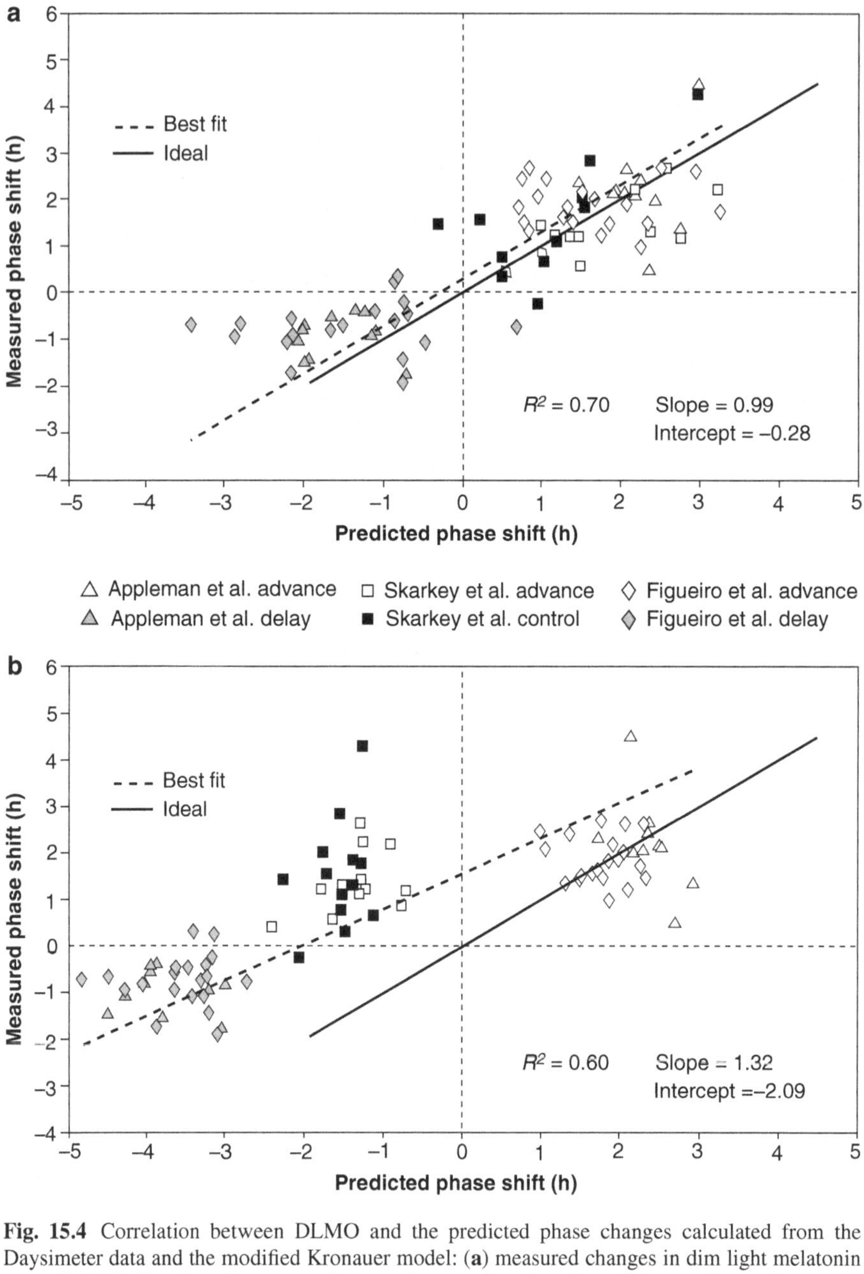

Fig. 15.4 Correlation between DLMO and the predicted phase changes calculated from the Daysimeter data and the modified Kronauer model: (**a**) measured changes in dim light melatonin onset (DLMO) from baseline to post-intervention are plotted on the ordinate, and circadian stimulus (CS)-oscillator model predictions based on actual measured light exposures during the intervention are plotted on the abscissa; (**b**) measured changes in DLMO from baseline to post-intervention are plotted on the ordinate, and CS-oscillator model predictions based solely on the treatment light exposures (i.e., not using light exposures measured throughout the day by the Daysimeter) are plotted on the abscissa. The ideal fit for both graphs was calculated using the least squares method, where the difference between the measured DLMO and the predicted DLMO was calculated

Future Trends in Lighted Environments

Since CS is defined in terms of the circadian system's input–output relationship (see section "Current Trends in Lighted Environments") and various field studies have shown that it is associated with better sleep, mood, and well-being, the future trends discussed in this section will use CS as the principal metric, with corresponding lux levels provided. In general, when occupied by day-active/night-resting people, the lighted environments should deliver a CS ≥ 0.3 at eye level and a CS < 0.1 starting 2 h prior to their desired bedtimes. The LRC has developed a CS calculator (https://www.lrc.rpi.edu/cscalculator/) to help lighting professionals select light sources and targeted photopic light levels that will increase the potential for proper circadian light exposure in buildings. This tool facilitates calculations of CL_A and CS for over 180 example light source spectra as well as user-supplied light source spectra.

Workplace Lighting

To promote circadian entrainment among daytime office workers, lighting systems and solutions must be tailored to meet the circadian requirements of the people who spend their workday in a particular office space while also addressing the unique characteristics of that space. Factors to be taken into account include (1) the workers' ages; (2) the type of work performed; (3) the presence and number of windows, as well as the amount of light they contribute to the space; (4) the direction in which the windows face, (5) the space's ceiling height; (6) the presence of obstructing features such as cabinets and partitions; (7) the reflectance values of the office furniture and fixtures; and (8) the amount of CS that the space's occupants receive at eye level and the times of day when they experience it throughout the entire workday.

As with any environment, morning lighting at work should provide plenty of CS to help synchronize the biological clock and promote circadian entrainment. In the simplest terms, this can be achieved by increasing the amount of light reaching the eye by sitting next to a window for 30 min, opening window shades, and/or increasing the office's ambient light level. Field research shows that relatively simple lighting strategies can achieve sufficiently high levels of CS (i.e., CS $\geq$ 0.3) using commonly available lighting systems [30, 60], such as the 2 × 4 troffer lighting fixture employed for the lighting schedule shown in Fig. 15.5. Ideally, to better promote entrainment, the lighting system's CS levels should be lowered toward the end of the day, especially if the office space is to be occupied through the early evening.

Healthcare Facility Lighting

Healthcare represents another sector that stands to benefit from circadian-effective lighting systems. The modern hospital is in effect a city in microcosm, which poses a formidable challenge to lighting design because the "city" is all under one roof.

The hospital functions around the clock and year-round, its patients can range in age from premature infants to the elderly, and the people in any given space can be very ill patients or workers and visitors who are in good health. Furthermore, many healthcare personnel work long hours that can include night shifts and rotating shifts and are exposed to different lighting characteristics at different circadian times.

For patients and their visitors, lighting schemes should more or less follow those outlined for daytime office workers in Fig. 15.5, since all patients must remain entrained to the hospital's daytime caregiving schedule and most visitors are usually present only before patients' bedtimes. The same may also be said for caregivers

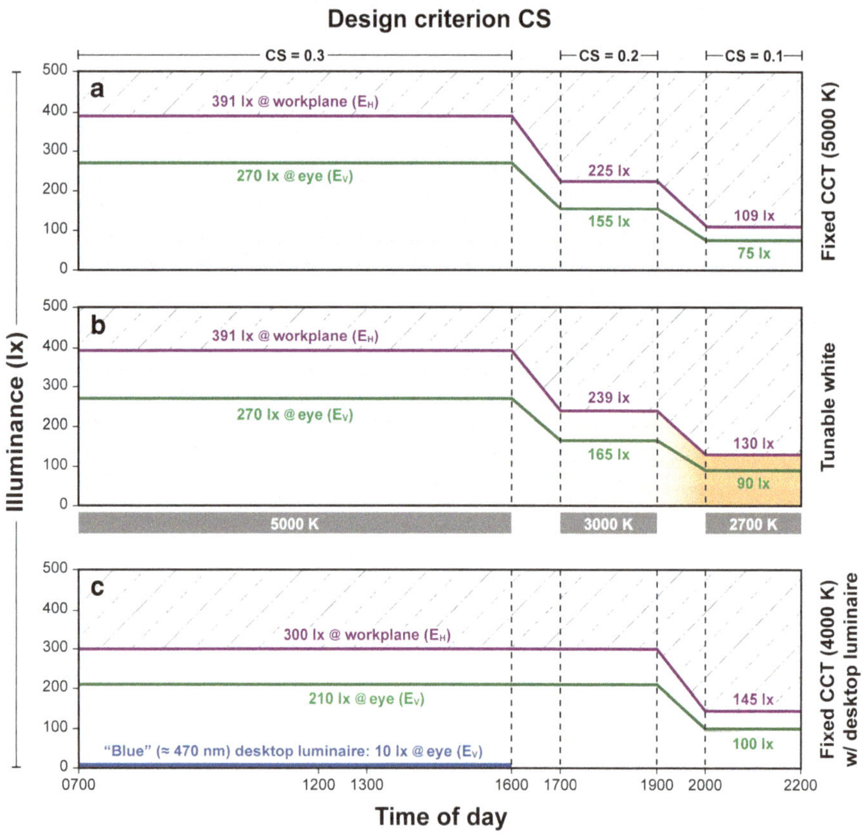

Fig. 15.5 Three lighting schemes that can be used to provide the minimum CS for promoting circadian entrainment (CS = 0.3) using typical 2 × 4 troffers in an open office environment: (**a**) fixed CCT (5000 K), (**b**) tunable white (5000–2700 K), and (**c**) fixed CCT (4000 K) supplemented by a personal desktop luminaire delivering 30 lx of saturated short-wavelength "blue" (≈ 470 nm) light at eye level. It is most desirable to specify lighting systems that vary in output and spectrum (if possible) throughout the day according to the circadian needs of the space's occupants. Though not shown here, during the night shift, systems can also be designed to deliver saturated long-wavelength "red" (≈ 630 nm) light to promote alertness without elevated CS levels and disruption to nighttime workers' melatonin rhythms [31]

Fig. 15.6 Rendering of a nurses' station during the day shift (left) and night shift (right). High CS to promote circadian entrainment and alertness is provided using higher photopic illuminance levels and saturated blue light from the desktop luminaires during the daytime. Illuminance and CS levels are lower at night, and alertness is promoted by saturated red light that does not affect the circadian system. The same desktop devices, in this case LED light therapy panels, can be configured to deliver both types of light shown here

who work exclusively during the day. Caregivers have dramatically different lighting needs during the night shift, however (Fig. 15.6), and circadian disruption (and its consequent health problems) has been demonstrated among those who work rotating shifts and, especially, among those who work throughout the night [61, 62]. Lighting tips for night-shift caregivers are provided in the sidebar.

Lighting Tips for Night-Shift Caregiver Entrainment

Night-shift caregivers are usually entrained to a day-shift lifestyle because they continue to follow daytime social and domestic activities, making adaptation to a nocturnal lifestyle both impractical and undesirable. For these caregivers, it would be best to:

- Minimize bright light (> 20–30 lx at the eye, CS > 0.05) during the night shift, starting at around 2300, taking care to avoid the use of self-luminous personal electronic devices.
- Receive saturated red (630 nm) light of at least 40–60 lx (CS = 0) at the eye in rest areas or workspaces (either intermittently or continuously), which will provide an alerting stimulus similar to drinking a cup of coffee throughout the night and will not affect melatonin levels.
- Use task lights to increase light levels on the workplane (i.e., horizontal surfaces such as desks, tables, workstations, etc.) and for specific critical tasks, such as insertion of an IV.
- Take public transportation to avoid falling asleep behind the wheel on the drive home.

Another alternative for nighttime caregivers is a "compromise solution" that uses light early in the night to delay feelings of sleepiness until after the shift is over, but not so late that it does not dramatically differ from their wake times on working days. In other words, the caregivers would remain entrained to a day-shift lifestyle but would become "night owls." This adaptation permits easier transitions between night and day shifts and can be achieved by caregivers if they:

- Receive high levels of illumination until 0300–0400 (at least 200–300 lx ($CS \geq 0.3$) at the eye from a white light source) in workspaces followed by dim white light (20–30 lx at the eye, $CS < 0.05$) until the end of the shift.
- Lower levels of saturated blue light (e.g., 30 lx at the eye of a 470 nm light) could also be added as a task light, but care should be taken to avoid compromising important aspects of visual performance (e.g., color rendering) required for carrying out visual tasks.
- Receive saturated red (630 nm) light of at least 40–60 lx ($CS = 0$) at the eye in rest areas or workspaces (either intermittently or continuously), which will provide an alerting stimulus similar to drinking a cup of coffee throughout the night and will not affect melatonin levels
- Use task lights to increase light levels on the workplane and for specific critical tasks such as insertion of an IV.
- Wear dark sunglasses on the way home after work to prevent outdoor light from affecting the night owl adaptation.
- Take public transportation to avoid falling asleep behind the wheel on the drive home.

Home Lighting

The home lighting environment is crucial for circadian entrainment since it is where people typically wake up to begin their day and prepare for sleep at the day's end. Given what we know about the circadian system, circadian-effective lighting systems must be adapted to follow humans *between* the spaces they occupy, from home to work, to retail and recreational spaces, and then back home again in the evening. Research by the LRC, in conjunction with Lund University and the Swedish Energy Agency, has made a crucial first step toward realizing an individualized lighting prescription system in the Swedish Healthy Home project. The system is built around the Daysimeter (or similar device) [13], which tracks personal light exposures and physical activity (actigraphy) throughout the 24-h day and transmits the data for storage on a smartphone or some other PED (see Fig. 15.3). Upon arrival at home at the end of the day, the data are then automatically transmitted to the home's central lighting control system, which then

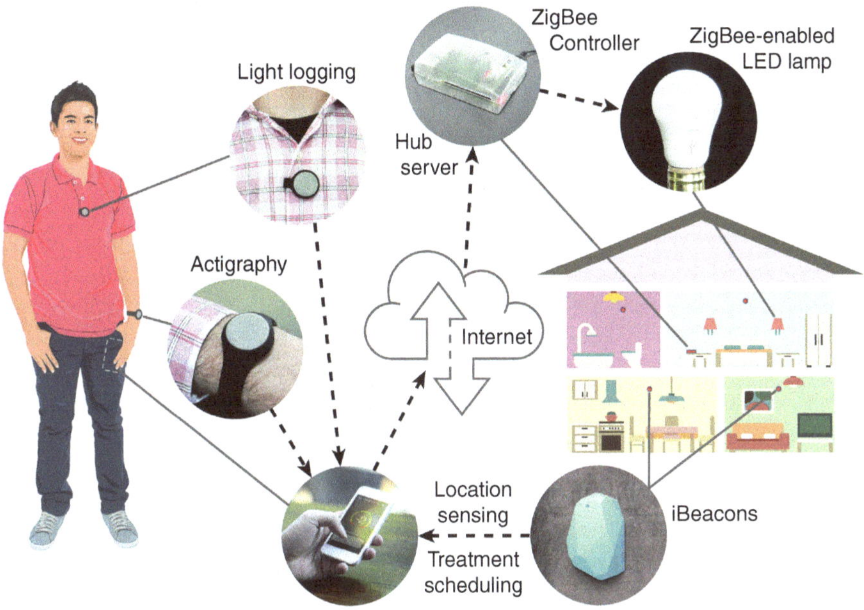

Fig. 15.7 Schematic of the Swedish Healthy Home lighting system. Light exposures and actigraphy are recorded using a Daysimeter worn as a pendant and on the wrist, and the data are transmitted via Bluetooth to a handheld electronic device such as a cellphone

makes any necessary adjustments to the lighting in any space a person happens to occupy (Fig. 15.7).

In future developments, we envision the system recognizing users' personal light exposures as they move between buildings or rooms and dynamically adjusting the ambient lighting to maintain entrainment, rather than simply making adjustments upon their arrival at home in the evening. If the automatic controls are not appropriate or practical at a given time, moreover, we also envision the system being overridden and configured to provide notifications of any lighting needs and measures that should be taken. The system could also send a notice advising users to use personal light "dosing" or filtering devices such as light goggles or spectrally filtered glasses should the system override not be appropriate for their personal needs. Because the system is dynamic and portable, it would also be easily translated to one's home, schoolroom, or workplace.

Conclusion

It is well established that a robust, regular 24-h light–dark pattern minimizes circadian disruption, which in turn minimizes negative health and performance outcomes. Circadian disruption can be observed and become an issue when people travel across multiple time zones, use self-luminous displays in the evening, spend most of their daytime hours in dim interiors, stay up late to view media, and move from building to building and space to space throughout the day. In other words, circadian disruption can occur when the 24-h light–dark pattern is no longer regular and predictable.

The challenge for lighting researchers and professionals is that they have been so closely tied to thinking about a particular building—that is, a single static place where one needs light for the performance of tasks and the perception of the space's ambience, instantaneously. Circadian hygiene is not instantaneous, but cumulative. Today, because people continually carry self-luminous displays and pursue active lives that profoundly influence their 24-h pattern of light and dark, they do not have a single lighting entity that is responsible for total 24-h light exposure patterns and therefore cannot adequately address 24-h light exposure issues.

The challenge for sleep clinicians who wish to use light as a non-pharmacological treatment for circadian sleep disorders is that they need to start "thinking outside the (light) box." Lighting technologies now exist that can incorporate precision medicine in the delivery of tailored lighting interventions to treat circadian sleep disorders. Light sources that can be tuned to maximally affect (or not) the biological clock while reducing glare can be specified and purchased. Measurement tools that can determine the real-time dose experienced by users are readily available for clinicians and researchers. As a next step, sleep clinicians should seek to collaborate with researchers to increase the number of applied research studies demonstrating the efficacy and feasibility of these new light therapy options. We have to let go of the past and take advantage of the latest lighting technology advances.

A new profession needs to emerge, such as personal light and health coaching, or new software applications need to be developed to keep track of light–dark exposures and provide recipes for maintaining entrainment or correcting circadian disruption. This can already be accomplished where users do not change their living space across the 24-h day (e.g., senior living facilities [44] and submarines [63]). Another area for real impact could be healthy lighting for schoolchildren, who have a regular daily routine, and the education of parents to better control light and thereby ensure adequate and consistent sleep. The next step would be to educate teachers and parents about the significance of a robust 24-h light–dark pattern. Office spaces pose a greater challenge, but one can begin to envision the use of a workplace "light oasis" (Fig. 15.8), where workers could receive their circadian light exposures during the daytime with the aid of an application that informs them about what they need to do and when they need to do it. The technology now exists for implementing some of these solutions and, perhaps, changing the lighted environment so that it indeed promotes circadian health.

Fig. 15.8 Light oases of varying configurations are practical solutions for providing circadian stimulus for office workers. Such oases could be used in conjunction with a personalized circadian application that informs workers about what light they might need and when they need it

References

1. Commission Internationale de l'Éclairage. Light as a true visual quantity: principles of measurement. Paris: Commission Internationale de l'Éclairage; 1978.
2. Boyce PR, Rea MS. Lighting and human performance II: beyond visibility models toward a unified human factors approach to performance. Electric Power Research Institute, Palo Alto, CA, National Electrical Manufacturers Association, VA, Environmental Protection Agency Office of Air and Radiation, Washington, DC; 2001.
3. Cajochen C. Alerting effects of light. Sleep Med Rev. 2007;11(6):453–64.
4. Souman JL, Tinga AM, Te Pas SF, van Ee R, Vlaskamp BNS. Acute alerting effects of light: a systematic literature review. Behav Brain Res. 2018;337:228–39.
5. Van de Werken M, Gimenez MC, de Vries B, Beersma DG, Gordijn MC. Short-wavelength attenuated polychromatic white light during work at night: limited melatonin suppression without substantial decline of alertness. Chronobiol Int. 2013;30(7):843–54.
6. Figueiro MG, Bierman A, Plitnick B, Rea MS. Preliminary evidence that both blue and red light can induce alertness at night. BMC Neurosci. 2009;10:105.
7. Plitnick B, Figueiro MG, Wood B, Rea MS. The effects of red and blue light on alertness and mood at night. Light Res Technol. 2010;42(4):449–58.
8. Sahin L, Figueiro MG. Alerting effects of short-wavelength (blue) and long-wavelength (red) lights in the afternoon. Physiol Behav. 2013;116–117:1–7.
9. Figueiro MG, Sahin L, Wood B, Plitnick B. Light at night and measures of alertness and performance: implications for shift workers. Biol Res Nurs. 2016;18(1):90–100.
10. Elliot AJ, Maier MA. Color psychology: effects of perceiving color on psychological functioning in humans. Annu Rev Psychol. 2014;65:95–120.
11. United Nations Department of Economic and Social Affairs Population Division. World urbanization prospects: the 2018 revision. United Nations, New York. 2018. https://population.un.org/wup/. Accessed 8 Jan 2019.
12. Bierman A, Klein TR, Rea MS. The Daysimeter: a device for measuring optical radiation as a stimulus for the human circadian system. Meas Sci Technol. 2005;16:2292–9.
13. Figueiro MG, Hamner R, Bierman A, Rea MS. Comparisons of three practical field devices used to measure personal light exposures and activity levels. Light Res Technol. 2013;45(4):421–34.
14. Rea MS, Figueiro MG, Bierman A, Bullough JD. Circadian light. J Circadian Rhythms. 2010;8(1):2.

15. Rea MS, Figueiro MG, Bullough JD, Bierman A. A model of phototransduction by the human circadian system. Brain Res Rev. 2005;50(2):213–28.
16. Rea MS, Figueiro MG, Bierman A, Hamner R. Modelling the spectral sensitivity of the human circadian system. Light Res Technol. 2012;44(4):386–96.
17. Nagare R, Plitnick B, Figueiro MG. Does the iPad Night Shift mode reduce melatonin suppression? Light Res Technol. 2019;51(3):373–83.
18. Sharkey KM, Carskadon MA, Figueiro MG, Zhu Y, Rea MS. Effects of an advanced sleep schedule and morning short wavelength light exposure on circadian phase in young adults with late sleep schedules. Sleep Med. 2011;12(7):685–92.
19. Rea MS, Brons JA, Figueiro MG. Measurements of light at night (LAN) for a sample of female school teachers. Chronobiol Int. 2011;28(8):673–80.
20. Leslie RP, Smith A, Radetsky LC, Figueiro MG, Yue L. Patterns to daylight schools for people and sustainability. Lighting Research Center, Rensselaer Polytechnic Institute, Troy, NY; 2010.
21. Miller D, Figueiro MG, Bierman A, Schernhammer E, Rea MS. Ecological measurements of light exposure, activity and circadian disruption. Light Res Technol. 2010;42(3):271–84.
22. Stevens RG. Circadian disruption and breast cancer: from melatonin to clock genes. Epidemiology. 2005;16(2):254–8.
23. Stevens RG, Rea MS. Light in the built environment: potential role of circadian disruption in endocrine disruption and breast cancer. Cancer Causes Control. 2001;12(3):279–87.
24. Stevens RG, Blask DE, Brainard GC, Hansen J, Lockley SW, Provencio I, et al. Meeting report: the role of environmental lighting and circadian disruption in cancer and other diseases. Environ Health Perspect. 2007;115(9):1357–62.
25. Wegrzyn LR, Tamimi RM, Rosner BA, Brown SB, Stevens RG, Eliassen AH, et al. Rotating night-shift work and the risk of breast cancer in the Nurses' health studies. Am J Epidemiol. 2017;185(5):532–40.
26. Kamdar BB, Tergas AI, Mateen FJ, Bhayani NH, Oh J. Night-shift work and risk of breast cancer: a systematic review and meta-analysis. Breast Cancer Res Treat. 2013;138(1):291–301.
27. Heckman CJ, Kloss JD, Feskanich D, Culnan E, Schernhammer ES. Associations among rotating night shift work, sleep, and skin cancer in Nurses' Health Study II participants. Occup Environ Med. 2017;74(3):169–75.
28. Gooley JJ, Chamberlain K, Smith KA, SBS K, SMW R, Van Reen E, et al. Exposure to room light before bedtime suppresses melatonin onset and shortens melatonin duration in humans. J Clin Endocrinol Metab. 2011;96(3):E463–72.
29. Boubekri M, Cheung IN, Reid KJ, Wang CH, Zee PC. Impact of windows and daylight exposure on overall health and sleep quality of office workers: a case-control pilot study. J Clin Sleep Med. 2014;10(6):603–11.
30. Figueiro MG, Steverson B, Heerwagen J, Kampschroer K, Hunter CM, Gonzales K, et al. The impact of daytime light exposures on sleep and mood in office workers. Sleep Health. 2017;3(3):204–15.
31. Figueiro MG, Kalsher M, Steverson BC, Heerwagen J, Kampschroer K, Rea MS. Circadian-effective light and its impact on alertness in office workers. Light Res Technol. 2019;51(2):171–83.
32. Figueiro MG. A proposed 24 h lighting scheme for older adults. Light Res Technol. 2008;40(2):153–60.
33. Figueiro MG, Hunter CM, Higgins PA, Hornick TR, Jones GE, Plitnick B, et al. Tailored lighting intervention for persons with dementia and caregivers living at home. Sleep Health. 2015;1(4):322–30.
34. Higuchi S, Lee SI, Kozaki T, Harada T, Tanaka I. Late circadian phase in adults and children is correlated with use of high color temperature light at home at night. Chronobiol Int. 2016;33(4):448–52.
35. Hysing M, Pallesen S, Stormark KM, Jakobsen R, Lundervold AJ, Sivertsen B. Sleep and use of electronic devices in adolescence: results from a large population-based study. BMJ Open. 2015;5(1):e006748.

36. Figueiro MG, Erdener B, Jayawardena A, Lesniak NZ, Reh R, Sahin L, et al. The impact of self-luminous electronic devices on melatonin suppression. Dig Tech Pap. 2011;42(1):408–11.
37. Figueiro MG, Wood B, Plitnick B, Rea MS. The impact of watching television on evening melatonin levels. J Soc Inf Disp. 2013;21(10):417–21.
38. Owens J. Insufficient sleep in adolescents and young adults: an update on causes and consequences. Pediatrics. 2014;134(3):e921–32.
39. Figueiro MG, Brons JA, Plitnick B, Donlan B, Leslie RP, Rea MS. Measuring circadian light and its impact on adolescents. Light Res Technol. 2011;43(2):201–15.
40. Figueiro MG, Rea MS. Short-wavelength light enhances cortisol awakening response in sleep-restricted adolescents. Int J Endocrinol. 2012;2012:301935.
41. Figueiro MG, Overington D. Self-luminous devices and melatonin suppression in adolescents. Light Res Technol. 2016;48(8):966–75.
42. Figueiro MG, Rea MS, Eggleston G. Light therapy and Alzheimer's disease. Sleep Review. 2003;4(1):24.
43. Hanford N, Figueiro MG. Light therapy and Alzheimer's disease and related dementia: past, present, and future. J Alzheimers Dis. 2013;33(4):913–22.
44. Figueiro MG, Plitnick BA, Lok A, Jones GE, Higgins P, Hornick TR, et al. Tailored lighting intervention improves measures of sleep, depression, and agitation in persons with Alzheimer's disease and related dementia living in long-term care facilities. Clin Interv Aging. 2014;9:1527–37.
45. Engwall M, Fridh I, Johansson L, Bergbom I, Lindahl B. Lighting, sleep and circadian rhythm: an intervention study in the intensive care unit. Intensive Crit Care Nurs. 2015;31(6):325–35.
46. Giménez MC, Geerdinck LM, Versteylen M, Leffers P, Meekes GJ, Herremans H, et al. Patient room lighting influences on sleep, appraisal and mood in hospitalized people. J Sleep Res. 2017;26(2):236–46.
47. Figueiro M, Plitnick B, Rea M. Research note: a self-luminous light table for persons with Alzheimer's disease. Light Res Technol. 2016;48(2):253–9.
48. Valdimarsdottir HB, Figueiro MG, Holden W, Lutgendorf S, Wu LM, Ancoli-Israel S, et al. Programmed environmental illumination during autologous stem cell transplantation hospitalization for the treatment of multiple myeloma reduces severity of depression: a preliminary randomized controlled trial. Cancer Med. 2018;7(9):4345–53.
49. Oldham MA, Lee HB, Desan PH. Circadian rhythm disruption in the critically ill: an opportunity for improving outcomes. Crit Care Med. 2016;44(1):207–17.
50. Choi JH, Beltran LO, Kim HS. Impacts of indoor daylight environments on patient average length of stay (ALOS) in a healthcare facility. Build Environ. 2012;50:65–75.
51. Reutrakul S, Knutson KL. Consequences of circadian disruption on Cardiometabolic health. Sleep Med Clin. 2015;10(4):455–68.
52. Smolensky MH, Hermida RC, Reinberg A, Sackett-Lundeen L, Portaluppi F. Circadian disruption: new clinical perspective of disease pathology and basis for chronotherapeutic intervention. Chronobiol Int. 2016;33(8):1101–19.
53. Wirz-Justice A, Graw P, Krauchi K, Sarrafzadeh A, English J, Arendt J, et al. 'Natural' light treatment of seasonal affective disorder. J Affect Disord. 1996;37(2–3):109–20.
54. Avery DH, Eder DN, Bolte MA, Hellekson CJ, Dunner DL, Vitiello MV, et al. Dawn simulation and bright light in the treatment of SAD: a controlled study. Biol Psychiatry. 2001;50(3):205–16.
55. Terman M, Terman J. Bright light therapy: side effects and benefits across the symptom spectrum. J Clin Psychiatry. 1999;60(11):799–808.
56. Kronauer RE, Forger DB, Jewett ME. Quantifying human circadian pacemaker response to brief, extended, and repeated light stimuli over the phototopic range. J Biol Rhythm. 1999;14(6):500–16.
57. Rea MS, Bierman A, Ward G, Figueiro MG. Field tests of a model of the human circadian oscillator. SLEEP 2014, the 28th Annual Meeting of the American Academy of Sleep Medicine; May 31 to June 4, 2014; Minneapolis, MN; 2014.

58. Appleman K, Figueiro MG, Rea MS. Controlling light-dark exposure patterns rather than sleep schedules determines circadian phase. Sleep Med. 2013;14(5):456–61.
59. Figueiro MG, Plitnick B, Rea MS. The effects of chronotype, sleep schedule and light/dark pattern exposures on circadian phase. Sleep Med. 2014;15(12):1554–64.
60. Figueiro MG, Steverson B, Heerwagen J, Rea MS, editors. Daylight in office buildings: impact of building design on personal light exposures, sleep and mood. 28th CIE Session; 2015 June 28 – July 4; Manchester, UK: Commission Internationale de l'Éclairage.
61. Haus EL, Smolensky MH. Shift work and cancer risk: potential mechanistic roles of circadian disruption, light at night, and sleep deprivation. Sleep Med Rev. 2013;17(4):273–84.
62. Boivin DB, Boudreau P. Impacts of shift work on sleep and circadian rhythms. Pathol Biol (Paris). 2014;62(5):292–301.
63. Young CR, Jones GE, Figueiro MG, Soutière SE, Keller MW, Richardson AM, et al. At-sea trial of 24-h-based submarine watchstanding schedules with high and low correlated color temperature light sources. J Biol Rhythm. 2015;30(2):144–54.

Index

R. R. Auger (ed.), *Circadian Rhythm Sleep-Wake Disorders*,
https://doi.org/10.1007/978-3-030-43803-6

Printed in Singapore by Markono Print Media Pte Ltd